Statistical
Data Mining
Using SAS
Applications

Second Edition

Chapman & Hall/CRC
Data Mining and Knowledge Discovery Series

SERIES EDITOR
Vipin Kumar

University of Minnesota
Department of Computer Science and Engineering
Minneapolis, Minnesota, U.S.A

AIMS AND SCOPE

This series aims to capture new developments and applications in data mining and knowledge discovery, while summarizing the computational tools and techniques useful in data analysis. This series encourages the integration of mathematical, statistical, and computational methods and techniques through the publication of a broad range of textbooks, reference works, and handbooks. The inclusion of concrete examples and applications is highly encouraged. The scope of the series includes, but is not limited to, titles in the areas of data mining and knowledge discovery methods and applications, modeling, algorithms, theory and foundations, data and knowledge visualization, data mining systems and tools, and privacy and security issues.

PUBLISHED TITLES

UNDERSTANDING COMPLEX DATASETS:
DATA MINING WITH MATRIX DECOMPOSITIONS
David Skillicorn

COMPUTATIONAL METHODS OF FEATURE
SELECTION
Huan Liu and Hiroshi Motoda

CONSTRAINED CLUSTERING: ADVANCES IN
ALGORITHMS, THEORY, AND APPLICATIONS
Sugato Basu, Ian Davidson, and Kiri L. Wagstaff

KNOWLEDGE DISCOVERY FOR
COUNTERTERRORISM AND LAW ENFORCEMENT
David Skillicorn

MULTIMEDIA DATA MINING: A SYSTEMATIC
INTRODUCTION TO CONCEPTS AND THEORY
Zhongfei Zhang and Ruofei Zhang

NEXT GENERATION OF DATA MINING
Hillol Kargupta, Jiawei Han, Philip S. Yu,
Rajeev Motwani, and Vipin Kumar

DATA MINING FOR DESIGN AND MARKETING
Yukio Ohsawa and Katsutoshi Yada

THE TOP TEN ALGORITHMS IN DATA MINING
Xindong Wu and Vipin Kumar

GEOGRAPHIC DATA MINING AND
KNOWLEDGE DISCOVERY, SECOND EDITION
Harvey J. Miller and Jiawei Han

TEXT MINING: CLASSIFICATION, CLUSTERING,
AND APPLICATIONS
Ashok N. Srivastava and Mehran Sahami

BIOLOGICAL DATA MINING
Jake Y. Chen and Stefano Lonardi

INFORMATION DISCOVERY ON ELECTRONIC
HEALTH RECORDS
Vagelis Hristidis

TEMPORAL DATA MINING
Theophano Mitsa

RELATIONAL DATA CLUSTERING: MODELS,
ALGORITHMS, AND APPLICATIONS
Bo Long, Zhongfei Zhang, and Philip S. Yu

KNOWLEDGE DISCOVERY FROM DATA STREAMS
João Gama

STATISTICAL DATA MINING USING SAS
APPLICATIONS, SECOND EDITION
George Fernandez

Chapman & Hall/CRC
Data Mining and Knowledge Discovery Series

Statistical Data Mining Using SAS Applications

Second Edition

George Fernandez

CRC Press
Taylor & Francis Group
Boca Raton London New York

CRC Press is an imprint of the
Taylor & Francis Group, an **informa** business

A CHAPMAN & HALL BOOK

CRC Press
Taylor & Francis Group
6000 Broken Sound Parkway NW, Suite 300
Boca Raton, FL 33487-2742

© 2010 by Taylor and Francis Group, LLC
CRC Press is an imprint of Taylor & Francis Group, an Informa business

No claim to original U.S. Government works

Printed in the United States of America on acid-free paper
10 9 8 7 6 5 4 3 2 1

International Standard Book Number: 978-1-4398-1075-0 (Hardback)

Library of Congress Cataloging-in-Publication Data

Fernandez, George, 1952-
   Statistical data mining using SAS applications / author, George Fernandez. -- 2nd ed.
       p. cm. -- (Chapman & Hall/CRC data mining and knowledge discovery series)
   First published under title: Data mining using SAS applications.
   Includes bibliographical references and index.
   ISBN 978-1-4398-1075-0 (hardcover : alk. paper)
       1. Commercial statistics--Computer programs. 2. SAS (Computer file) 3. Data mining.
   I. Fernandez, George, 1952- Data mining using SAS applications II. Title. III. Series.

   HF1017.F476 2010
   006.3'12--dc22   2010011768

Visit the Taylor & Francis Web site at
http://www.taylorandfrancis.com

and the CRC Press Web site at
http://www.crcpress.com

Contents

Preface

Objective

The objective of the second edition of this book is to introduce statistical data mining concepts, describe methods in statistical data mining from sampling to decision trees, demonstrate the features of user-friendly data mining SAS tools and, above all, allow the book users to download compiled data mining SAS (Version 9.0 and later) macro files and help them perform complete data mining. The user-friendly SAS macro approach integrates the statistical and graphical analysis tools available in SAS systems and provides complete statistical data mining solutions without writing SAS program codes or using the point-and-click approach. Step-by-step instructions for using SAS macros and interpreting the results are emphasized in each chapter. Thus, by following the step-by-step instructions and downloading the user-friendly SAS macros described in the book, data analysts can perform complete data mining analysis quickly and effectively.

Why Use SAS Software?

The SAS Institute, the industry leader in analytical and decision support solutions, offers a comprehensive data mining solution that allows you to explore large quantities of data and discover relationships and patterns that lead to intelligent decision-making. Enterprise Miner, SAS Institute's data mining software, offers an integrated environment for businesses that need to conduct comprehensive data mining. However, if the Enterprise Miner software is not licensed at your organization, but you have license to use other SAS BASE, STAT, and GRAPH modules, you could still use the power of SAS to perform complete data mining by using the SAS macro applications included in this book.

Including complete SAS codes in the data mining book for performing comprehensive data mining solutions is not very effective because a majority of business and statistical analysts are not experienced SAS programmers. Quick results from data mining are not feasible since many hours of code modification and debugging program errors are required if the analysts are required to work with SAS program codes. An alternative to the point-and-click menu interface modules is the user-friendly SAS macro applications for performing several data mining tasks, which are included in this book. This macro approach integrates statistical and graphical tools available in the latest SAS systems (version 9.2) and provides user-friendly data analysis tools, which allow the data analysts to complete data mining tasks quickly, without writing SAS programs, by running the SAS macros in the background. SAS Institute also released a learning edition (LE) of SAS software in recent years and the readers who have no access to SAS software can buy a personal edition of SAS LE and enjoy the benefits of these powerful SAS macros (See Appendix 3 for instructions for using these macros with SAS EG and LE).

Coverage:

The following types of analyses can be performed using the user-friendly SAS macros.

- Converting PC databases to SAS data
- Sampling techniques to create training and validation samples
- Exploratory graphical techniques:
  - Univariate analysis of continuous response
  - Frequency data analysis for categorical data
- Unsupervised learning:
  - Principal component
  - Factor and cluster analysis
  - *k*-mean cluster analysis
  - Biplot display
- Supervised learning: Prediction
  - Multiple regression models
    - Partial and VIF plots, plots for checking data and model problems
    - Lift charts
    - Scoring
    - Model validation techniques
  - Logistic regression
    - Partial delta logit plots, ROC curves false positive/negative plots
    - Lift charts
- Model validation techniques
Supervised learning: Classification
  - Discriminant analysis
    - Canonical discriminant analysis—biplots
    - Parametric discriminant analysis
    - Nonparametric discriminant analysis
    - Model validation techniques
  - CHAID—decisions tree methods
    - Model validation techniques

Why Do I Believe the Book Is Needed?

During the last decade, there has been an explosion in the field of data warehousing and data mining for knowledge discovery. The challenge of understanding data has led to the development of new data mining tools. Data-mining books that are currently available mainly address data-mining principles but provide no instructions and explanations to carry out a data-mining project. Also, many existing data analysts are interested in expanding their expertise in the field of data-mining and are looking for how-to books on data mining by using the power of the SAS STAT and GRAPH modules. Business school and health science instructors teaching in MBA programs or MPH are currently incorporating data mining into their curriculum and are looking for how-to books on data mining using the available software. Therefore, this second edition book on statistical data mining, using SAS macro applications, easily fills the gap and complements the existing data-mining book market.

Key Features of the Book

No SAS programming experience required: This is an essential how-to guide, especially suitable for data analysts to practice data mining techniques for knowledge discovery. Thirteen very unique user-friendly SAS macros to perform statistical data mining are described in the book. Instructions are given in the book in regard to downloading the compiled SAS macro files, macro-call file, and running the macro from the book's Web site. No experience in modifying SAS macros or programming with SAS is needed to run these macros.

Complete analysis in less than 10 min.: After preparing the data, complete predictive modeling, including data exploration, model fitting, assumption checks, validation, and scoring new data, can be performed on SAS datasets in less than 10 min.

SAS enterprise minor not required: The user-friendly macros work with the standard SAS modules: BASE, STAT, GRAPH, and IML. No additional SAS modules or the SAS enterprise miner is required.

No experience in SAS ODS required: Options are available in the SAS macros included in the book to save data mining output and graphics in RTF, HTML, and PDF format using SAS new ODS features.

More than 150 figures included in this second edition: These statistical data mining techniques stress the use of visualization to thoroughly study the structure of data and to check the validity of statistical models fitted to data. This allows readers to visualize the trends and patterns present in their database.

Textbook or a Supplementary Lab Guide

This book is suitable for adoption as a textbook for a statistical methods course in statistical data mining and research methods. This book provides instructions and tools for quickly performing a complete exploratory statistical method, regression analysis, logistic regression multivariate methods, and classification analysis. Thus, it is ideal for graduate level statistical methods courses that use SAS software.

Some examples of potential courses:

- Biostatistics
- Research methods in public health
- Advanced business statistics
- Applied statistical methods
- Research methods
- Advanced data analysis

What Is New in the Second Edition?

- *Active internet connection is no longer required to run these macros*: After downloading the compiled SAS macros and the mac-call files and installing them in the C:\ drive, users can access these macros directly from their desktop.
- *Compatible with version 9*: All the SAS macros are compatible with SAS version 9.13 and 9.2 Windows (32 bit and 64 bit).
- *Compatible with SAS EG*: Users can run these SAS macros in SAS Enterprise Guide (4.1 and 4.2) code window and in SAS learning Edition 4.1 by using the special macro-call files and special macro files included in the downloadable zip file. (See Appendixes 1 and 3 for more information.)
- *Convenient help file location*: The help files for all 13 included macros are now separated from the chapter and included in Appendix 2.
- *Publication quality graphics*: Vector graphics format such as EMF can be generated when output file format TXT is chosen. Interactive ActiveX graphics can be produced when Web output format is chosen.
- *Macro-call error check*: The macro-call input values are copied to the first 10 title statements in the first page of the output files. This will help to track the macro input errors quickly.

Additionally the following new features are included in the SAS-specific macro application:

I. Chapter 2

a. Converting PC data files to SAS data (EXLSAS2 macro)
  - All numeric (m) and categorical variables (n) in the Excel file are converted to X_1-X_m and C_1-C_n, respectively. However, the original column names will be used as the variable labels in the SAS data. This new feature helps to maximize the power of the user-friendly SAS macro applications included in the book.

– Options for renaming any X_1-X_n or C_1-C_n variables in a SAS data step are available in EXLSAS2 macro application.
– Using SAS ODS graphics features in version 9.2, frequency distribution display of all categorical variables will be generated when WORD, HTML, PDF, and TXT format are selected as output file formats.

b. Randomly splitting data (RANSPLIT2)
– Many different sampling methods such as simple random sampling, stratified random sampling, systematic random sampling, and unrestricted random sampling are implemented using the SAS SURVEYSELECT procedure.

II. Chapter 3

a. Frequency analysis (FREQ2)
– For one-way frequency analysis, the Gini and Entropy indexes are reported automatically.
– Confidence interval estimates for percentages in frequency tables are automatically generated using the SAS SURVEYFREQ procedure. If survey weights are specified, then these confidence interval estimates are adjusted for survey sampling and design structures.

b. Univariate analysis (UNIVAR2)
– If survey weights are specified, then the reported confidence interval estimates are adjusted for survey sampling and design structures using SURVEYMEAN procedure.

III. Chapter 4

a. PCA and factor analysis (FACTOR2)
– PCA and factor analysis can be performed using the covariance matrix.
– Estimation of Cronbach coefficient alpha and their 95% confidence intervals when performing latent factor analysis.
– Factor pattern plots (New 9.2: statistical graphics feature) before and after rotation.
– Assessing the significance and the nature of factor loadings (New 9.2: statistical graphics feature).
– Confidence interval estimates for factor loading when ML factor analysis is used.

b. Disjoint cluster analysis (DISJCLUS2)

IV. Chapter 5

a. Multiple linear regressions (REGDIAG2)
– Variable screening step using GLMSELECT and best candidate model selection using AICC and SBC.

- Interaction diagnostic plots for detecting significant interaction between two continuous variables or between a categorical and continuous variable.
- Options are implemented to run the ROBUST regression using SAS ROBUSTREG when extreme outliers are present in the data.
- Options are implemented to run SURVEYREG regression using SAS SURVEYREG when the data is coming from a survey data and the design weights are available.

b. Logistic regression (LOGIST2)
- Best candidate model selection using AICC and SBC criteria by comparing all possible combination of models within an optimum number of subsets determined by the sequential step-wise selection using AIC.
- Interaction diagnostic plots for detecting significant interaction between two continuous variables or between a categorical and continuous variable.
- LIFT charts for assessing the overall model fit are automatically generated.
- Options are implemented to run survey logistic regression using SAS PROC SURVEYLOGISTIC when the data is coming from a survey data and the design weights are available.

V. Chapter 6

CHAID analysis (CHAID2)
- Large data (>1000 obs) can be used.
- Variable selection using forward and stepwise selection and backward elimination methods.
- New SAS SGPLOT graphics are used in data exploration.

Potential Audience

- This book is suitable for SAS data analysts, who need to apply data mining techniques using existing SAS modules for successful data mining, without investing a lot of time in buying new software products, or spending time on additional software learning.
- Graduate students in business, health sciences, biological, engineering, and social sciences can successfully complete data analysis projects quickly using these SAS macros.
- Big business enterprises can use data mining SAS macros in pilot studies involving the feasibility of conducting a successful data mining endeavor before investing big bucks on full-scale data mining using SAS EM.
- Finally, any SAS users who want to impress their boss can do so with quick and complete data analysis, including fancy reports in PDF, RTF, or HTML format.

Additional Resources

Book's Web site: A Web site has been setup at http://www.cabnr.unr.edu/gf/dm. Users can find information in regard to downloading the sample data files used in the book, and additional reading materials. Users are also encouraged to visit this page for information on any errors in the book, SAS macro updates, and links for additional resources.

Acknowledgments

I am indebted to many individuals who have directly and indirectly contributed to the development of this book. I am grateful to my professors, colleagues, and my former and present students who have presented me with consulting problems over the years that have stimulated me to develop this book and the accompanying SAS macros. I would also like to thank the University of Nevada–Reno and the Center for Research Design and Analysis faculty and staff for their support during the time I spent on writing the book and in revising the SAS macros.

I have received constructive comments about this book from many CRC Press anonymous reviewers, whose advice has greatly improved this edition. I would like to acknowledge the contribution of the CRC Press staff from the conception to the completion of this book. I would also like to thank the SAS Institute for providing me with an opportunity to continuously learn about this powerful software for the past 23 years and allowing me to share my SAS knowledge with other users.

I owe a great debt of gratitude to my family for their love and support as well as their great sacrifice during the last 12 months while I was working on this book. I cannot forget to thank my late dad, Pancras Fernandez, and my late grandpa, George Fernandez, for their love and support, which helped me to take challenging projects and succeed. Finally, I would like to thank the most important person in my life, my wife Queency Fernandez, for her love, support, and encouragement that gave me the strength to complete this book project within the deadline.

George Fernandez
University of Nevada-Reno
gcjf@unr.edu

About the Author

George Fernandez, Ph.D., is a professor of applied statistical methods and serves as the director of the Reno Center for Research Design and Analysis, University of Nevada. His publications include an applied statistics book, a CD-Rom, 60 journal papers, and more than 30 conference proceedings. Dr. Fernandez has more than 23 years of experience teaching applied statistics courses and SAS programming.

He has won several best-paper and poster presentation awards at regional and international conferences. He has presented several invited full-day workshops on applications of user-friendly statistical methods in data mining for the American Statistical Association, including the joint meeting in Atlanta (2001); Western SAS* users conference in Arizona (2000), in San Diego (2002) and San Jose (2005); and at the 56th Deming's conference, Atlantic City (2003). He was keynote speaker and workshop presenter for the 16th Conference on Applied Statistics, Kansas State University, and full-day workshop presenter at the 57th session of the International Statistical Institute conference at Durbin, South Africa (2009). His recent paper, "A new and simpler way to calculate body's Maximum Weight Limit–BMI made simple," has received worldwide recognition.

---

* This was originally an acronym for statistical analysis system. Since its founding and adoption of the term as its trade name, the SAS Institute, headquartered in North Carolina, has considerably broadened its scope.

Chapter 1

Data Mining: A Gentle Introduction

1.1 Introduction

Data mining, or knowledge discovery in databases (KDD), is a powerful information technology tool with great potential for extracting previously unknown and potentially useful information from large databases. Data mining automates the process of finding relationships and patterns in raw data and delivers results that can either be utilized in an automated decision support system or assessed by decision makers. Many successful enterprises practice data mining for intelligent decision making.[1] Data mining allows the extraction of nuggets of knowledge from business data that can help enhance customer relationship management (CRM)[2] and can help estimate the return on investment (ROI).[3] Using powerful advanced analytical techniques, data mining enables institutions to turn raw data into valuable information and thus gain a critical competitive advantage.

With data mining, the possibilities are endless. Although data mining applications are popular among forward-thinking businesses, other disciplines that maintain large databases could reap the same benefits from properly carried out data mining. Some of the potential applications of data mining include characterizations of genes in animal and plant genomics, clustering and segmentations in remote sensing of satellite image data, and predictive modeling in wildfire incidence databases.

The purpose of this chapter is to introduce data mining concepts, provide some examples of data mining applications, list the most commonly used data mining techniques, and briefly discuss the data mining applications available in the

SAS software. For a thorough discussion of data mining concept, methods, and applications, see the following publications.[4-6]

1.2 Data Mining: Why it is Successful in the IT World

In today's world, we are overwhelmed with data and information from various sources. Advances in the field of IT make the collection of data easier than ever before. A business enterprise has various systems such as transaction processing system, HR management system, accounting system, and so on, and each of these systems collects huge piles of data everyday. Data mining is an important part of business intelligence that deals with how an organization uses, analyzes, manages, and stores data it collects from various sources to make better decisions. Businesses that have already invested in business intelligence solutions will be in a better position to undertake right measures to survive and continue its growth. Data mining solutions provide an analytical insight into the performance of an organization based on historical data, but the economic impact on an organization is linked to many issues and, in many cases, to external forces and unscrupulous activities. The failure to predict this does not undermine the role of data mining for organizations, but on the contrary, makes it more important, especially for regulatory bodies of governments, to predict and identify such practices in advance and take necessary measures to avoid such circumstances in future. The main components of data mining success are described in the following subsections.

1.2.1 Availability of Large Databases: Data Warehousing

Data mining derives its name from the fact that analysts search for valuable information in gigabytes of huge databases. For the past two decades, we have seen a dramatic increase—at an explosive rate—in the amount of data being stored in electronic format. The increase in the use of electronic data-gathering devices such as point-of-sale, Web logging, or remote sensing devices has contributed to this explosion of available data. The amount of data accumulated each day by various businesses and scientific and governmental organizations around the world is daunting. With data warehousing, business enterprises can collect data from any source within or outside the organization, reorganize the data, and place it in new dynamic storage for efficient utilization. Business enterprises of all kinds now computerize all their business activities and their abilities to manage their valuable data resources. One hundred gigabytes of databases are now common, and terabyte (1000 GB) databases are now feasible in enterprises. Data warehousing techniques enable forward-thinking businesses to collect, save, maintain, and retrieve data in a more productive way.

Data warehousing (DW) collects data from many different sources, reorganizes it, and stores it within a readily accessible repository that DW should support relational, hierarchical, and multidimensional database management systems, and is designed specifically to meet the needs of data mining. A DW can be loosely defined as any centralized data repository that makes it possible to extract archived operational data and overcome inconsistencies between different data formats. Thus, data mining and knowledge discovery from large databases become feasible and productive with the development of cost-effective data warehousing.

A successful data warehousing operation should have the potential to integrate data wherever it is located and whatever its format. It should provide the business analyst with the ability to quickly and effectively extract data tables, resolve data quality problems, and integrate data from different sources. If the quality of the data is questionable, then business users and decision makers cannot trust the results. In order to fully utilize data sources, data warehousing should allow you to make use of your current hardware investments, as well as provide options for growth as your storage needs expand. Data warehousing systems should not limit customer choices, but instead should provide a flexible architecture that accommodates platform-independent storage and distributed processing options.

Data quality is a critical factor for the success of data warehousing projects. If business data is of an inferior quality, then the business analysts who query the database and the decision makers who receive the information cannot trust the results. The quality of individual records is necessary to ensure that the data is accurate, updated, and consistently represented in the data warehousing.

1.2.2 Price Drop in Data Storage and Efficient Computer Processing

Data warehousing became easier, more efficient, and cost-effective as the cost of data processing and database development dropped. The need for improved and effective computer processing can now be met in a cost-effective manner with parallel multiprocessor computer technology. In addition to the recent enhancement of exploratory graphical statistical methods, the introduction of new machine-learning methods based on logic programming, artificial intelligence, and genetic algorithms have opened the doors for productive data mining. When data mining tools are implemented on high-performance parallel-processing systems, they can analyze massive databases in minutes. Faster processing means that users can automatically experiment with more models to understand complex data. High speed makes it practical for users to analyze huge quantities of data.

1.2.3 New Advancements in Analytical Methodology

Data mining algorithms embody techniques that have existed for at least 10 years, but have only recently been implemented as mature, reliable, understandable tools that consistently outperform older methods. Advanced analytical models and algorithms, including data visualization and exploration, segmentation and clustering, decision trees, neural networks, memory-based reasoning, and market basket analysis, provide superior analytical depth. Thus, quality data mining is now feasible with the availability of advanced analytical solutions.

1.3 Benefits of Data Mining

For businesses that use data mining effectively, the payoffs can be huge. By applying data mining effectively, businesses can fully utilize data about customers' buying patterns and behavior, and can gain a greater understanding of customers' motivations to help reduce fraud, forecast resource use, increase customer acquisition, and halt customer attrition. After a successful implementation of data mining, one can sweep through databases and identify previously hidden patterns in one step. An example of pattern discovery is the analysis of retail sales data to identify seemingly unrelated products that are often purchased together. Other pattern discovery problems include detecting fraudulent credit card transactions and identifying anomalous data that could represent data entry keying errors. Some of the specific benefits associated with successful data mining are listed here:

- Increase customer acquisition and retention.
- Uncover and reduce frauds (determining if a particular transaction is out of the normal range of a person's activity and flagging that transaction for verification).
- Improve production quality, and minimize production losses in manufacturing.
- Increase *upselling* (offering customers a higher level of services or products such as a gold credit card versus a regular credit card) and *cross-selling* (selling customers more products based on what they have already bought).
- Sell products and services in combinations based on *market-basket analysis* (by determining what combinations of products are purchased at a given time).

1.4 Data Mining: Users

A wide range of companies have deployed successful data mining applications recently.[1] While the early adopters of data mining belong mainly to information-intensive industries such as financial services and direct mail marketing, the technology is applicable to any institution looking to leverage a large data warehouse to extract information that can be used in intelligent decision making. Data mining applications reach across industries and business functions. For example, telecommunications, stock exchanges, credit card, and insurance companies use data mining to detect fraudulent use of their services; the medical industry uses data mining to predict the effectiveness of surgical procedures, diagnostic medical tests, and medications; and retailers use data mining to assess the effectiveness of discount coupons and sales' promotions. Data mining has many varied fields of application, some of which are listed as follows:

- *Retail/Marketing*: An example of pattern discovery in retail sales is to identify seemingly unrelated products that are often purchased together. Market-basket analysis is an algorithm that examines a long list of transactions in order to determine which items are most frequently purchased together. The results can be useful to any company that sells products, whether it is in a store, a catalog, or directly to the customer.
- *Banking*: A credit card company can leverage its customer transaction database to identify customers most likely to be interested in a new credit product. Using a small test mailing, the characteristics of customers with an affinity for the product can be identified. Data mining tools can also be used to detect patterns of fraudulent credit card use, including detecting fraudulent credit card transactions and identifying anomalous data that could represent data entry keying errors. It identifies "loyal" customers, predicts customers likely to change their credit card affiliation, determines credit card spending by customer groups, finds hidden correlations between different financial indicators, and can identify stock trading rules from historical market data. It also finds hidden correlations between different financial indicators and identifies stock trading rules from historical market data.
- *Insurance and health care*: It claims analysis—that is, which medical procedures are claimed together. It predicts which customers will buy new policies, identifies behavior patterns of risky customers, and identifies fraudulent behavior.
- *Transportation*: State departments of transportation and federal highway institutes can develop performance and network optimization models to predict the life-cycle cost of road pavement.
- *Product manufacturing companies*: They can apply data mining to improve their sales process to retailers. Data from consumer panels, shipments, and competitor activity can be applied to understand the reasons for brand and store switching. Through this analysis, manufacturers can select promotional strategies that best reach their target customer segments. The distribution schedules among outlets can be determined, loading patterns can be analyzed, and the distribution schedules among outlets can be determined.
- *Health care and pharmaceutical industries*: Pharmaceutical companies can analyze their recent sales records to improve their targeting of high-value physicians and determine which marketing activities will have the greatest impact in the next few months. The ongoing, dynamic analysis of the data warehouse allows the best practices from throughout the organization to be applied in specific sales situations.
- *Internal Revenue Service (IRS) and Federal Bureau of Investigation (FBI)*: The IRS uses data mining to track federal income tax frauds. The FBI uses data mining to detect any unusual pattern or trends in thousands of field reports to look for any leads in terrorist activities.

1.5 Data Mining: Tools

All data mining methods used now have evolved from the advances in computer engineering, statistical computation, and database research. Data mining methods are not considered to replace traditional statistical methods but extend the use of statistical and graphical techniques. Once it was thought that automated data mining tools would eliminate the need for statistical analysts to build predictive models. However, the value that an analyst provides cannot be automated out of existence. Analysts will still be needed to assess model results and validate the plausibility of the model predictions. Since data mining software lacks the human experience and intuition to recognize the difference between a relevant and irrelevant correlation, statistical analysts will remain in great demand.

1.6 Data Mining: Steps

1.6.1 Identification of Problem and Defining the Data Mining Study Goal

One of the main causes of data mining failure is not defining the study goals based on short- and long-term problems facing the enterprise. The data mining specialist should define the study goal in clear and sensible terms of what the enterprise hopes to achieve and how data mining can help. Well-identified study problems lead to formulated data mining goals, and data mining solutions geared toward measurable outcomes.[4]

1.6.2 Data Processing

The key to successful data mining is using the right data. Preparing data for mining is often the most time-consuming aspect of any data mining endeavor. A typical data structure suitable for data mining should contain observations (e.g., customers and products) in rows and variables (demographic data and sales history) in columns. Also, the measurement levels (interval or categorical) of each variable in the dataset should be clearly defined. The steps involved in preparing the data for data mining are as follows:

> *Preprocessing*: This is the data-cleansing stage, where certain information that is deemed unnecessary and may slow down queries is removed. Also, the data is checked to ensure that a consistent format (different types of formats used in dates, zip codes, currency, units of measurements, etc.) exists. There is always the possibility of having inconsistent formats in the database because the data is drawn from several sources. Data entry errors and extreme outliers should be removed from the dataset since influential outliers can affect the modeling results and subsequently limit the usability of the predicted models.

Data integration: Combining variables from many different data sources is an essential step since some of the most important variables are stored in different data marts (customer demographics, purchase data, and business transaction). The uniformity in variable coding and the scale of measurements should be verified before combining different variables and observations from different data marts.

Variable transformation: Sometimes, expressing continuous variables in standardized units, or in log or square-root scale, is necessary to improve the model fit that leads to improved precision in the fitted models. Missing value imputation is necessary if some important variables have large proportions of missing values in the dataset. Identifying the response (target) and the predictor (input) variables and defining their scale of measurement are important steps in data preparation since the type of modeling is determined by the characteristics of the response and the predictor variables.

Splitting database: Sampling is recommended in extremely large databases because it significantly reduces the model training time. Randomly splitting the data into "training," "validation," and "testing" is very important in calibrating the model fit and validating the model results. Trends and patterns observed in the training dataset can be expected to generalize the complete database if the training sample used sufficiently represents the database.

1.6.3 Data Exploration and Descriptive Analysis

Data exploration includes a set of descriptive and graphical tools that allow exploration of data visually both as a prerequisite to more formal data analysis and as an integral part of formal model building. It facilitates discovering the unexpected as well as confirming the expected. The purpose of data visualization is pretty simple: let the user understand the structure and dimension of the complex data matrix. Since data mining usually involves extracting "hidden" information from a database, the understanding process can get a bit complicated. The key is to put users in a context they feel comfortable in, and then let them poke and prod until they understand what they did not see before. Understanding is undoubtedly the most fundamental motivation to visualizing the model.

Simple descriptive statistics and exploratory graphics displaying the distribution pattern and the presence of outliers are useful in exploring continuous variables. Descriptive statistical measures such as the mean, median, range, and standard deviation of continuous variables provide information regarding their distributional properties and the presence of outliers. Frequency histograms display the distributional properties of the continuous variable. Box plots provide an excellent visual summary of many important aspects of a distribution. The box plot is based on the 5-number summary plot that is based on the median, quartiles, and extreme values. One-way and multiway frequency tables of categorical data are useful in summarizing group distributions, relationships between groups, and checking for rare events. Bar charts show frequency information for categorical variables and display differences among the different groups in them. Pie charts compare the levels or classes of a categorical variable to each other and to the whole. They use the size of pie slices to graphically represent the value of a statistic for a data range.

1.6.4 Data Mining Solutions: Unsupervised Learning Methods

Unsupervised learning methods are used in many fields under a wide variety of names. No distinction between the response and predictor variable is made in unsupervised learning methods. The most commonly practiced unsupervised methods are latent variable models (principal component and factor analyses), disjoint cluster analyses, and market-basket analysis.

- *Principal component analysis (PCA)*: In PCA, the dimensionality of multivariate data is reduced by transforming the correlated variables into linearly transformed uncorrelated variables.
- *Factor analysis (FA)*: In FA, a few uncorrelated hidden factors that explain the maximum amount of common variance and are responsible for the observed correlation among the multivariate data are extracted.
- *Disjoint cluster analysis (DCA)*: It is used for combining cases into groups or clusters such that each group or cluster is homogeneous with respect to certain attributes.
- *Association and market-basket analysis*: Market-basket analysis is one of the most common and useful types of data analysis for marketing. Its purpose is to determine what products customers purchase together. Knowing what products consumers purchase as a group can be very helpful to a retailer or to any other company.

1.6.5 Data Mining Solutions: Supervised Learning Methods

The supervised predictive models include both classification and regression models. Classification models use categorical response, whereas regression models use continuous and binary variables as targets. In regression, we want to approximate the regression function, while in classification problems, we want to approximate the probability of class membership as a function of the input variables. Predictive modeling is a fundamental data mining task. It is an approach that reads training data composed of multiple input variables and a target variable. It then builds a model that attempts to predict the target on the basis of the inputs. After this model is developed, it can be applied to new data that is similar to the training data, but that does not contain the target.

- *Multiple linear regressions* (*MLRs*): In MLR, the association between the two sets of variables is described by a linear equation that predicts the continuous response variable from a function of predictor variables.
- *Logistic regressions:* It allows a binary or an ordinal variable as the response variable and allows the construction of more complex models rather than straight linear models.
- *Neural net* (*NN*) *modeling*: It can be used for both prediction and classification. NN models enable the construction of train and validate multiplayer feed-forward network models for modeling large data and complex interactions with many predictor variables. NN models usually contain more parameters than a typical statistical model, and the results are not easily interpreted and no explicit rationale is given for the prediction. All variables are treated as numeric, and all nominal variables are coded as binary. Relatively more training time is needed to fit the NN models.
- *Classification and regression tree* (*CART*): These models are useful in generating binary decision trees by splitting the subsets of the dataset using all predictor variables to create two child nodes repeatedly, beginning with the entire dataset. The goal is to produce subsets of the data that are as homogeneous as possible with respect to the target variable. Continuous, binary, and categorical variables can be used as response variables in CART.
- *Discriminant function analysis*: This is a classification method used to determine which predictor variables discriminate between two or more naturally occurring groups. Only categorical variables are allowed to be the response variable, and both continuous and ordinal variables can be used as predictors.
- *CHAID decision tree* (*Chi-square Automatic Interaction Detector*): This is a classification method used to study the relationships between a categorical response measure and a large series of possible predictor variables, which may interact among one another. For qualitative predictor variables, a series of chi-square analyses are conducted between the response and predictor variables to see if splitting the sample based on these predictors leads to a statistically significant discrimination in the response.

1.6.6 Model Validation

Validating models obtained from training datasets by independent validation datasets is an important requirement in data mining to confirm the usability of the developed model. Model validation assess the quality of the model fit and protect against overfitted or underfitted models. Thus, it could be considered as the most important step in the model-building sequence.

1.6.7 Interpret and Make Decisions

Decision making is one of the most critical steps for any successful business. No matter how good you are at making decisions, you know that making an intelligent decision is difficult. The patterns identified by the data mining solutions can be interpreted into knowledge, which can then be used to support business decision making.

1.7 Problems in the Data Mining Process

Many of the so-called data mining solutions currently available on the market today either do not integrate well, are not scalable, or are limited to one or two modeling techniques or algorithms. As a result, highly trained quantitative experts spend more time trying to access, prepare, and manipulate data from disparate sources, and less time modeling data and applying their expertise to solve business problems. And the data mining challenge is compounded even further as the amount of data and complexity of the business problems increase. It is usual for the database to often be designed for purposes different from data mining, so properties or attributes that would simplify the learning task are not present, nor can they be requested from the real world.

Data mining solutions rely on databases to provide the raw data for modeling, and this raises problems in that databases tend to be dynamic, incomplete, noisy, and large. Other problems arise as a result of the adequacy and relevance of the information stored. Databases are usually contaminated by errors, so it cannot be assumed that the data they contain is entirely correct. Attributes, which rely on subjective or measurement judgments, can give rise to errors in such a way that some examples may even be misclassified. Errors in either the values of attributes or class information are known as *noise*. Obviously, where possible, it is desirable to eliminate noise from the classification information as this affects the overall accuracy of the generated rules. Therefore, adopting a software system that provides a complete data mining solution is crucial in the competitive environment.

1.8 SAS Software the Leader in Data Mining

SAS Institute,[7] the industry leader in analytical and decision-support solutions, offers a comprehensive data mining solution that allows you to explore large quantities of data and discover relationships and patterns that lead to proactive decision making. The SAS data mining solution provides business technologists and quantitative experts the necessary tools to obtain the enterprise knowledge for helping their organizations to achieve a competitive advantage.

1.8.1 SEMMA: The SAS Data Mining Process

The SAS data mining solution is considered a process rather than a set of analytical tools. The acronym SEMMA[8] refers to a methodology that clarifies this process. Beginning with a statistically representative sample of your data, SEMMA makes it easy to apply exploratory statistical and visualization techniques, select and transform the most significant predictive variables, model the variables to predict outcomes, and confirm a model's accuracy. The steps in the SEMMA process include the following:

Sample your data by extracting a portion of a large dataset big enough to contain the significant information, and yet small enough to manipulate quickly.

Explore your data by searching for unanticipated trends and anomalies in order to gain understanding and ideas.

Modify your data by creating, selecting, and transforming the variables to focus on the model selection process.

Model your data by allowing the software to search automatically for a combination of data that reliably predicts a desired outcome.

Assess your data by evaluating the usefulness and reliability of the findings from the data mining process.

By assessing the results gained from each stage of the SEMMA process, you can determine how to model new questions raised by the previous results, and thus proceed back to the exploration phase for additional refinement of the data. The SAS data mining solution integrates everything you need for discovery at each stage of the SEMMA process: These data mining tools indicate patterns or exceptions and mimic human abilities for comprehending spatial, geographical, and visual information sources. Complex mining techniques are carried out in a totally code-free environment, allowing you to concentrate on the visualization of the data, discovery of new patterns, and new questions to ask.

1.8.2 SAS Enterprise Miner for Comprehensive Data Mining Solution

Enterprise Miner,[9,10] SAS Institute's enhanced data mining software, offers an integrated environment for businesses that need to conduct comprehensive data mining. Enterprise Miner combines a rich suite of integrated data mining tools, empowering users to explore and exploit huge databases for strategic business advantages. In a single environment, Enterprise Miner provides all the tools needed to match robust data mining techniques to specific business problems, regardless of the amount or source of data, or complexity of the business problem. However, many small business, nonprofit institutions, and academic universities are still currently not using the SAS Enterprise Miner, but they are licensed to use SAS BASE, STAT, and GRAPH modules. Thus, these user-friendly SAS macro applications for data mining are targeted at this group of customers. Also, providing the complete SAS codes for performing comprehensive data mining solutions is not very effective because a majority of the business and statistical analysts are not experienced SAS programmers. Quick results from data mining are not feasible since many hours of code modification and debugging program errors are required if the analysts are required to work with SAS program code.

1.9 Introduction of User-Friendly SAS Macros for Statistical Data Mining

As an alternative to the point-and-click menu interface modules, the user-friendly SAS macro applications for performing several data mining tasks are included in this book. This macro approach integrates the statistical and graphical tools available in SAS systems and provides user-friendly data analysis tools that allow the data analysts to complete data mining tasks quickly without writing SAS programs by running the SAS macros in the background. Detailed instructions and help files for using the SAS macros are included in each chapter. Using this macro approach, analysts can effectively and quickly perform complete data analysis and spend more time exploring data and interpreting graphs and output rather than debugging their program errors, etc. The main advantages of using these SAS macros for data mining are as follows:

- Users can perform comprehensive data mining tasks by inputting the macro parameters in the macro-call window and by running the SAS macro.
- SAS code required for performing data exploration, model fitting, model assessment, validation, prediction, and scoring are included in each macro. Thus, complete results can be obtained quickly by using these macros.
- Experience in SAS output delivery system (ODS) is not required because options for producing SAS output and graphics in RTF, WEB, and PDF are included within the macros.
- Experience in writing SAS programs code or SAS macros is not required to use these macros.
- SAS-enhanced data mining software *Enterprise Miner* is not required to run these SAS macros.
- All SAS macros included in this book use the same simple user-friendly format. Thus, minimum training time is needed to master the usage of these macros.
- Regular updates to the SAS macros will be posted in the book Web site. Thus, readers can always use the updated features in the SAS macros by downloading the latest versions.

1.9.1 Limitations of These SAS Macros

These SAS macros do not use SAS Enterprise Miner. Thus, SAS macros are not included for performing neural net, CART, and market-basket analysis since these data mining tools require the SAS special data mining software SAS Enterprise Miner.

1.10 Summary

Data mining is a journey—a continuous effort to combine your enterprise knowledge with the information you extracted from the data you have acquired. This chapter briefly introduces the concept and applications of data mining techniques; that is, the secret and intelligent weapon that unleashes the power in your data. The SAS institute, the industry leader in analytical and decision support solutions, provides the powerful software called *Enterprise Miner* to perform complete data mining solutions. However, many small business and academic institutions do not have the license to use the application, but they have the license for SAS BASE, STAT, and GRAPH. As an alternative to the point-and-click menu interface modules, user-friendly SAS macro applications for performing several statistical data mining tasks are included in this book. Instructions are given in the book for downloading and applying these user-friendly SAS macros for producing quick and complete data mining solutions.

References

1. SAS Institute Inc., Customer success stories at http://www.sas.com/success/ (last accessed 10/07/09).
2. SAS Institute Inc., Customer relationship management (CRM) at http://www.sas.com/solutions/crm/index.html (last accessed 10/07/09).
3. SAS Institute Inc., SAS Enterprise miner product review at http://www.sas.com/products/miner/miner_review.pdf (last accessed 10/07/09).
4. Two Crows Corporation, *Introduction to Data Mining and Knowledge Discovery*, 3rd ed., 1999 at http://www.twocrows.com/intro-dm.pdf.
5. Berry, M. J. A. and Linoff, G. S. *Data Mining Techniques: For Marketing, Sales, and Customer Support*, John Wiley & Sons, New York, 1997.
6. Berry, M. J. A. and Linoff, G. S., *Mastering Data Mining: The Art and Science of Customer Relationship Management*, Second edition, John Wiley & Sons, New York, 1999.
7. SAS Institute Inc., *The Power to Know* at http://www.sas.com.
8. SAS Institute Inc., *Data Mining Using Enterprise Miner Software: A Case Study Approach*, 1st ed., Cary, NC, 2000.
9. SAS Institute Inc., The Enterprise miner, http://www.sas.com/products/miner/index.html (last accessed 10/07/09).
10. SAS Institute Inc., The Enterprise miner standalone tutorial, http://www.cabnr.unr.edu/gf/dm/em.pdf (last accessed 10/07/09).

Chapter 2

Preparing Data for Data Mining

2.1 Introduction

Data is the backbone of data mining and knowledge discovery. However, real-world data is usually not available in data-mining-ready form. Therefore, the biggest challenge for data miners is preparing data suitable for modeling. Many enterprises maintain central data storage and access facilities called *data warehouses*. Data warehousing is defined as a process of centralized data management and allows analysts to access, update, and maintain the data for analysis and reporting. Thus, data warehouse technology improves the efficiency of extracting and preparing data for data mining. Popular data warehouses use relational databases (e.g., Oracle, Informix, and Sybase) and PC data formats (spreadsheets and MS Access). Roughly 70% of data mining operation time is spent on preparing the data obtained from different sources.[1] Therefore, careful consideration in time and effort should be given to preparing data tables suitable for data mining projects.

2.2 Data Requirements in Data Mining

Summarized data is not suitable for data mining, because information about individual records or products is not available. For example, to identify profitable customers, individual customer records, including demographic information, are necessary to profile or cluster customers based on their purchasing patterns. Similarly, to identify the characteristics of profitable customers in a predictive modeling target (outcome or response) and input (predictor), variables should be included. Therefore, for solving specific data mining objectives, extract suitable data from data warehouses or collect new data that meets the data mining requirements.

2.3 Ideal Structures of Data for Data Mining

The rows (observations or cases) and columns (variables) format, similar to a spreadsheet worksheet file, is required for data mining. The rows usually contain information regarding individual records or consumer products. The columns describe the attributes (variables) of individual cases. The scales of the variables can be continuous or categorical. Total sales or product, number of units purchased by each customer, and annual income per customer are some examples of continuous variables. Gender, race, and age group are considered categorical variables. Knowledge of the possible maximum and minimum values for the continuous variables can help to identify and exclude extreme outliers from the data. Similarly, knowledge of the possible levels for categorical variables can help to detect data entry errors and anomalies in the data.

Constant values in continuous (e.g., zip code) or categorical fields (state code) should not be included in any predictive or descriptive data mining modeling, since these values are unique for each case and do not help to discriminate or group individual cases. Similarly, unique information about customers such as phone numbers and social security numbers should also be excluded from predictive data mining. However, these unique value variables can be used as ID variables to identify individual cases and to exclude extreme outliers. Also, avoid including highly correlated (correlation coefficient > 0.95) continuous predictor variables in predictive modeling since they can produce unstable predictive models that work only with the sample used.

2.4 Understanding the Measurement Scale of Variables

The measurement scale of the target and input variables determines the type of modeling technique that is appropriate for a specific data mining project. Therefore, understanding the nature of the measurement scale of variables used in modeling is an important data mining requirement. The variables can be generally classified into continuous or categorical:

A *Continuous* variable is a numeric variable that describes a quantitative attribute of the cases and has a continuous scale of measurement. Means and standard deviations are commonly used to quantify the central tendency and dispersion. Total sales per customer and total manufacturing costs per product are examples of interval scales. Interval scale target variable is a requirement for multiple regression and neural net modeling.

Categorical variables can be further classified into

Nominal—A categorical variable with more than two levels. Mode is the pre-ferred estimate for measuring central tendency, and frequency analysis is the common form of descriptive technique. Different kinds of accounts in banking, telecommunication services, and insurance policies are some examples for a nominal variable. Discriminant analysis and decision tree methods are suitable for modeling nominal target variables.

Binary—A categorical variable with only two levels. Sale versus no sale and good versus bad credit are some examples of a binary variable. Logistic regression is suitable for modeling binary target-dependent variables.

Ordinal—A categorical or discrete rank variable with more than two levels. Ordinal logistic regression is suitable for modeling ordinal-dependent variables.

2.5 Entire Database or Representative Sample

To find trends and patterns in data, data miners can use the entire database or randomly selected sample from it. Although using the entire database is currently feasible with the current high-powered computing environment, using randomly selected representative samples in model building is more attractive for the following reasons:

Using random samples allows the modeler to develop the model from the *training* or *calibration* sample, validate it with a holdout *validation* dataset, and test it with another independent *test sample*.

Mining a representative random sample is easier, more efficient, and can produce accurate results similar to those produced when using the entire database.

When samples are used, data exploration and visualization help to gain insights, leading to faster and more accurate models.

Representative samples require a relatively shorter time to cleanse, explore, develop, and validate models for, and are therefore more cost-effective than using entire databases.

2.6 Sampling for Data Mining

The sample used in modeling should represent the entire database since the main goal in data mining is to make predictions about the entire database. The size and other characteristics of the selected sample determine whether the sample used in modeling is a good representation of the entire database. The following types of sampling are commonly practiced in data mining[1]:

Simple random sampling: The most common sampling method in data mining. Each observation or case in the database has an equal chance of being included in the sample.

Cluster sampling: The database is divided into clusters at the first stage of sample selection and a predetermined number of clusters are randomly selected based on simple random sampling. All the records from those randomly selected clusters are included in the study.

Stratified random sampling: The database is divided into mutually exclusive strata or subpopulations. Random samples are then taken from each stratum proportional to its population size.

2.6.1 Sample Size

The number of input variables, the functional form of the model (linear, nonlinear, models with interactions, etc.), and the size of the databases can influence the sample size requirement in data mining. By default, the SAS Enterprise Miner software takes a simple random sample of 2000 cases from the data table and divides it into training (40%), validation (30%), and test (30%) datasets.[2] If the number of cases is less than 2000, the whole database is used in the model building. Data analysts can use these sampling proportions as a guideline in determining the sample sizes. However, depending on the data mining objectives and the nature of the database, data miners can modify sample-size proportions.

2.7 User-Friendly SAS Applications Used in Data Preparation

SAS software has many powerful features available for extracting data from different DBMSs. Some of these features are described in the following section. The readers are expected to have basic knowledge in using SAS to perform the following operations. The *Little SAS Book*[3] can be used as an introductory SAS guide to become familiar with SAS systems and SAS programming.

2.7.1 Preparing PC Data Files before Importing into SAS Data

MS Excel, Access, dBase, and tab- and comma-delimited are some of the popular PC data file formats used in data mining. These file types can be easily converted to SAS datasets by using the PROC ACCESS or PROC IMPORT procedures in SAS. In addition, in SAS version 9.2, the PROC IMPORT wizard is also capable of importing JMP (jmp), SPSS (sav), and Stata (dta) files[4] (Figure 2.1). A graphical user interface (GUI)-based IMPORT WIZARD is also available in SAS to convert a single PC file type to SAS dataset. However, before converting the PC file types, the following points should be considered:

■ The maximum number of rows and columns allowed in a EXCEL 2003 worksheet is 65536 × 246. Although Excel 2007 can store 1 million rows and 16,000

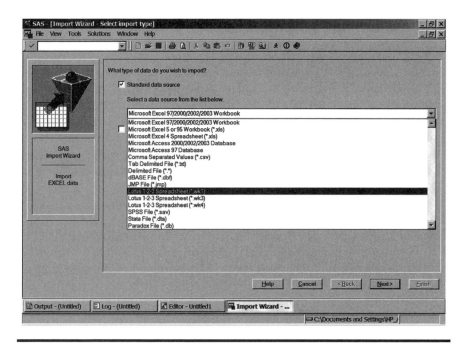

Figure 2.1 **Screen copy of SAS IMPORT (Version 9.2) showing the valid PC file types that can be imported to SAS datasets.**

columns, importing Excel 2007 files is not implemented in SAS 9.2 (phase 1). However, this feature will become available in future releases of SAS.

- Check whether the first row of your worksheet contains the names of the variables stored in the columns. Select names that are valid SAS variable names (one word). Also, remove any blank rows in the worksheet.
- Save only one data table per worksheet if you are going to use the EXLSAS2 macro. Name your data table to "Sheet1" if you are importing an MS Access table.
- SAS cannot read the worksheet file that you have currently opened in Excel. This will cause a sharing violation error. Be sure to close your Excel file before trying to convert it in SAS.
- If you want to create a permanent SAS data file, you could assign a LIBNAME before you import the PC file into a SAS dataset. For information on the LIBNAME statement and making permanent SAS data files, refer to *The Little SAS Book*.[3]
- Make sure that each column in a worksheet contains either numeric or character variables. Do not mix numeric and character values in the same column. The results of most Excel formulas should import into SAS without a problem.

2.7.2 Converting PC Data Files to SAS Datasets Using the SAS Import Wizard

The SAS Import Wizard available in the SAS/ACCESS module can be used to import or export Excel 4, 5, 7, 95, 98, 2000, and 2003 files, as well as Microsoft Access files (2000, 2002, and 2003). The GUIs in the Import Wizard lead through menus and provide step-by-step instructions for transferring data between external data sources and SAS datasets.[5] The types of files that can be imported depend on your operating system and the SAS/ACCESS engines you have installed. The following steps are involved in using the Import Wizard for importing a PC file:

- *Step 1. Selecting the PC file type*: The Import Wizard can be activated by using the pull-down menu to select FILE and then clicking IMPORT. For a list of available data sources to choose from, click the drop-down arrow (Figure 2.1). Select the file format in which your data is stored. To read an Excel file, click the black triangle, and choose the type of Excel file (4.0, 5.0, 7.0, 95, 97, and 2000–2003 spreadsheets). You can also select other PC file types such as MS Access (2000–2003 tables), dBASE (5.0, IV, III+, and III files), and text files such as tab-delimited and comma-separated files. After selecting the file type, click the NEXT button to continue.
- *Step 2. Selecting the PC file location*: In the Import Wizard, select file window, type the full path for your file, or click BROWSE to find your file. Then click the NEXT button to go to the next screen. On the second screen after you choose the Excel file, the OPTIONS button becomes active. The OPTIONS button allows you to choose which worksheet you want to read (if your file has multiple sheets), to specify whether the first row of the spreadsheet contains the variable names, and to choose the range of the worksheet that you want to read. Generally, you can ignore these options.
- *Step 3. Selecting the temporary or permanent SAS dataset name*: The third screen prompts for the SAS data file name. Select the LIBRARY (the alias name for the folder) and member (SAS dataset name) you want for your SAS data file. For example, to create temporary data file called "fraud," choose WORK for the LIBRARY and "fraud" as a valid SAS dataset name for the member. When you are ready, click FINISH, and SAS will convert the Excel spreadsheet that you chose into a SAS data file (Figure 2.2).
- *Step 4. Final check*: Check the LOG window for a message indicating that SAS has successfully converted the Excel file to a SAS dataset. Also, compare the number of observations and variables in the SAS dataset with the source Excel file to make sure that the imported SAS file is identical to the source PC data file.

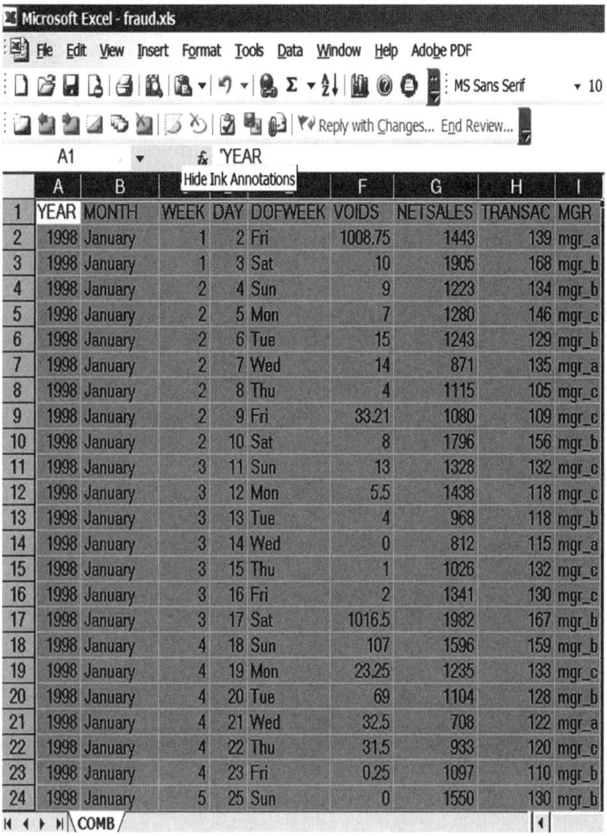

	A	B	C	D	E	F	G	H	I
1	YEAR	MONTH	WEEK	DAY	DOFWEEK	VOIDS	NETSALES	TRANSAC	MGR
2	1998	January	1	2	Fri	1008.75	1443	139	mgr_a
3	1998	January	1	3	Sat	10	1905	168	mgr_b
4	1998	January	2	4	Sun	9	1223	134	mgr_b
5	1998	January	2	5	Mon	7	1280	146	mgr_c
6	1998	January	2	6	Tue	15	1243	129	mgr_b
7	1998	January	2	7	Wed	14	871	135	mgr_a
8	1998	January	2	8	Thu	4	1115	105	mgr_c
9	1998	January	2	9	Fri	33.21	1080	109	mgr_c
10	1998	January	2	10	Sat	8	1796	156	mgr_b
11	1998	January	3	11	Sun	13	1328	132	mgr_c
12	1998	January	3	12	Mon	5.5	1438	118	mgr_c
13	1998	January	3	13	Tue	4	968	118	mgr_b
14	1998	January	3	14	Wed	0	812	115	mgr_a
15	1998	January	3	15	Thu	1	1026	132	mgr_c
16	1998	January	3	16	Fri	2	1341	130	mgr_c
17	1998	January	3	17	Sat	1016.5	1982	167	mgr_b
18	1998	January	4	18	Sun	107	1596	159	mgr_b
19	1998	January	4	19	Mon	23.25	1235	133	mgr_c
20	1998	January	4	20	Tue	69	1104	128	mgr_b
21	1998	January	4	21	Wed	32.5	708	122	mgr_a
22	1998	January	4	22	Thu	31.5	933	120	mgr_c
23	1998	January	4	23	Fri	0.25	1097	110	mgr_b
24	1998	January	5	25	Sun	0	1550	130	mgr_b

Figure 2.2 Screen copy of MS Excel 2003 worksheet fraud.xls opened in Office 2003 showing the required structure of the PC spreadsheet.

2.7.3 EXLSAS2 SAS Macro Application to Convert PC Data Formats to SAS Datasets

The EXLSAS2 macro application can be used as an alternative to the SAS Import Wizard to convert PC file types to SAS datasets. The SAS procedure PROC IMPORT is the main tool if the EXLSAS2 macro is used with the post-SAS version 9.0. PROC IMPORT can import a wide variety of types and versions of PC files. The enhanced features included in the EXLSAS2 SAS macro application are the following:

- Multiple PC files can be converted in single operations.
- *New*: Outputting the macro input values specified by the user in the last run of EXLSAS2 macro submission.

- A sample printout of the first 10 observations is produced in the output file.
- The characteristics of the numeric and character variables and number of observations in the converted SAS data file are reported in the output file.
- Descriptive statistics of all the numeric variables and the frequency information of all character variables are reported in the output file.
- *New*: All numeric (m) and categorical variables (n) in the Excel file are converted to X_1 - X_m and C_1 - C_n, respectively. However, the original column names will be used as the variable labels in the SAS data. This new feature helps to maximize the power of the user-friendly SAS macro applications included in the book. However, to keep the Excel column names as the SAS variable names, use the SAS import wizard to import the Excel data file.
- *New*: Options for renaming any X_1 - X_m or C_1 - C_n variables in a SAS data step is available in the EXLSAS2 macro application. Refer to the EXLSAS2 help file in Appendix 2.
- *New*: Using SAS ODS graphics features in version 9.2, the frequency distribution display of all categorical variables will be generated when WORD, HTML, PDF, and TXT format are selected as the output file format.
- Options for saving the output tables in WORD, HTML, PDF, and TXT formats are available.

Software requirements for using the EXLSAS2 macro:

- SAS/BASE and SAS/ACCESS interface to PC file formats must be licensed and installed at your site.
- The EXLSAS2 macro has only been tested in the Windows (Windows XP and later) environment. However, to import DBF, CSV, and tab-delimited files in the Unix platform, the EXLSAS2 macro could be used.
- SAS version 9.0 and above is required for full utilization.

2.7.4 Steps Involved in Running the EXLSAS2 Macro

Step 1. Prepare the PC data file by following the recommendations given in Section 2.7.1.

Step 2. Open the EXLSAS2.SAS macro-call file from the \dmsas2e\maccal\ folder into the SAS EDITOR window in the SAS display manager. Instructions are given in Appendix 1 regarding downloading the macro-call and sample data files from this book's Web site. Click the RUN icon to submit the macro-call file EXLSAS2.SAS to open the MACRO–CALL window called EXLSAS2 (Figure 2.3).

Special note to SAS Enterprise Guide (EG) CODE window users: Because the user-friendly SAS macro application included in this book uses SAS WINDOW/

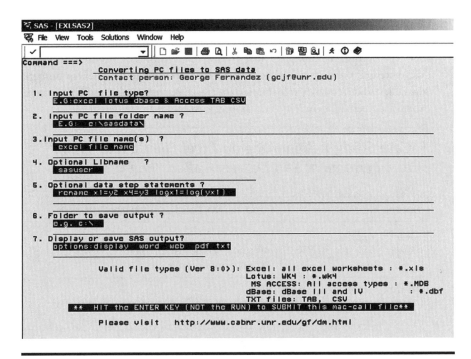

Figure 2.3 Screen copy of EXLSAS2 macro-call window showing the macro-call parameters required to import PC file types to SAS datasets.

DISPLAY commands, and these commands are not compatible with SAS EG, open the traditional EXLSAS2 macro-call file included in the \dmsas2e\mac-cal\nodisplay\ into the SAS editor. Read the instructions given in Appendix 3 regarding using the traditional macro-call files in the SAS EG/SAS Learning Edition (LE) code window.

Step 3: Input the appropriate parameters in the macro-call window by following the instructions provided in the EXLSAS2 macro help file in Appendix 2. After inputting all the required macro parameters, check whether the cursor is in the last input field, and then hit the ENTER key (not the RUN icon) to submit the macro.

Step 4: Examine the LOG window (only in DISPLAY mode) for any macro execution errors. If you see any errors in the LOG window, activate the EDITOR window, resubmit the EXLSAS2 macro-call file, check your macro input values, and correct if you see any input errors.

Step 5: Save the output files. If no errors are found in the LOG window, activate the EDITOR window, resubmit the EXLSAS2 macro-call file, and change the macro input value from DISPLAY to any other desirable format.

The PC file will be imported to a temporary (If the macro input #4 is blank or WORK) or permanent (If a LIBNAME is specified in the macro input #4) SAS dataset. The output, including the first 10 observations of the imported SAS data, characteristics of numeric and character variables, simple statistics for numeric variables, and frequency information for the character variables, will be saved as the user-specified format in the user-specified folder as a single file.

2.7.5 Case Study 1: Importing an Excel File Called "Fraud" to a Permanent SAS Dataset Called "Fraud"

Source file	fraud.xls. MS Excel sheet 2003
Variables	Daily retail sales, number of transactions, net sales, voids, and manager on duty in a small convenience store
Number of observations	923

- Open the Excel file "fraud" and check whether all the specified data requirements reported in Section 2.7.1 are satisfied. The screen copy of the Excel file with the required format is shown in Figure 2.2. Close the "fraud" worksheet file, and exit from Excel.
- Open the EXLSAS2 macro-call window in SAS (see Figure 2.3), and input the appropriate macro-input values by following the suggestions given in the help file in Appendix 2. Submit the EXLSAS2 macro to import the "fraud" Excel sheet to a SAS dataset called "fraud."
- A printout of the first 10 observations, including all variables in the SAS data file "fraud," is displayed (Table 2.1). Examine the printout to see whether SAS imported all the variables from the Excel worksheet correctly.
- Examine the PROC CONTENTS display of all the variables in the SAS dataset called "fraud." The characteristics of all numeric and character variables are presented in Tables 2.2 and 2.3, respectively.
- Examine the simple descriptive statistics for all the numeric variables (Table 2.4). Note that the variables YEAR, WEEK, and DAY are treated as numeric. The total number of observations in the dataset is 923. Confirm that three observations in VOIDS and TRANSAC, and two observations in NETSALES are missing in the Excel file. Also, examine the minimum and maximum numbers for all the numeric variables, and verify that no unusual or extreme values are present.
- Examine the frequency information (Tables 2.5 and 2.6) for all the character variables. Make sure that character variable levels are entered consistently. SAS systems treat uppercase and lowercase data values differently. For example, "April," "april," and "APRIL" are considered different data values. The frequency information for MGR (manager on duty) indicated that managers

Table 2.1   List of the First 10 Observations of the Imported SAS Data Using PROC PRINT in Macro EXLSAS2

X1	C1	X2	X3	C2	VOIDS	X5	X6	C3
1998	January	1	2	Fri	1008.75	1443	139	mgr_a
1998	January	1	3	Sat	10.00	1905	168	mgr_b
1998	January	2	4	Sun	9.00	1223	134	mgr_b
1998	January	2	5	Mon	7.00	1280	146	mgr_c
1998	January	2	6	Tue	15.00	1243	129	mgr_b
1998	January	2	7	Wed	14.00	871	135	mgr_a
1998	January	2	8	Thu	4.00	1115	105	mgr_c
1998	January	2	9	Fri	33.21	1080	109	mgr_c
1998	January	2	10	Sat	8.00	1796	156	mgr_b
1998	January	3	11	Sun	13.00	1328	132	mgr_c

Mgr-a and Mgr-e were on duty relatively fewer times than the other three managers (Figure 2.4). This information should be considered in modeling.

2.7.6 SAS Macro Applications—RANSPLIT2: Random Sampling from the Entire Database

The RANSPLIT2 macro can be used to draw training, validation, and test samples from the entire database. SAS Data step and the RANUNI function are the main tools in the RANSPLIT2 macro. The advantages of using the RANSPLIT2 macro are

Table 2.2   Numeric Variable Descriptions Using PROC CONTENTS Procedure in Macro EXLSAS2

NAME	LABEL	NOBS	LENGTH
VOIDS	VOIDS	923	8
X1	YEAR	923	8
X2	WEEK	923	8
X3	DAY	923	8
X5	NETSALES	923	8
X6	TRANSAC	923	8

Table 2.3 Character Variable Descriptions Using PROC CONTENTS Procedure in Macro EXLSAS2

Obs	NAME	LABEL	NOBS	LENGTH	FORMAT	INFORMAT
1	C1	MONTH	923	9	$	$
2	C2	DOFWEEK	923	3	$	$
3	C3	MGR	923	5	$	$

The distribution pattern among the training, validation, and test samples for user-specified numeric variables could be examined graphically by box plots to confirm that all three-sample distributions are similar.

New: The attributes of numeric and categorical variables and the macro input values specified by the user in the last run of RANSPLIT2 macro submission, added to the output.

New: Many different sampling methods such as simple random sampling, stratified random sampling, systematic random sampling and unrestricted random sampling, are implemented using the SAS SURVEYSELECT procedure.

A sample printout of the first 10 observations can be examined from the TRAINING sample.

Options for saving the output tables and graphics in WORD, HTML, PDF, and TXT formats are available.

Software requirements for using the RANSPLIT2 macro are:

SAS/BASE, SAS/STAT, and SAS/GRAPH must be licensed and installed at your site.

SAS version 9.0 and above is required for full utilization.

2.7.7 Steps Involved in Running the RANSPLIT2 Macro

Step 1: Open a copy of the temporary SAS data, and examine the variables.

Step 2: Open the RANSPLIT2.sas macro-call file into the SAS EDITOR window. Instructions are given in Appendix 1 regarding downloading the macro-call and sample data files from this book's Web site. Click the RUN icon to submit the macro-call file RANSPLIT2.SAS to open the MACRO–CALL window called RANSPLIT2 (Figure 2.5).

Special note to SAS Enterprise Guide (EG) CODE window users: Because the user-friendly SAS macro application included in this book uses SAS WINDOW/DISPLAY commands, and these commands are not compatible

Table 2.4 Simple Statistics: Numeric Variables Using PROC MEANS within Macro EXLSAS2

Variable	Label	N	Mean	Median	Minimum	Maximum	Quartile Range	Skewness	Kurtosis
X1	YEAR	923	1998.875	1999.000	1998.0	2000.000	2.000	0.223	−1.353
X2	WEEK	923	3.027	3.000	1.0	6.000	2.000	0.041	−1.039
X3	DAY	923	15.794	16.000	1.0	31.000	15.000	0.006	−1.175
VOIDS	VOIDS	920	69.660	15.000	0.0	1752.450	59.250	5.508	35.562
X5	NETSALES	921	1324.328	1233.000	7.0	4114.000	606.750	1.181	2.886
X6	TRANSAC	920	132.258	127.000	10.0	259.000	39.000	0.896	1.593

Table 2.5 Frequencies of Character Variable: Month Using PROC FREQ within Macro EXLSAS2

C1—MONTH	Frequency
April	89
August	88
December	57
February	83
January	83
July	87
June	88
March	92
May	88
November	53
October	58
September	57

Table 2.6 Frequency of Character Variable: Day of the WEEK Using PROC FREQ Employing Macro EXLSAS2

C2—Day of the Week	Frequency
Fri	133
Mon	133
Sat	130
Sun	137
Thu	128
Tue	129
Wed	133

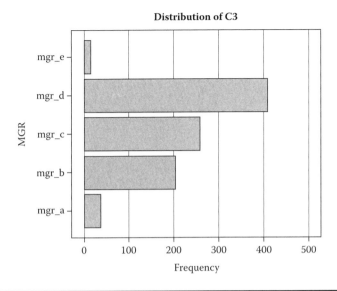

Figure 2.4 Frequency distribution of categorical variable MGR generated by running the SAS macro EXLSAS2.

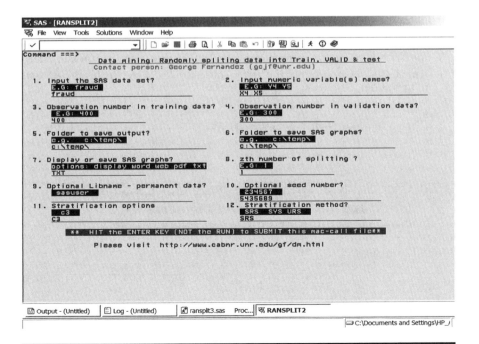

Figure 2.5 Screen copy of RANSPLIT2 macro-call window showing the macro-call parameters required to split the database into training, validation, and test samples.

with SAS EG, open the traditional RANSPLIT2 macro-call file included in the \dmsas2e\maccal\nodisplay\ into the SAS editor. Read the instructions given in Appendix 3 regarding using the traditional macro-call files in the SAS EG/SAS Learning Edition (LE) code window.

Step 3: Input the appropriate parameters in the macro-call window by following the instructions provided in the RANSPLIT2 macro help file in Appendix 2. After inputting all the required macro parameters, check whether the cursor is in the last input field, and then hit the ENTER key (not the RUN icon) to submit the macro.

Step 4: Examine the LOG window (only in DISPLAY mode) for any macro execution errors. If you see any errors in the LOG window, activate the EDITOR window, resubmit the RANSPLIT2 macro-call file, check your macro input values, and correct if you see any input errors.

Step 5: Save the output files. If no errors are found in the LOG window, activate the EDITOR window, resubmit the RANSPLIT2 macro-call file, and change the macro input value from DISPLAY to any other desirable output format. If sample size input for validation sample is blank, a random sample with user-specified sample size will be saved as **Training**, and the left-over observations in the database will be saved as **Validation** datasets. If sample sizes are specified for both **Training** and **Validation** input, random samples with user-specified sample size will be saved as **Training** and **Validation** samples, and the left-over observations will be saved as the **Test** sample. The new SAS datasets will be saved as temporary (if the macro input #9 is blank or WORK) or permanent files (if a LIBNAME is specified in the macro input #9). The printout of the first 10 observations of the **Training** SAS data and box plots illustrating distribution pattern among the training, validation, and test samples for user-specified numeric variables can be saved in a user-specified format in the user-specified folder.

2.7.8 Case Study 2: Drawing Training (400), Validation (300), and Test (All Left-Over Observations) Samples from the SAS Data Called "Fraud"

Data Requirement:

Source file	SAS dataset "fraud"
Variables	Daily retail sales, number of transactions, net sales, voids, and manager on duty in a small convenience store
Number of observations	923

- Open the RANSPLIT2 macro-call window in SAS (see Figure 2.5), input the appropriate macro-input values by following the suggestions given in the help file (Appendix 2). Submit the RANSPLIT2 macro. SAS will draw stratified random samples using *C3* as the stratification variable, and split the entire database into three samples. It will save these **train** (400 obs), **valid** (300 obs), and **test** (left-over observations in the database) as permanent SAS datasets in the LIBRARY called GF.
- The output file shows a list of first 10 observations from the train dataset (Table 2.7). This dataset will be used in calibrating or training the models. Examine the contents and characteristics of the variables in the SAS dataset called "fraud-train."
- The distribution pattern among the training, validation, and test samples for one of the numeric variables, NETSALES, can be graphically examined by the box plot (Figure 2.6) created by the RANSPLIT2 SAS macro. A box plot shows the distribution pattern and the central tendency of the data. The line between

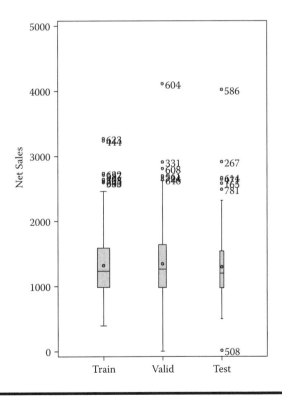

Figure 2.6   A box plot illustrating the distribution pattern among the training, validation, and test samples for the continuous variable NETSALES generated by running the SAS macro RANSPLIT2.

Table 2.7 Macro RANSPLIT2—PROC PRINT Output: The First 10 Observations—Training Data

Obs	MGR	YEAR	MONTH	WEEK	DAY	DOFWEEK	VOIDS	NETSALES	TRANSAC	ID
1	mgr_b	1998	January	1	3	Sat	10.00	1905	168	2
2	mgr_c	1998	January	2	9	Fri	33.21	1080	109	8
3	mgr_c	1998	January	3	11	Sun	13.00	1328	132	10
4	mgr_c	1998	January	3	12	Mon	5.50	1438	118	11
5	mgr_c	1998	January	3	16	Fri	2.00	1341	130	15
6	mgr_b	1998	January	3	17	Sat	1016.50	1982	167	16
7	mgr_c	1998	January	4	19	Mon	23.25	1235	133	18
8	mgr_a	1998	January	4	21	Wed	32.50	708	122	20
9	mgr_c	1998	January	4	22	Thu	31.50	933	120	21
10	mgr_b	1998	January	4	23	Fri	0.25	1097	110	22

the lowest adjacent limit and the bottom of the box represents one-fourth of the data. One-fourth of the data falls between the bottom of the box and the median, and another one-fourth between the median and the top of the box. The line between the top of the box and the upper adjacent limit represents the final one-fourth of the observations. For more information about interpreting the box plot, see Chapter 3. The box plot confirmed that the distribution showed a similar pattern for NETSALES among training, validation, and test samples, and confirmed that the random sampling was successful.

2.8 Summary

Data mining and knowledge discovery is driven by massive amount of data. Sizes of business databases are growing at exponential rates because of the multitude of data that exist. Today, organizations are accumulating vast and growing amounts of data in different formats and different databases. Dynamic data access is critical for data navigation applications, and the ability to store large databases is critical to data mining. The data may exist in a variety of formats such as relational databases, mainframe systems, or flat files. Therefore, in data mining it is common to work with data from several different sources. Roughly, 70% of the time in data mining is spent preparing the data.

The methods of extracting and preparing suitable data for data mining are covered in this chapter. The importance of calibrating the prediction model by a training sample, validating the model using a validation sample, and fine-tuning the model by the test data is briefly addressed. Steps involved in working with the user-friendly SAS macro applications for importing the PC worksheet files to SAS dataset and randomly splitting the entire database into training, validation, and test data are shown by using a small business dataset called "fraud."

References

1. SAS Institute Inc., Data Mining and the Case for Sampling. Solving Business problems Using SAS Enterprise Miner Software. *SAS Institute Best Practice Paper*. Cary, NC. (http://www.cabnr.unr.edu/gf/dm/sasdm.pdf).
2. SAS Institute Inc., *Data Mining Using Enterprise Miner Software: A Case Study Approach*, 1st ed., Cary, NC, 2000.
3. Delwiche, L. D., and Slaughter, S. J., *The Little SAS Book: A Primer*, 4th ed., SAS Institute, Cary, NC. SAS Institute Inc., 2008.
4. SAS Institute Inc., SAS/ACCESS® 9.2 Interface to PC Files SAS Institute Inc., Cary, NC, 2009.
5. Cunningham, P. G. Importing data from Microsoft Excel Worksheets into SAS Documentation from USDA Forest Service, PNW Research Station, http://www.cabnr.unr.edu/gf/dm/Excel-SASImport.pdf.
6. http://support.sas.com/documentation/cdl/en/acpcref/61891/PDF/default/acpcref.pdf.

Chapter 3

Exploratory Data Analysis

3.1 Introduction

The goal of exploratory data analysis (EDA) is to examine the underlying structure of the data and learn about the systematic relationship among many variables. EDA includes a set of descriptive and graphical tools for exploring data visually, both as a prerequisite to more formal data analysis and as an integral part of formal model building. It facilitates discovering the unexpected, as well as confirming the expected patterns and trends. Although the two terms are used almost interchangeably, EDA is not identical to statistical graphical analysis. As an important step in data mining, EDA employs graphical and descriptive statistical techniques to obtain insights into a dataset, detect outliers and anomalies, and to test the underlying model assumptions. Thus, thorough data exploration is an important prerequisite for any successful data mining project. For additional information on EDA, see Chambers et al.[1] and Cleveland and McGill.[2]

3.2 Exploring Continuous Variables

Simple descriptive statistics and exploratory graphics displaying the distribution pattern and the presence of outliers are useful in exploring continuous variables. Commonly used descriptive statistics and exploratory graphics suitable for analyzing continuous variables are described next.

3.2.1 Descriptive Statistics

Simple descriptive statistics of continuous variables are useful in summarizing central tendency, quantifying variability, detecting extreme outliers, and checking for distributional assumptions. SAS procedures MEANS, SUMMARY, and UNIVARIATE provide a wide range summary and exploratory statistics. For additional information on statistical theory, formulas, and computational details, the readers should refer to Schlotzhauer and Little[3] and SAS Institute.[4]

3.2.1.1 Measures of Location or Central Tendency

- *Arithmetic mean*: The arithmetic mean is the most commonly used measure of central tendency of a continuous variable. The mean is equal to the sum of the variable values (totals) divided by the number of observations. However, the mean is not a robust estimate because it can be heavily influenced by a few extreme values or outliers in the tails of a distribution.
- *Median*: The median is the middle value of a ranked continuous variable and the number that separates the bottom 50% of the data from the top 50%. Thus, half of the values in a sample will have values that are equal to or larger than the median, and half will have values that are equal to or smaller than the median. The median is less sensitive to extreme outliers than the mean and, therefore, it is a better measure than the mean for highly skewed distributions. For example, the median salary is usually more informative than the mean salary when summarizing average salary. The mean value is higher than the median in positively skewed distributions and lower than the median in negatively skewed distributions.
- *Mode*: The mode is the most frequent observation in a distribution. This is the most commonly used measure of central tendency with nominal data.
- *Geometric mean*: The geometric mean is an appropriate measure of central tendency when averages of rates or index numbers are required. It is the nth root of the product of a positive variable. For example, to estimate the average rate of return of a 3-year investment that earns 10% the first year, 50% the second year, and 30% the third year, the geometric mean of these three rates should be used.
- *Harmonic mean*: The harmonic mean is the reciprocal of the mean of the reciprocals. The harmonic mean of N positive numbers, $(x_1, x_2, ..., x_n)$ is equal to $N/(1/x_1 + 1/x_2 + \cdots + 1/x_n)$. The harmonic mean is used to estimate the mean of sample sizes and rates. For example, when averaging rate of speed, which is measured by miles per hour, harmonic mean is the appropriate measure rather than arithmetic mean in averaging the rate.

3.2.1.2 Robust Measures of Location

- *Winsorized mean*: The Winsorized mean compensates for the presence of extreme values in the mean computation by setting the tail values equal to a certain percentile value. For example, when estimating a 95% Winsorized mean, first the bottom 2.5% of the values are set equal to the value corresponding to the 2.5th percentile, while the upper 2.5% of the values are set equal to the value corresponding to the 97.5th percentile, and then the arithmetic mean is computed.

■ *Trimmed mean*: The trimmed mean is calculated by excluding a given percentage of the lowest and highest values and then computing the arithmetic mean of the remaining values. For example, by excluding the lower and upper 2.5% of the values and taking the mean of the remaining scores, a 5% trimmed mean is computed. The median is considered as the mean trimmed 100% and the arithmetic mean is the mean trimmed 0%. A trimmed mean is not as affected by extreme outliers as an arithmetic mean. Trimmed means are commonly used in sports ratings to minimize the effects of extreme ratings, possibly caused by biased judges.

3.2.1.3 Five-Number Summary Statistics

The five-number summary of a continuous variable consists of the minimum value, the first quartile, the median, the third quartile, and the maximum value. The median or the second quartile is the middle value of the sorted data. The first quartile is the 25th percentile, and the third quartile is the 75th percentile of the sorted data. The range between the first and third quartiles includes half of the data. The difference between the third quartile and the first quartile is called the interquartile range (IQR). Thus, these five numbers display the full range of variation (from minimum to maximum), the common range of variation (from first to third quartile), and a typical value (the median).

3.2.1.4 Measures of Dispersion

■ *Range*: The range is the difference between the maximum and minimum values. It is easy to compute since only two values, the minimum and maximum, are used in the estimation. However, a great deal of information is ignored, and the range is greatly influenced by outliers.

■ *Variance*: The variance is the average measure of the variation. It is computed as the average of the square of the deviation from the average. However, because variance relies on the squared differences of a continuous variable from the mean, a single outlier has greater impact on the size of the variance than does a single value near the mean.

■ *Standard deviation*: The standard deviation is the square root of the variance. In a normal distribution, about 68% of the values fall within one standard deviation of the mean, and about 95% of the values fall within two standard deviations of the mean. Both variance and standard deviation measurements take into account the difference between each value and the mean. Consequently, these measures are based on a maximum amount of information.

■ *Interquartile range*: The interquartile range is a robust measure of dispersion. It is the difference between the 75th percentile (Q3) and the 25th percentile (Q1). The interquartile range (IQR) is hardly affected by extreme scores; therefore, it is a good measure of spread for skewed distributions. In normally distributed data, the IQR is approximately equal to 1.35 times the standard deviation.

3.2.1.5 Standard Errors and Confidence Interval Estimates

■ *Standard error*: The standard error equals the standard deviation of the sampling distribution of a given statistic in repeated sampling. Standard errors show the amount of sampling fluctuation that exists in the estimated statistics in repeated sampling. Confidence interval estimation and statistical significance testing are dependent on the magnitude of the standard errors. The standard error of a statistic depends on the sample size. In general, the larger the sample size, the smaller the standard error.

■ *Confidence interval*: The confidence interval is an interval estimate that quantifies the uncertainty caused by sampling error. It provides a range of values, which are likely to include an unknown population parameter, since the estimated range is being calculated from a given set of sample data. If independent samples are taken repeatedly from the same population, and a confidence interval is calculated for each sample, then a certain percentage of the intervals will include the unknown population parameter. The width of the confidence interval provides some idea about the uncertainty of the unknown parameter estimates. A very wide interval may indicate that more data should be collected before making inferences about the parameter.

3.2.1.6 Detecting Deviation from Normally Distributed Data

■ *Skewness*: The skewness is a measure that quantifies the degree of asymmetry of a distribution. A distribution of a continuous variable is symmetric if it looks the same to the left and right of the center point. Data from positively skewed (skewed to the right) distributions have values that are clustered together below the mean, but have a long tail above the mean. Data from negatively skewed (skewed to the left) distributions have values that are clustered together above the mean, but have a long tail below the mean. The skewness estimate for a normal distribution equals zero. A negative skewness estimate indicates that the data is skewed left (the left tail is heavier than the right tail), and a positive skewness estimate indicates that the data is skewed right (the right tail is heavier than the left tail).

■ *Kurtosis*: The kurtosis is a measure that quantifies whether the data are peaked or flat relative to a normal distribution. Datasets with large kurtosis have a distinct peak near the mean, decline rather rapidly, and have heavy tails. Datasets with low kurtosis have a flat top near the mean rather than a sharp peak. Kurtosis can be both positive and negative. Distributions with positive kurtosis have typically heavy tails. Kurtosis and skewness estimates are very sensitive to the presence of outliers. These estimates may be influenced by a few extreme observations in the tails of the distribution. Therefore, these statistics are not a robust measure of nonnormality. The Shapiro–Wilks test[5] and the D Agostino–Pearson omnibus tests[6] are commonly used for detecting nonnormal distributions.

3.2.2 Graphical Techniques Used in EDA of Continuous Data

Graphical techniques convert complex and messy information in large databases into meaningful effective visual displays. There are no quantitative analogues that will give the same insight as well-chosen graphics in data exploration. SAS/GRAPH procedures GCHART and GPLOT, SGPLOT, SGSCATTER; SAS/BASE procedure UNIVARIATE; and SAS/QC procedure SHEWHART and the new ODS STAT GRAPHICS provide many types of graphical displays to explore continuous variables.[7] This section provides a brief description of some useful graphical techniques used in EDA of continuous data:

■ *Frequency histogram*: The frequency histogram of a continuous variable displays the class intervals on the horizontal axis and the frequencies of the classes on the vertical axis (see Figure 3.1 for an example of a histogram). The frequency of each class is represented by a vertical bar, which has a height equal to the frequency of that class, and each value in the dataset must belong to one and only one class. A desirable, but not essential, requirement is that the classes have the same width. Unlike in a bar chart, the class intervals are drawn immediately adjacent to each other.

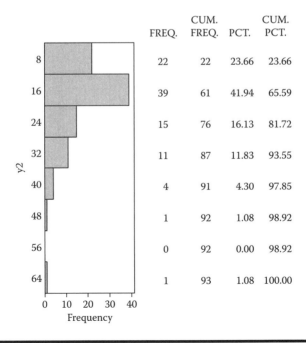

	FREQ.	CUM. FREQ.	PCT.	CUM. PCT.
8	22	22	23.66	23.66
16	39	61	41.94	65.59
24	15	76	16.13	81.72
32	11	87	11.83	93.55
40	4	91	4.30	97.85
48	1	92	1.08	98.92
56	0	92	0.00	98.92
64	1	93	1.08	100.00

Figure 3.1 Frequency histogram illustrating the distribution pattern of car mid-price generated using the UNIVAR2 SAS macro.

■ *Box plot*: The box plot provides an excellent visual summary of many important aspects of a distribution. The box plot is based on the 5-number summary plot that is based on the median, quartiles, and extreme values. The box stretches from the lower hinge (1st quartile) to the upper hinge (the 3rd quartile) and therefore contains the middle half of the scores in the distribution. The median is shown as a line across the box (see Figure 3.2 for an example of a box plot). Therefore, 1/4 of the distribution is between this line

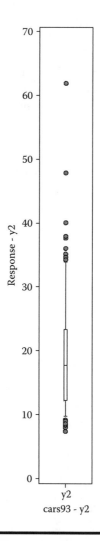

Figure 3.2   A box plot display illustrating the 5-number summary statistics of car midprice generated using the UNIVAR2 SAS macro.

and the top of the box, and 1/4 of the distribution is between this line and the bottom of the box. A box plot may be useful in detecting skewness to the right or to the left.

■ *Normal probability plot*: The normal probability plot is a graphical technique for assessing whether or not the continuous variable is approximately normally distributed. The data are plotted against a theoretical normal distribution in such a way that the points should form an approximate straight line. Departures from this straight line indicate departures from normality. A normal probability plot, also known as a normal Q-Q plot or normal quantile-quantile plot, is the plot of the ordered data values (*Y* axis) against the associated quantiles of the normal distribution (*X* axis). For data from a normal distribution, the points of the plot should lie close to a straight line. Normal probability plots may also be useful in detecting skewness to the right or left (see Figure 3.3 for an example of a normal probability plot). If outliers are present, the normality test may reject the null hypothesis that the distribution is normal even when the remainders of the data do in fact come from a normal distribution. Often, the effect of an assumption violation on the normality test result depends on the extent of the violation. Some small violations may have little practical effect on the analysis, while serious violations may render the normality test result incorrect or uninterpretable.

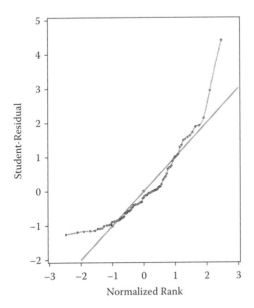

Figure 3.3   Normal probability displays illustrating the right-skewed distribution of car midprice generated using the UNIVAR2 SAS macro.

3.3 Data Exploration: Categorical Variable

One-way and multiway frequency tables of categorical data are useful in summarizing group distributions, relationships between groups, and checking for rare events. SAS procedure FREQ provides wide-range frequency tables and exploratory statistics. For additional information on statistical theory, formulas, and computational details, the reader is recommended to refer to the following SAS documentation.[8]

3.3.1 Descriptive Statistical Estimates of Categorical Variables

- *Gini index:* A Gini index measures heterogeneity of one-way frequencies of a categorical variable. This index can range from 0 (zero heterogeneity [perfect homogeneity—when all the frequencies are concentrated within one level of a categorical variable]) to 1 (maximum heterogeneity [when total frequencies are equally distributed among all the levels of a categorical variable]).[9] When a categorical variable has k levels and p_i is the frequency of the ith categorical level, the Gini index (G) is equal to $g/([(k-1)/k]$, where

$$g = 1 - \text{sum of } (p_1^2 + p_2^2 + \cdots + p_k^2).$$

- *Entropy:* An entropy measurement is another measure of heterogeneity; it ranges between 0 (perfect homogeneity) and 1 (maximum heterogeneity).[9] When a categorical variable has k levels and p_i is the frequency of ith categorical level, the entropy index (E) is equal to $e / log\ e$, where

$$e = - \text{sum of } [(p_i * \log p_i) + \ldots + (p_k * \log p_k)].$$

- *Cross-tabulation:* This is a two-way table showing the frequencies for each level in one categorical variable across the levels of other categorical variables. One of the categorical variables is associated with the columns of the contingency table; the other categorical variable is associated with the rows of the contingency table. This table is commonly used to display the correlation between two categorical variables.
- *Pearson's chi-square test for independence:* For a contingency table, this tests the null hypothesis that the row classification factor and the column classification factor are independent by comparing observed and expected frequencies. The expected frequencies are calculated by assuming the null hypothesis is true. The chi-square test statistic is the sum of the squares of the differences between the observed and expected frequencies, with each squared difference divided by the corresponding expected frequency.

3.3.2 Graphical Displays for Categorical Data

The graphical techniques employed in this chapter to display categorical data are quite simple, consisting of bar, block, and pie charts. SAS/GRAPH procedure GCHART provides many types of graphical displays to explore categorical variables.[7] This section provides a brief description of some simple graphical techniques used in EDA of categorical data. For advanced methods in exploring categorical data, see Reference 10.

- *Bar charts*: The bar charts display a requested statistic (frequency, percentage) based on the values of one or more categorical variables. They are useful for displaying exact magnitudes emphasizing differences among the charted values, and comparing a number of discontinuous values against the same scale. Thus, bar charts allow us to see the differences between categories, rather than displaying trends. Stacked bar and block charts are effective in showing relationships between two-way and three-way tables. See Figures 3.4 and 3.5 for examples of stacked bar and block charts.
- *Pie charts*: Compare the levels or classes of a categorical variable to each other and to the whole. They use the size of pie slices to graphically represent the value of a statistic for a data range. Pie charts are useful for examining how the values of a variable contribute to the whole and for comparing the values of several variables. Donut charts, which are modified pie charts, are useful in displaying differences between groups in two-way data (see Figure 3.6 for a sample donut chart).

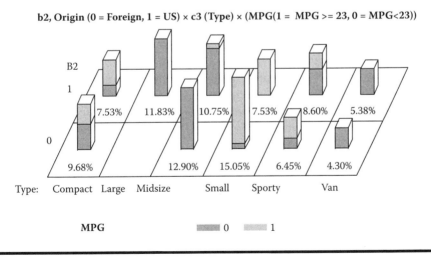

Figure 3.4 Stacked block chart illustrating the three-way relationship between car type, car origin, and the fuel efficiency (MPG) generated using FREQ2 SAS macro.

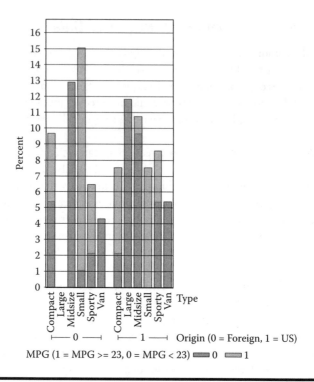

Figure 3.5 Stacked vertical bar chart illustrating the three-way relationship between car type, car origin, and fuel efficiency (MPG), generated using the FREQ2 SAS macro.

3.4 SAS Macro Applications Used in Data Exploration

SAS software has many statistical and graphical features for exploring numeric and categorical variables in large datasets. Some of the features are described in the following section. Readers are expected to have basic knowledge of using SAS to perform the following operations. *The Little SAS Book*[11] can be used as an introductory SAS guide to become familiar with SAS systems and SAS programming.

3.4.1 Exploring Categorical Variables Using the SAS Macro FREQ2

The FREQ2 macro application is an enhanced version of SAS PROC FREQ with graphical capabilities for exploring categorical data. Since the release of SAS version 8.0, many additional statistical capabilities have been made available for data exploration in the PROC FREQ SAS procedure.[8] The enhanced features included in the FREQ2 SAS macro application are:

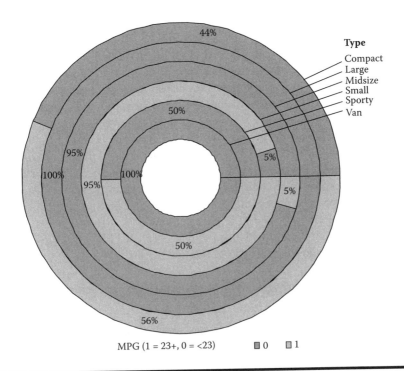

Figure 3.6 Donut chart illustrating the relationship between car type and fuel efficiency (MPG), generated using the FREQ2 SAS macro.

- *New*: Data description table, including all numerical and categorical variable names and labels, and the macro input values specified by the user in the last run of FREQ2 macro submission are included in the output.
- Vertical, horizontal, block, and pie charts for exploring one-way, two-way, and three-way frequency tables are automatically produced.
- *New*: For one-way frequency analysis, Gini and entropy index are reported automatically.
- *New*: Confidence interval estimates for percentages in frequency tables are automatically generated using SAS SURVEYFREQ procedure. If survey weights are specified, then these confidence interval estimates are adjusted for survey sampling and design structures.[12]
- Options for saving the output tables and graphics in WORD, HTML, PDF, and TXT formats are available.

Software requirements for using the FREQ2 macro:

- SAS/BASE and SAS/GRAPH must be licensed and installed at your site.
- FREQ2 macro has only been tested in the WINDOWS (Windows XP and later) environment.
- SAS versions 9 and above are required for full utilization.

3.4.1.1 Steps Involved in Running the FREQ2 Macro

- Create or open a temporary SAS data file.
- Open the FREQ2.sas macro-call file from the \dmsas2e\maccal\ folder into the SAS EDITOR window in the SAS display manager. Instructions are given in Appendix 1 regarding downloading the macro-call and sample data files from this book's Web site. (*Special note to SAS Enterprise Guide (EG) CODE window users:* Because the user-friendly SAS macro application included in this book uses SAS WINDOW/ DISPLAY commands, and these are not compatible with SAS EG, open the traditional FREQ2 macro-call file included in the \dmsas2e\maccal\nodisplay\ into the SAS editor. Read the instructions given in Appendix 3 regarding using the traditional macro-call files in the SAS EG/SAS Learning Edition (LE) code window.)
- Click the RUN icon to submit the macro-call file FREQ2.sas to open the MACRO window called FREQ2 (Figure 3.7).

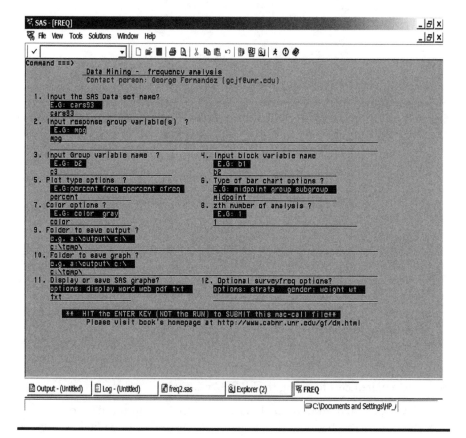

Figure 3.7 Screen copy of FREQ2 macro-call window, showing the macro-call parameters required for exploring the categorical variable.

- Input the appropriate parameters in the macro-call window by following the instructions provided in the FREQ2 macro help file in Appendix 2. After inputting all the required macro parameters, check whether the cursor is in the last input field (#12), and then hit the ENTER key (not the RUN icon) to submit the macro.
- Examine the LOG window only in the DISPLAY mode for any macro execution errors.
- If you see any errors in the LOG window, activate the EDITOR window, resubmit the FREQ2.sas macro-call file, check your macro input values, and correct if you see any input errors.
- Otherwise, activate the EDITOR window, resubmit the FREQ2.sas macro-call file, and change the macro input (#11) value from DISPLAY to any other desirable format (see Appendix 2). The output, including exploratory graphics and frequency statistics, will be saved in the user-specified format in the user-specified folder as a single file if you select one of these three file formats: WORD, WEB, or PDF. If you select TXT as the file format in #11 macro input field, SAS output and graphics files will be saved as separate files.

3.4.2 Case Study 1: Exploring Categorical Variables in a SAS Dataset

Source file	*cars93*
Categorical variables	**MPG** (0: Fuel efficiency below 26 mile/gallon; 1: Fuel efficiency > 26 miles/gallon);
	b2 (Origin of cars: 0: Foreign; 1: American);
	c3 (type of vehicle: Compact, Large, Midsize, Small, Sporty, and Van)
Number of observations	93
Data source[13]	Lock, R. H. (1993)

Open the FREQ2 macro-call window in SAS (Figure 3.7), and input the appropriate macro-input values following the suggestions given in the help file (Appendix 2). Input MPG (miles per gallon) as the target categorical variable in macro input field #2. Input b2 (origin) as the group variable in the macro input field #3. To account for the differences in car types, input c3 (car type) as the block variable in the macro input field #4. After inputting other graphical and file saving parameters, submit the FREQ2 macro-call window, and SAS will output frequency statistics and exploratory charts for MPG categorical variables by car origin and car type. Only selected output and graphics generated by the FREQ2 macro are described in the following text.

Table 3.1   Frequency, Percentage, and 95% Confidence Interval (CI) Values for Origin: Using SAS Proc SURVEYFREQ in Macro FREQ2

Origin (b2)	Frequency	Percent	95% Lower CI for Percent	95% Upper CI for Percent
Foreign (0)	45	48.38	38.04	58.73
Domestic (1)	48	51.61	41.26	61.96

The one-way frequency, percentage, and 95% confidence interval statistics for car origin and car type are presented in Tables 3.1 and 3.2. Additionally, for one-way frequencies, the following statistics are generated: Group level with the highest frequency, number of levels, highest frequency count, highest frequency percentage, exact lower confidence interval, binomial proportion, exact upper confidence levels, testing the binomial proportion = 0.5, two-sided *p*-value, Gini, and entropy estimates. For the proportion of fuel-inefficient cars, the 95% confidence intervals and exact confidence intervals are given in Table 3.3. The hypothesis test that the proportion of fuel-inefficient cars in the database is not equal to 0.5 could be rejected at 5% level (*p*-value = 0.0.0488 in Table 3.3). Both Gini and entropy indices are greater than 0.95, showing that a very high level of heterogeneity exists for the frequency distribution of fuel efficiency.

Two-way percentage statistics for car type and MPG are illustrated in a donut chart (Figure 3.6). A two-way frequency table, including 95% confidence interval estimates for car type × MPG group, is presented in Table 3.4. Zero frequencies were observed for car type equals large and van and fuel-efficient equals 1. The variation in frequency distribution by car type × car origin × MPG group is illustrated

Table 3.2   Frequency, Percentage, and 95% Confidence Interval (CI) Values for Car Type: Using SAS Proc SURVEYFREQ in Macro FREQ2

C3 (Type)	Frequency	Percent	95% Lower CI for Percent	95% Upper CI for Percent
Compact	16	17.20	9.39	25.02
Large	11	11.83	5.14	18.51
Midsize	22	23.66	14.86	32.45
Small	21	22.58	13.92	31.24
Sporty	14	15.05	7.65	22.46
Van	9	9.68	3.56	15.80

Table 3.3   Binomial 95% CI, *p*-values and Gini and Entropy Statistics for Categorical Variable MPG: One-Way Analysis Output Using Macro FREQ2

FREQ-VAR	Highest Frequency Level	Number of Categories	Highest Frequency Count	Highest Frequency %	Exact Lower CL, Binomial Proportion	Exact Upper CL, Binomial Proportion	p-value, Binomial P (two-sided)	Gini	Entropy
mpg	0	2	56	60.22	0.50	0.70	0.0488	0.96	0.97

in a stacked block chart (Figure 3.4) and in a stacked vertical bar chart (Figure 3.5). No large car is found among the 44 foreign-made cars. Irrespective of the origin, a majority of the compact and small cars are more fuel-efficient than the midsize, sporty, large, and van-type vehicles.

The null hypothesis that car type and fuel efficiency (MPG) are independent is not feasible when some of the categories in the two-way table have zero frequencies (Table 3.4). However, the null hypothesis that car origin and fuel efficiency (MPG) are independent is not rejected at the 5% level based on the chi-square test derived from the SAS SURVEYFREQ procedure (*p*-value = 0.1916 in Table 3.5).

3.4.3 EDA Analysis of Continuous Variables Using SAS Macro UNIVAR2

The UNIVAR2 macro is a powerful SAS application for exploring and visualizing continuous variables. SAS procedures UNIVARIATE, GCHART, GPLOT, SHEWHART, and the enhanced ODS Statistical Graphs introduced in version 9.2, are the main tools utilized in the UNIVAR2 macro. The enhanced features included in the UNIVAR2 SAS macro application are:

- *New*: Data description table, including all numerical and categorical variable names and labels and the macro input values specified by the user in the last run of UNIVAR2 macro submission are generated.
- Additional central tendency estimates such as geometric and the harmonic means are reported.
- Test statistics and p-values for testing significant skewness and kurtosis are reported.
- Outliers are detected based on (1) greater than $1.5 \times IQR$ and (2) STUDENT value greater than 2.5 criteria. New datasets are also created after excluding the outliers based on the $1.5 \times IQR$ criterion.
- Separate output and exploratory graphs for all observations, 5% trimmed data, and outlier excluded data are created for each specified continuous variable.

Table 3.4 Two-Way Frequency Table for Car Type × MPG: Using SAS Proc SURVEYFREQ in Macro FREQ2

C3 (Type)	MPG (1=mpg>=23, 0=mpg<23)	Frequency	Percent	Std. Err. of Percent	95% Lower CI for Percent	85% Upper CI for Percent
Compact	0	7	7.52	2.75	2.06	12.98
	1	9	9.67	3.08	3.55	15.79
	Total	16	17.20	3.93	9.38	25.01
Large	0	11	11.82	3.36	5.14	18.51
	1	0	.	.	.	.
	Total	11	11.82	3.36	5.14	18.51
Midsize	0	21	22.58	4.35	13.92	31.23
	1	1	1.07	1.07	0.00	3.21
	Total	22	23.65	4.43	14.85	32.45
Small	0	1	1.07	1.07	0.00	3.21
	1	20	21.50	4.28	12.99	30.01
	Total	21	22.58	4.35	13.92	31.23
Sporty	0	7	7.52	2.75	2.06	12.98
	1	7	7.52	2.75	2.06	12.98
	Total	14	15.05	3.72	7.64	22.45
Van	0	9	9.67	3.08	3.55	15.79
	1	0	.	.	.	.
	Total	9	9.67	3.08	3.55	15.79
Total	0	56	60.21	5.10	50.08	70.34
	1	37	39.78	5.10	29.65	49.91
	Total	93	100.00			

Table 3.5 Hypothesis Testing That Car Origin and MPG are Independent Using a Chi-Square Test—Derived from PROC SURVEYFREQ Output Using Macro FREQ2

Wald Chi-Square Test	
Chi-Square	1.7307
F Value	1.7307
Num DF	1
Den DF	92
Pr > F	0.1916
Sample Size	93

- *New*: If survey weights are specified, then the reported confidence interval estimates are adjusted for survey sampling and design structures using the SURVEYMEAN procedure.[12]
- Options for saving the output tables and graphics in WORD, HTML, PDF, and TXT formats are available.

Software requirements for using the UNIVAR2 macro:

- SAS/BASE, SAS/GRAPH, and SAS QC must be licensed and installed at your site.
- SAS versions 9.13 and above are required for full utilization.

3.4.3.1 Steps Involved in Running the UNIVAR2 Macro

- Create or open a temporary SAS dataset containing at least one numerical variable.
- Open the UNIVAR2.SAS macro-call file from the dmsas2e\maccal\ folder into the SAS EDITOR window. Instructions are given in Appendix 1 regarding downloading the macro-call and sample data files from this book's Web site.
- *Special note to SAS Enterprise Guide (EG) CODE window users*: Because the user-friendly SAS macro application included in this book uses SAS WINDOW/ DISPLAY commands, and these are not compatible with SAS EG, open the traditional UNIVAR2 macro-call file included in the \dmsas2e\ maccal\nodisplay\ into the SAS editor. Please read the instructions given in Appendix 3 regarding using the traditional macro-call files in the SAS EG/ SAS Learning Edition (LE) code window.

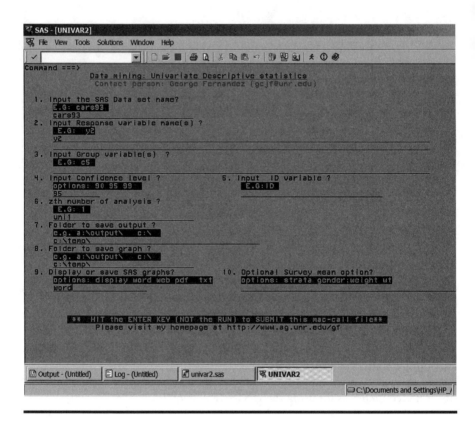

Figure 3.8   Screen copy of the UNIVAR2 macro-call window, showing the macro-call parameters required for exploring continuous variables.

- Click the RUN icon to submit the macro-call file UNIVAR2.sas to open the MACRO–CALL window called UNIVAR2 (Figure 3.8).
- Input the appropriate parameters in the macro-call window by following the instructions provided in the UNIVAR2 macro help file in Appendix 2. After inputting all the required macro parameters, check whether the cursor is in the last input field, and then hit the ENTER key (not the RUN icon) to submit the macro.
- Examine the LOG window (only in DISPLAY mode) for any macro execution errors. If you see any errors in the LOG window, activate the EDITOR window, resubmit the UNIVAR2.SAS macro-call file, check your macro input values, and correct if you see any input errors.
- Save the output files. If no errors are found in the LOG window, activate the EDITOR window, resubmit the UNIVAR2.SAS macro-call file, and change the macro input value from DISPLAY to any other desirable format. The printout of all descriptive statistics and exploratory graphs can be saved as a user-specified format file in the user-specified folder.

3.4.4 Case Study 2: Data Exploration of a Continuous Variable Using UNIVAR2

Data Description	
Data name	SAS dataset CARS93
Continuous variable	Y2 Midprice
Number of observations	93
Data source[13]	Lock, R. H. (1993)

Open the UNIVAR2 macro-call window in SAS (Figure 3.8), and input the appropriate macro-input values by following the suggestions given in the help file (see Appendix 2). Input Y2 (Midprice) as the response variable (#2). Leave the #3 (group variable) field blank because data exploration on midprice is performed for the entire dataset. Submit the UNIVAR2 macro, and SAS will output descriptive statistics, exploratory graphs, and create a new dataset in the WORK folder after excluding the outliers. Only selected output and graphics generated by the UNIVAR2 macro are now described.

The histogram of the midprice value clearly shows a right-skewed distribution (Figure 3.1). More than 50% of the midcar prices are below 18,000. The variation in midprice is clearly illustrated by the control chart (Figure 3.9). The control chart

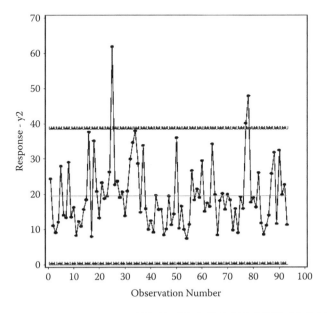

Figure 3.9 Control chart illustrating the variation in car midprice (Y2), generated using the UNIVAR2 SAS macro.

Table 3.6   Five Different Central Tendency Measures of Midprice Computed Using SAS Macro UNIVAR2

Mean	Median	Most Frequent Value	Geometric Mean	Harmonic Mean	Sample Size
19.5097	17.7	15.9	17.5621	15.9329	93

displays the variation in continuous response variables by the observation number. The mean and mean ± 2 standard deviation boundary lines are useful in detecting extreme observations and their location in the dataset. The five different mean values for midprice of the 93 cars are given in Table 3.6. A large positive difference between the mean and the median and mean and the geometric mean clearly shows that the midprice data is right-skewed. Excluding a total of 5% extreme observations (4 observations out of 93) at both ends reduced the mean price from 19.50 to 18.98 (Table 3.7). However, the median price is not affected by deleting the extreme observations. The box plot display illustrates the 5-number summary statistics graphically and shows the presence of outliers (Figure 3.2). The 5-number summary statistics for midcar prices before and after 5% trimming is given in Tables 3.8 and 3.9, respectively. The median midcar price is 17.7 thousand, and 50% of the midcar prices varies between 12.5 and 23.3 thousand. Excluding the 5% extreme values at both ends reduced the range and the standard deviation but not the IQR of the midcar price (Tables 3.10 and 3.11). The 95% Winsorized mean computed after replacing the high extreme values with 97.5th percentile values and the low extreme values with 2.5th percentile values for midprice are reported in Table 3.12. The Winsorized mean (19.05) value falls between the arithmetic mean (19.5) and the 5% trimmed mean (18.97).

Confidence intervals (95%) for mean and estimates of variation for midcar price assuming normality are presented in Table 3.13. Confidence intervals assuming normality and distribution-free confidence intervals for all quantile values are reported in Table 3.14. Because the midprice distribution is significantly right-skewed, distribution-free confidence intervals provide more reliable information on the interval estimates.

Table 3.7   Three Different Central Tendency Measures of 5% Trimmed Midprice Computed Using SAS Macro UNIVAR2

Mean	Median	Most Frequent Value	Sample Size
18.98	17.7	15.9	89

Table 3.8   Five-Number Summary Statistics of Midprice Computed Using SAS Macro UNIVAR2

Smallest Value	Lower Quartile	Median	Upper Quartile	Largest Value
7.4	12.2	17.7	23.3	61.9

Table 3.9   Five-Number Summary Statistics of 5% Trimmed Midprice Computed Using SAS Macro UNIVAR2

Smallest Value	Lower Quartile	Median	Upper Quartile	Largest Value
8.3	12.5	17.7	22.7	40.1

Table 3.10   Measures of Variation of Midprice Computed Using SAS Macro UNIVAR2

Range	Interquartile Range	Standard Deviation
54.5	11.1	9.66

Table 3.11   Measures of Variation of 5% Trimmed Midprice Computed Using the SAS Macro UNIVAR2

Range	Interquartile Range	Standard Deviation
31.8	10.2	8.03

Table 3.12   Winsorized Mean and Related Statistics of Midprice Computed Using SAS Macro UNIVAR2

Percent Winsorized in Tail	Number Winsorized in Tail	Winsorized Mean	Std Error Winsorized Mean	95% Lower CI	95% Upper CI
5.38	5	19.05	0.953	17.15	20.94

Table 3.13 Confidence Interval (95%) Estimates of Midprice Assuming Normality Computed Using SAS Macro UNIVAR2

Parameter	Estimate	95% Lower CI	95% Upper CI
Mean	19.50	17.52	21.49
Std Deviation	9.65	8.44	11.28
Variance	93.30	71.27	127.44

Observation numbers 45 and 21 are identified as large extreme values based on both IQR and STUDENT statistics (Tables 3.15 and 3.16). In addition, observation number 43 is also identified as extreme based on the IQR criterion (Table 3.15). The presence of large extreme values for midprice is the main cause for significant skewness and kurtosis, and confirms that the distribution is not normal based on the D'Agostino–Pearson omnibus normality test (Table 3.17). The upward curvature in the normal probability plot of the STUDENT $((Y_i\text{-}Y_{mean})/Y_{std}))$ value confirms that the distribution is right-skewed (Figure 3.3).

Table 3.14 Confidence Intervals (95%) for Quantile Values of Midprice Computed Using SAS Macro UNIVAR2

Quantile	Estimate	95% Confidence Limits (Lower-Upper) Assuming Normality		95% Confidence Limits (Lower-Upper) Distribution Free	
100 (Max)	61.9				
99	61.9	38.61	46.33	40.1	61.9
95	37.7	32.68	38.82	33.9	61.9
90	33.9	29.47	34.86	28.0	37.7
75 (Q3)	23.3	23.96	28.38	19.9	28.7
50 (Median)	17.7	17.52	21.49	15.8	19.5
25 (Q1)	12.2	10.63	15.04	11.1	14.9
10	9.8	4.15	9.54	8.4	11.1
5	8.4	0.19	6.33	7.4	9.8
1	7.4	−7.31	0.40	7.4	8.3
0 (Min)	7.4	—	—	—	—

Table 3.15   Extreme Observations in Midprice Based on IQR Criterion Computed Using SAS Macro UNIVAR2

ID	Midprice
21	47.9
43	40.1
45	61.9

Table 3.16   Extreme Observations in Midprice Based on STUDENT Criterion Computed Using SAS Macro UNIVAR2

Id	Midprice	Student Residual	Outlier
21	47.9	2.95506	*
45	61.9	4.41228	****

*Outlier.

****Extreme outlier.

Table 3.17   Test Statistics Checking For Normal Distribution of Midprice Computed Using SAS Macro UNIVAR2

Skewness	p-value for Skewness	Kurtosis Statistic	p-value for Kurtosis	Chi-Square Value for Kurtosis	p-value—D' Agostino–Pearson Omnibus Normality Test
1.508	<0.0001	6.183	0.0005	36.310	<.0001

3.4.5 Case Study 3: Exploring Continuous Data by a Group Variable Using UNIVAR2

3.4.5.1 Data Descriptions

Data name	SAS dataset CARS93
Continuous variables	Y2 Midprice
Number of observations	93
Group variable	B2 (Origin of cars: 0: Foreign, 1: Domestic (American make)
Data source[13]	Lock, R. H. (1993)

Open the UNIVAR2 macro-call window in SAS (Figure 3.10), and input the appropriate macro-input values by following the suggestions given in the help file in Appendix 2. Input the Y2 (midprice) as the response variable in macro-call field #2. Because data exploration on midprice is performed by the origin of vehicle, input b2 (origin) as the group variable in the macro-call field #3. Submit the UNIVAR2 macro, and SAS will generate descriptive statistics, exploratory graphs, and create new SAS datasets by excluding the outliers by the group variable origin. Only the features of selected output and graphics are described in the following text.

Approximately 50% of the 93 cars in the dataset are foreign-made. The mean, median, and geometric mean estimates for the foreign cars are greater than that of domestic cars (Table 3.18). The five-number summary statistics, variation, and the presence of outliers in both domestic and the foreign cars are illustrated in a box plot display (Figure 3.11). The observed lowest price is 7.4 thousand for an American-made car, and the price of the most expensive car is 61.9 thousand for a foreign-made car (Table 3.19). The median price of foreign cars is approximately 3000 more than that of domestic cars.

The variation in midcar price for foreign cars is greater than the variation for American cars, as illustrated by the combined control chart and box plots in Figure 3.12 and by the range, IQR, and standard deviation estimates (Table 3.20). One observation is identified as the outlier in both foreign and domestic car types by the 3-sigma cutoff value in the individual control charts. However, many extreme values are detected based on individual IQR criterion in the box plots (Figure 3.12). Observation number 23, the most expensive car in the database, is identified as an outlier among the foreign cars based on 2.5-student and 1.5-IQR criteria. Among the domestic cars, two observations are identified as outliers based on the 2.5-student criterion (Table 3.21). Since the IQR value for the domestic cars

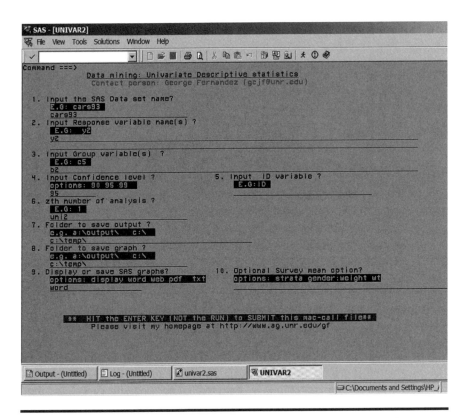

Figure 3.10 Screen copy of the UNIVAR2 macro call window, showing the macro-call parameters required for exploring continuous variables by a group variable in the macro-call parameter 3.

Table 3.18 Different Mean Statistics for Midprice by Origin Computed Using SAS Macro UNIVAR2

Origin Foreign (0)					
Mean	*Median*	*Most Frequent Value*	*Geometric Mean*	*Harmonic Mean*	*Sample Size*
20.50	19.1	8.4	17.95	15.84	45

Origin Domestic (1)					
Mean	*Median*	*Frequent Value*	*Geometric Mean*	*Harmonic Mean*	*Sample Size*
18.57	16.3	11.1	17.19	16.01	48

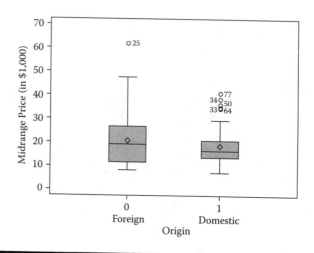

Figure 3.11 A box plot display comparing the 5-number summary statistics of car midprice by car origin, generated using the UNIVAR2 SAS macro.

is relatively smaller than that for foreign cars, the 1.5-IQR criterions detect many observations as outliers (Table 3.22). The distributional pattern differences between the foreign and the domestic car types are illustrated in Figure 3.13. The normal probability plots of student residual for both foreign and domestic cars clearly indicate the nature of right-skewed distribution for midprice (Figure 3.14). Other statistics, such as confidence interval estimates, trimmed mean statistics, Winsorized mean estimates, extreme outliers, and normality check statistics, generated by the UNIVAR2 macro are not shown.

Table 3.19   Five-Number Summary Statistics for Midprice by Origin Computed Using the SAS Macro UNIVAR2

Origin Foreign (0)				
Smallest Value	Lower Quartile	Median	Upper Quartile	Largest Value
8	11.6	19.1	26.7	61.9

Origin Domestic (1)				
Smallest Value	Lower Quartile	Median	Upper Quartile	Largest Value
7.4	13.45	16.3	20.75	40.1

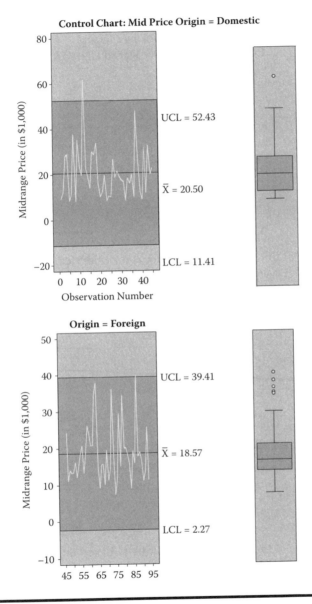

Figure 3.12 A combination of control chart and box plot display of comparing the variation and 5-number summary statistics of car midprice by car origin, generated using the UNIVAR2 SAS macro.

Table 3.20 Estimates of Dispersion Statistics for Midprice by Origin Computed Using SAS Macro UNIVAR2

Origin Foreign (0)		
Range	*Interquartile Range*	*Standard Deviation*
53.9	15.1	11.30
Origin Domestic (1)		
Range	*Interquartile Range*	*Standard Deviation*
32.7	7.3	7.81

Table 3.21 List of Outliers for Midprice by Origin Computed Using SAS Macro UNIVAR2

Origin Foreign (0)			
ID	*Midprice*	*Studentized Residual*	*Outlier*
23	61.9	3.70211	***
Origin Domestic (1)			
ID	*Midprice*	*Studentized Residual*	*Outlier*
82	38	2.51156	*
66	40.1	2.78305	*

* Outlier
** Extreme outlier
*** Most extreme outlier

Table 3.22 List of Outliers Based on 1.5-IQR for Midprice in Domestic Cars Computed Using SAS Macro UNIVAR2

ID	*Midprice*
47	36.1
66	40.1
82	38
90	34.3
91	34.7

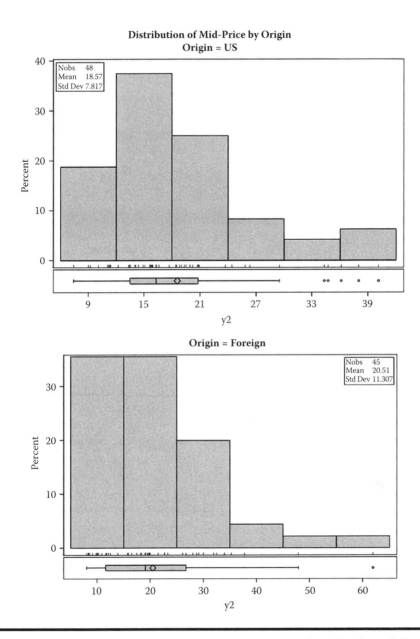

Figure 3.13 Display of frequency histograms comparing the right-skewed distribution of car midprice by car origin, generated using the UNIVAR2 SAS macro.

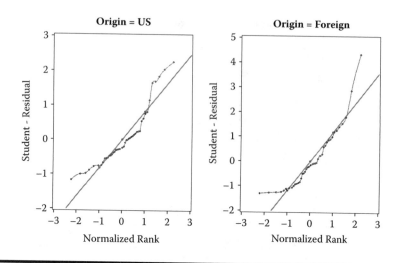

Figure 3.14 **Display of normal probability plots comparing the right-skewed distribution of car midprice by car origin, generated using the UNIVAR2 SAS macro.**

3.5 Summary

The methods of exploring continuous and categorical data using user-friendly SAS macro applications FREQ2 and UNIVAR2 are covered in this chapter. Both descriptive summary statistics and graphical analysis are used to explore continuous variables to learn about the data structure, detect outliers, and test the distributional assumptions. Frequency analysis, confidence interval estimates, and one-, two-, and multiway graphical charts are used to explore the relationships between categorical variables. Steps involved in using the user-friendly SAS macro applications FREQ2 and UNIVAR2 for exploring categorical and continuous variables are shown by using the cars93 dataset.

References

1. Chambers, J. M., Cleveland, W. S., Kleiner, B., and Tukey, P. A., *Graphical Methods for Data Analysis,* Duxbury Press, Boston, 1983.
2. Cleveland, W. and McGill, R., Graphical perception and graphical methods for analyzing scientific data. *Science* 229:828–833. 1985.
3. Schlotzhauer, S. D. and Littel, R. C., *SAS System for Elementary Statistical Analysis,* Cary, NC, SAS Institute Inc., 1990.
4. SAS Institute Inc., Statistical Computation—UNIVARIATE procedure Version 8 Online documentation Cary, NC: SAS Institute Inc., http://support.sas.com/91doc/getDoc/procstat.hlp/univariate_index.htm (last accessed 03-21-09).

5. Shapiro, S. S. and Wilk, M. B., An analysis of variance test for normality (complete samples). *Biometrika*, 591–611. 1965.

6. D Agostino, R. B., Belanger, A., and D Agostino, R. B. Jr., A suggestion for using powerful and informative tests of Normality. *The American Statistician*, 44:316–321. 1990.

7. SAS Institute Inc., Online documentation, version 9.2, SAS Institute Inc., Cary, NC. http://support.sas.com/documentation/onlinedoc/91pdf/index.html (last accessed 03-21-09).

8. SAS Institute Inc., Statistical Computation—FREQ procedure Version 8 Online documentation Cary, NC: SAS Institute Inc. http://support.sas.com/91doc/getDoc/procstat.hlp/freq_index.htm (last accessed 03-21-09).

9. Paolo, G., *Applied Data Mining—Statistical Methods for Business and Industry*, John Wiley & Sons, England, 2003.

10. Friendly, M. Visualizing Categorical Data: Data, Stories, and Pictures. SAS Users Group International, 25th Annual Conference. http://www.math.yorku.ca/SCS/vcd/vcdstory.pdf (last accessed 03-21-09).

11. Delwiche, L. D. and Slaughter, S. J., *The Little SAS Book: A Primer*, 3rd ed., Cary, NC, SAS Institute Inc., 2003.

12. SAS Institute Inc., SAS/STAT® 9.2 User's Guide, The SURVEYFREQ Procedure (Book Excerpt) documentation, Cary, NC, SAS Institute Inc. (http://support.sas.com/documentation/cdl/en/statugsurveyfreq/61835/PDF/default/statugsurveyfreq.pdf) (last accessed 03-21-09).

13. Lock, R. H., New CAR DATA, *Journal of Statistics Education*, 1, No. 1, 1993. (http://www.amstat.org/publications/jse/v1n1/datasets.lock.html) (last accessed 03-21-09).

Chapter 4

Unsupervised Learning Methods

4.1 Introduction

Analysis of multivariate data plays a key role in data mining and knowledge discovery. Multivariate data consists of many different attributes or variables recorded for each observation. If there are p variables in a database, each variable could be regarded as constituting a different dimension, in a p-dimensional hyperspace. This multidimensional hyperspace is often difficult to visualize, and thus, the main objectives of unsupervised learning methods are to reduce dimensionality, scoring all observations based on a composite index and clustering similar observations together based on multiattributes. Also, summarizing multivariate attributes by two or three dimensions that can be displayed graphically with minimal loss of information is useful in knowledge discovery.

The main difference between supervised and unsupervised learning methods is the underlying model structure. In supervised learning, relationships between input (independent, predictor) and the target (dependent, response) variables are being established. However, in unsupervised learning, no variable is defined as a target or response variable. In fact, for most types of unsupervised learning, the targets are same as the inputs. All the variables are assumed to be influenced by a few hidden or latent factors in supervised learning. Because of this feature, it is better to explore large complex models with unsupervised learning than with supervised learning methods.

Unsupervised learning methods are used in many fields under a wide variety of names. The most commonly practiced unsupervised methods are latent factor models (principal component and factor analyses) and disjoint cluster analyses. In principal component analysis, dimensionality of multivariate data is reduced by transforming the correlated variables into linearly transformed uncorrelated variables. In factor analysis, a few uncorrelated hidden factors that explain the maximum amount of common variance and are responsible for the observed correlation among the multivariate attributes are extracted. The relationship between the multiattributes and the extracted hidden factors is then investigated. Disjoint cluster analysis is used for combining cases (records, observation) into groups or clusters such that each group or cluster is homogeneous with respect to certain attributes. Each cluster is also assumed to be different from other groups with respect to the similar characteristics. This implies that observations belonging to the same cluster are as similar to each other as possible, while observations belonging to different clusters are as dissimilar as possible.

A brief account on nonmathematical description and application of these unsupervised learning methods is given in this chapter. For a mathematical account of principal component analysis, factor analysis, and disjoint cluster analysis, readers are encouraged to refer to Sharma[1] and Johnson and Wichern.[2]

4.2 Applications of Unsupervised Learning Methods

The applications of unsupervised learning methods in data mining have increased tremendously in recent years. Many of the application examples are multifaceted, since more than one unsupervised learning method can be used to solve one specific objective. To clearly differentiate the purpose of performing different unsupervised learning methods, one specific example for each unsupervised learning method application is described in this section.

Principal component analysis (PCA). A business analyst is interested in ranking 2000 mutual funds based on monthly performance over the last 2 years on 20 financial indicators and ratios. It would be very difficult to score each mutual funds based on 20 indicators and interpret the findings. Thus, the analyst performed PCA on a standardized 2000 × 20 data matrix and extracted the first two PCs. The first two components accounted for 74% of the variation contained in the 20 variables. Thus, the analyst used the first two PCs to develop scorecards and ranked the mutual funds.

Exploratory factor analysis. An online major book company collects and maintains a large database on annual purchase patterns on different categories of books, audio CDs, DVDs, and CD-ROMs for individual customers. The CEO for marketing is interested in using this database to find patterns among different consumer purchasing patterns so that different advertising strategies can be implemented for different groups of purchasing pattern. A business analyst used exploratory factor analysis and extracted hidden factors responsible for the observed correlation among the various purchase patterns. Using the correlation structure between the original purchase patterns and the hidden factors, the marketing CEO designed different advertising strategies. Also, the customers are also ranked based on the factor scores, and different levels of sales promotion are adopted for each customer based on the factor score values.

Disjoint cluster analysis. A major bank collects and maintains a large database on customer banking patterns on different bank services, including checking, saving, certificates of deposits, loans, and credit card. Based on the banking attributes, the bank CEO would like to segment bank customers into very active, moderate, and passive groups based on the last 3 years' data on individual customers banking pattern. The bank's business analyst performed a disjoint cluster analysis on the customer × banking indicator data and extracted nonoverlapping disjoint cluster groups. The marketing division used the segmented customer group information and tried different advertising strategies for different cluster groups.

4.3 Principal Component Analysis

Because it is hard to visualize multidimensional space, principal component analysis (PCA), a popular multivariate technique, is mainly used to reduce the dimensionality of p multiattributes to two or three dimensions. PCA summarizes the variation in a correlated multiattribute to a set of uncorrelated components, each of which is a particular linear combination of the original variables. The extracted uncorrelated components are called *principal components* (PCs) and are estimated from the eigenvectors of the covariance or correlation matrix of the original variables. Therefore, the objective of PCA is to achieve parsimony and reduce dimensionality by extracting the smallest number of components that account for most of the variation in the original multivariate data and to summarize the data with little loss of information.

In PCA, uncorrelated PCs are extracted by linear transformations of the original variables so that the first few PCs contain most of the variations in the original dataset. These PCs are extracted in decreasing order of importance so that the first PC accounts for as much of the variation as possible, and each successive component accounts for a little less. Following PCA, the analyst tries to interpret the first few PCs in terms of the original variables, and thereby arrive at a greater understanding of the data. To reproduce the total system variability of the original p variables, we need all p number of PCs. However, if the first few PCs account for a large proportion of the variability (80%–90%), we have achieved our objective of dimension reduction. Because the first PC accounts for the covariation shared by all attributes, this may be a better estimate than simple or weighted averages of the original variables. Thus, PCA can be useful when there is an unusually high degree of correlation present in the multiattributes.

In PCA, PCs can be extracted using either original multivariate attribute data or using the covariance or the correlation matrix if the original dataset is not available. In deriving PCs, the correlation matrix is commonly used when different variables in the dataset are measured using different units (annual income, educational level, and number of cars owned per family) or if different variables have different variances. Using the correlation matrix is equivalent to standardizing the variables to zero mean and unit standard deviation. The statistical theory, methods, and the computational aspects of PCA are presented in detail elsewhere.[3]

4.3.1 PCA Terminology

Eigenvalues. They measure the amount of the variation explained by each PC and will be largest for the first PC and smaller for the subsequent PCs. An eigenvalue greater than 1 indicates that PCs account for more variance than accounted for by one of the original variables in standardized data. This is commonly used as a cutoff point for which PCs are retained.

Eigenvectors. They provide the weights to compute the uncorrelated PCs, which are the linear combination of the centered standardized or centered unstandardized original variables.

PC scores. These are the derived composite scores computed for each observation in the dataset weighted by the eigenvectors for each PC. The means of PC scores are equal to zero, as these are the linear combination of the centered variables. These uncorrelated PC scores can be used in subsequent analyses, to check for multivariate normality,[4] to detect multivariate outliers,[4] or as a remedial measure in regression analysis with severe multicollinearity.[5]

Estimating the number of PCs. Several criteria are available for determining the number of PCs to be extracted, but these are just empirical guidelines rather than definite solutions. In practice, we seldom use a single criterion to decide on the number of PCs to extract. Some of the most commonly used guidelines are the Kaiser–Guttman rule, the scree and parallel analysis plot, and interpretability.[6]

– *Kaiser–Guttman rule.* This rule states that the number of PCs to be extracted should be equal to the number of PCs having an eigenvalue greater than 1.0.

– *Scree test.* Plotting the eigenvalues against the corresponding PC produces a screen plot that illustrates the rate of change in the magnitude of the eigenvalues for the PC. The rate of decline tends to be fast first, then levels off. The "elbow," or the point at which the curve bends, is considered to indicate the maximum number of PCs to extract. One less PC than the number at the elbow might be appropriate if you are concerned about getting an overly defined solution. However, scree plots may not give such a clear indication of the number of PCs at all times.

- *Parallel analysis.* To aid the decision making in the selection of number of PCs extracted in standardized data, another graphical method known as *parallel analysis* is suggested to enhance the interpretation of the scree plot. (See Sharma[7] for a description of the computational details of performing parallel analysis.) In parallel analysis, eigenvalues are extracted in repeated sampling from totally independent multivariate data with an exact dimension as the data of interest. Because the variables are not correlated in this simulated dataset, all the extracted eigenvalues should have a value equal to 1. However, due to sampling error, the first half of the PC will have eigenvalues greater than 1, and the second half of the PC will have eigenvalues less than 1. The average eigenvalues for each PC computed from the repeated sampling is overlaid on the same scree plot of the actual data. The optimum number of PCs is selected at the cutoff point where the scree plot and the parallel analysis curve intersect. An example of the scree–parallel analysis plot is presented in Figure 4.3.
- *Interpretability.* Another very important criterion for determining the number of PC is the interpretability of the PC extracted. The number of PC extracted should be evaluated not only according to empirical criteria but also to the criterion of meaningful interpretation.

PC loading. They are correlation coefficients between the PC scores and the original variables. They measure the degree of association of each variable in accounting for the variability in the PC. It is often possible to interpret the first few PCs in terms of *overall* effect or a *contrast* between groups of variables based on the structures of PC loadings. A high positive correlation between PC1 and a variable indicates that the variable is associated with the direction of the maximum amount of variation in the dataset. More than one variable might have a high correlation with PC1. A strong correlation between a variable and PC2 indicates that the variable is responsible for the next largest variation in the data perpendicular to PC1, and so on. Conversely, if a variable does not correlate to any PC axis, or correlates only with the last PC, or the one before the last PC, this usually suggests that the variable makes little or no contribution to the variation in the dataset. Therefore, PCA may often indicate which variables in a dataset are important and which ones may be of little consequence in accounting for the total variation in the data. Some of these low-performance variables might therefore be removed from consideration in order to simplify the overall analyses.

4.4 Exploratory Factor Analysis

Factor analysis is a multivariate statistical technique concerned with the extraction of a small number of hidden factors that are responsible for the correlation among the set of observable variables. It is assumed that observed variables are correlated because they share one or more underlying factors. Basically, factor analysis tells us what variables can be grouped or goes together. It transforms a correlation matrix or a highly correlated multivariate data matrix into a few major factors, so that the variables within the factors are more highly correlated with each other than with variables in the other factors.

Exploratory factor analysis (EFA) is the most common form of factor analysis; it seeks to uncover the underlying structure of a relatively large set of variables. An a priori assumption is that any variable may be associated with any factor. There is no prior theory regarding factor structures. We use estimated factor loadings to intuit the factor structure of the data.

If the extracted hidden factors account for most of the variation in the data matrix, the partial correlations among the observed variables will be close to zero. Thus, the hidden factors determine the values of the observed variables. Each observed variable could be expressed as a weighted composite of a set of latent factors. Given the assumption that the residuals are uncorrelated across the observed variables, the correlations among the observed variables are accounted for by the factors. Any correlation between a pair of observed variables can be explained in terms of their relationships with the latent factors. The statistical theory, methods, and the computational aspects of EFA are presented in detail elsewhere.[8,9]

4.4.1 Exploratory Factor Analysis versus Principal Component Analysis

Many different methods of factor analysis are available in the SAS system, and PCA is one of the common methods in factor analysis. While the two analyses are functionally very similar and are used for data reduction and summarization, they are quite different in terms of the underlying assumptions. In EFA, the variance of a single variable can be partitioned into common and unique variances. The common variance is considered shared by other variables included in the model, and the unique variance that includes the error component is unique to that particular variable. Thus, EFA analyzes only the common variance of the observed variables, whereas PCA summarizes the total variance and makes no distinction between common and unique variance.

The selection of PCA over the EFA is dependent on the objective of the analysis and assumptions about the variation in the original variables. EFA and PCA are considered similar since the objectives of both analyses are to reduce the original variables into fewer components called *factors* or *principal components*. However, they are also different since the extracted components serve different purposes. In EFA, a small number of factors are extracted to account for the intercorrelations among the observed variables and to identify the latent factors that explain why the variables are correlated with each other. However, the objective of PCA is to account for the maximum portion of the variance present in the original variables with a minimum number of PCs.

If the observed variables are measured relatively error free (customer age, number of years of education, or number of family members), or if it is assumed that the error and unique variance represent a small portion of the total variance in the original set of the variables, then PCA is appropriate. However, if the observed variables are only indicators of the latent factor, or if the error variance represents a significant portion of the total variance, then the appropriate technique is EFA.

4.4.2 *Exploratory Factor Analysis Terminology*

4.4.2.1 *Communalities and Uniqueness*

The proportion of variance in the observed variable that is attributed to the latent factors is called *communality*, and the proportion of variance in the observed variable that is not accounted by factors is called *uniqueness*. Thus, communalities and uniqueness sum to one. There are many different methods available to estimate communalities. When the communalities are assumed to be equal to 1.0, that is, when all the variables are completely predicted by the factors, and then this factor analysis is equivalent to PCA. However, the objective of PCA is dimension reduction rather than explaining observed correlations with underlying factors. A second method for estimating prior communalities is to use the squared multiple correlation (SMC) in a multiple regression model. In estimating the SMC, each variable is treated as the response in a regression model in which all the other variables are considered predictor or independent variables. The estimated R^2 from this multiple regression is used as the prior communality estimate for the SMC and is used in factor extraction. Similarly, the SMCs for all observed variables are estimated and used as the prior communality estimate. After the factor analysis is completed, the actual communality values are reestimated and reported as the final communality estimates.

4.4.2.2 *Heywood Case*

Communalities are squared correlations and, therefore, they should always lie between 0 and 1. However, the final communality estimates might be greater than 1 in common factor model. If communality equals 1, the condition is referred to as a Heywood case, and if communality exceeds 1, it is an ultra-Heywood case.[17] An ultra-Heywood case may result when some unique component has negative variance, a clear indication that something is wrong. The maximum likelihood factor analysis method is especially sensitive to quasi- or ultra-Heywood cases. During the iterative process, a variable with high communality is given a high weight that tends to increase its communality and subsequently increases its weight. This could be true if the common factor model fits the data perfectly, but it is not generally the case with real data. A final communality estimate that is less than the squared multiple correlation can, therefore, indicate poor fit, possibly due to not specifying enough factors. PCA, unlike common factor analysis, has none of these problems if the covariance or correlation matrix is computed correctly from a dataset with no missing values. However, severe rounding of the correlations with few decimals can produce negative eigenvalues in principal components. Possible causes include the following:

- Incorrect prior communality estimates
- Many common factors
- Few common factors
- Small sample size to provide stable estimates
- Inappropriate common factor model

4.4.2.3 Cronbach Coefficient Alpha

Analyzing latent factor such as customer satisfaction requires indicators to accurately measure them. Cronbach's coefficient alpha estimates the reliability of this type of scale by determining the internal consistency of the attribute or the average correlation of attributes within the factor. The Cronbach's coefficient alpha is a lower bound for the reliability coefficient when the multiattributes measure a common entity. Positive correlation is required for the alpha coefficient because the variables positively influence a common scale. Because the variances of some variables vary widely, you should use the standardized score to estimate reliability. The standardized alpha coefficient provides information about how each variable reflects the reliability of the scale with standardized variables. Cronbach's coefficient alpha of 0.7 is commonly considered a cutoff value. If the standardized Cronbach's alpha decreases after removing a variable from the construct, then this variable is strongly correlated with other variables in the scale. However, if the standardized alpha increases after removing a variable from the construct, then removing this variable from the scale makes the construct more reliable.

4.4.2.4 Factor Analysis Methods

A variety of different factor extraction methods, namely, principal component, principal factor, iterative principal factor, unweighted least-squares factor, maximum-likelihood factor, alpha factor, image analysis, and Harris component analysis, are available in the SAS PROC FACTOR procedure. The two most commonly employed factor analytic techniques are principal factor and maximum-likelihood factor analyses. Different FA techniques employ different criteria for extracting factors. Discussions on choosing different methods of factor extraction can be found in Sharma.[8]

4.4.2.5 Sampling Adequacy Check in Factor Analysis

Kaiser–Meyer–Olkin (KMO) statistics predicts if data is likely to factor well, based on correlation and partial correlation among the variables. A KMO statistic is reported for each variable, and for the overall variable matrix. KMO varies from 0 to 1.0, and overall KMO should be 0.6 or higher to proceed with successful factor analysis. If the overall KMO statistic is less than 0.6, then drop the variables with the lowest individual KMO statistic values, until the overall KMO rises above 0.6. To compute the overall KMO statistic, find the numerator, which is the sum of squared correlations of all variables in the analysis (except for the 1, self-correlations of variables with themselves). Then calculate the denominator, which is the sum of squared correlations plus the sum of squared partial correlations of each ith variable with each jth variable, controlling for others in the analysis. The partial correlation values should not be very large if one is to expect successful factor extraction.

4.4.2.6 Estimating the Number of Factors

Methods described for extracting the optimum number of PCs could be used in factor analysis also. Some of the most commonly used guidelines in estimating the number of factors are the modified Kaiser–Guttman rule, percentage of variance, the scree test, size of the residuals, and interpretability.[8]

- *Modified Kaiser–Guttman rule.* This rule states that the number of factors to be extracted should equal the number of factors having an eigenvalue greater than 1.0. The modified Kaiser–Guttman rule is adjusted downward when the common factor model is chosen, and the eigenvalue criterion should be lower and around the average of the initial communality estimates.
- *Percentage of variance.* Another criterion, related to the eigenvalue, is the percentage of the common variance (defined by the sum of communality estimates) that is explained by successive factors. For example, if you set the cutoff value at 75% of the common variance, then factors will be extracted until the sum of eigenvalues for the retained factors exceeds 75% of the common variance, defined as the sum of initial communality estimates.
- *Scree/parallel analysis plot.* Similar to PC analysis, scree plot and parallel analysis could be used to detect the optimum number of factors for standardized data. However, the parallel analysis suggested under PC is not feasible for the maximum-likelihood-based EFA method.
- *Chi-square test in maximum likelihood factors analysis method.* This test comprises two separate hypotheses tests. The first test, labeled "Test of H0: No common factors," tests the null hypothesis that no common factors can sufficiently explain the intercorrelations among the variables included in the analysis. You want this test to be statistically significant ($p < .05$). A nonsignificant value for this test statistic suggests that your intercorrelations may not be

strong enough to warrant performing a factor analysis since the results from such an analysis could probably not be replicated. The second chi-square test statistic, labeled "Test of H0: N number of factors are sufficient," is the test of the null hypothesis that N common factors are sufficient to explain the inter-correlations among the variables, where N is the number of factors you specify. This test is useful for testing the hypothesis that a given number of factors is sufficient to account for your data. In this instance, your goal is a small chi-square value relative to its degrees of freedom. This outcome results in a large p-value ($p > 0.05$). One downside of this test is that the chi-square test is very sensitive to sample size; given large degrees of freedom, this test will normally reject the null hypothesis of the residual matrix being a null matrix, even when the factor analysis solution is very good. Therefore, be careful in interpreting this test's significance value. Some datasets do not lend themselves to good factor solutions, regardless of the number of factors extracted.

■ *A priori hypothesis.* It can provide a criterion for deciding the number of factor to be extracted. If a theory or previous research suggests a certain number of factors and the analyst wants to confirm the hypothesis or replicate the previous study, then a factor analysis with the prespecified number of factor can be run. Ultimately, the criterion for determining the number of factors should be the repeatability of the factor solution. It is important to extract only factors that can be expected to replicate themselves when a new sample (validation sample) of subjects is employed.

■ *Interpretability:* Another very important criterion for determining the number of factors is the interpretability of the factors extracted. Factor solutions should be evaluated not only according to empirical criteria but also according to the criterion of theoretical meaningfulness.

4.4.2.7 Eigenvalues

They measure the amount of variation in the total sample accounted for by each factor and reveal the explanatory importance of the factors with respect to the variables. If a factor has a low eigenvalue, then it is only contributing a little to the explanation of variances in the variables and may be excluded as redundant. Note that the eigenvalue is not the percentage of variance explained but rather a measure of "amount" used for comparison with other eigenvalues. A factor's eigenvalue may be computed as the sum of its squared factor loadings for all the variables. Note that the eigenvalues associated with the unrotated and rotated solution will differ, though the sum of all eigenvalues will be the same.

4.4.2.8 Factor Loadings

They are the basis for assigning a label or name to the different factors, and they represent the correlation or linear association between the original variable and the latent factors. Factor loadings or a $p \times k$ matrix of correlations between the original variables and their factors, where p is the number of variables, and k is the number of factors retained. Factor loadings greater than 0.40 in absolute value are frequently used when making decisions regarding significant loading. As the sample size and the number of variables increases, the criterion may need to be adjusted slightly downward; and as the number of factors increases, it may need to be adjusted upward. The procedure described next outlines the steps of interpreting a factor matrix. Once all significant loadings are identified, we can assign some meaning to the factors based on the factor loading patterns. First, examine the significant loadings for each factor. In general, the larger the absolute size of the factor loading for a variable, the more important the variable is in interpreting the factor. The sign of the loadings also needs to be considered in labeling the factors. Considering all the variables' loading on a factor, including the size and sign of the loading, we can determine what the underlying factor may represent. The squared factor loading is the percentage of variance in that variable accounted by the factor. To get the percentage of variance in all the variables accounted for by each factor, add the sum of the squared factor loadings for that factor and divide by the number of variables. Note that the number of variables equals the sum of their variances, as the variance of a standardized variable is 1. This is the same as dividing the factor's eigenvalue by the number of variables. The ratio of the squared factor loadings for a given variable shows the relative importance of the different factors in explaining the variance of the given variable.

4.4.2.9 Factor Rotation

The idea of simple structure and ease of interpretation form the basis for rotation. The goal of factor rotation is to rotate the factors simultaneously in order to have as many zero loadings on each factor as possible. The sum of eigenvalues is not affected by rotation, but rotation will alter the eigenvalues of particular factors. The rotated factor pattern matrix is calculated by postmultiplying the original factor pattern matrix by the orthogonal transformation matrix. The simplest case of rotation is an orthogonal rotation in which the angle between the reference axes of factors is maintained at 90°. More complicated forms of rotation allow the angle between the reference axes to be other than a right angle; that is, factors can be correlated with each other. These types of rotational procedures are referred to as *oblique rotations*. Orthogonal rotation procedures are more commonly used than oblique rotation procedures. In some situations, theory may mandate that underlying latent factors be uncorrelated with each other and, therefore, oblique rotation procedures will not be appropriate. In other situations, when the correlations between the underlying factors are not assumed to be zero, oblique rotation procedures may yield simpler and more interpretable factor patterns. In all cases, interpretation is easiest if we achieve what is called simple structure. In simple structure, each variable is highly associated with one and only one factor. If that is the case, we can name factors for the observed variables highly associated with them.

The VARIMAX is the most widely used orthogonal rotation method, and PROMAX is the most popular oblique rotation method. VARIMAX rotation produces factors that have high correlations with one smaller set of variables and little or no correlation with another set of variables. Each factor will tend to have either large or small loadings of particular variables on it. A VARIMAX solution yields results that make it as easy as possible to identify each variable with a single factor.

PROMAX rotation is a nonorthogonal rotation method, which is computationally faster, and therefore is recommended for very large datasets. PROMAX rotation begins with a VARIMAX rotation, and it makes the larger loadings closer to 1.0 and the smaller loadings closer to 0, resulting in an easy-to-interpret, simple factor structure. When an oblique rotation method is performed, the output also includes a factor pattern matrix, which is a matrix of standardized regression coefficients for each of the original variables on the rotated factors. The meaning of the rotated factors is inferred from the variables significantly loaded on their factors. One downside of an oblique rotation method is that if the correlations among the factors are substantial, then it is sometimes difficult to distinguish among factors by examining the factor loadings. In such situations, investigate the factor pattern matrix, which displays the variance explained by each factor and the final communality estimates.

4.4.2.10 Confidence Intervals and the Significance of Factor Loading Converge

The traditional approach to determining significant factor loadings (loadings that are considered large in absolute values) employs rules of thumb such as 0.3 or 0.4. However, this estimate does not use the statistical evidence efficiently. A new COVER option is available in PROC FACTOR (version 9.2) to assess the significance and to estimate the confidence intervals of the factor loadings when the ML factor analysis method is used with multivariate normally distributed data.[10] The interpretation of the coverage adapted from the SAS documentation[17] is presented in Table 4.1.

4.4.2.11 Standardized Factor Scores

These are the scores of all the cases on all the factors, where cases are the rows and factors are the columns. Factor scores can quantify individual cases on a latent factor using a z-score scale, which ranges from approximately −3.0 to +3.0. The SAS FACTOR procedure can provide the estimated scoring confidents, which are then used in PROC SCORE to produce a matrix of estimated factor scores. You can then output these scores into a SAS dataset for further analysis.[11]

Table 4.1 Interpretations of the Factor Loading 95% Confidence Interval Specified Coverage Displays = 0.45

Positive Loadings	Negative Loadings	COVER = 0.45 Specified	Interpretation of Converge Display
[0]*	*[0]		The loadings are not significantly different from zero, and the CI covers a region of values that are smaller in magnitude than the loading coverage = 0.45. There is strong statistical evidence for positive or negative association between the variable and the extracted factor.
0[ ]*	*[ ]0		The loading estimate is significantly different from zero, but the CI covers an interval of values that are smaller in magnitude than the loading coverage = 0.45. There is strong statistical evidence for nonsignificant variable–factor relationship.
[0*]	[*0]	[0.45]	The loading estimate is not significantly different from zero or the coverage estimate = 0.45 value. The population loadings might be larger or smaller in magnitude than the COVER = 0.45 value. There is strong statistical evidence for a nonsignificant variable–factor relationship.
0[*]	[*]0		The loading estimate is significantly different from zero but not from the COVER estimate = 0.45. There is marginal statistical evidence for a significant variable–factor relationship.
0*[ ]	[ ]*0	0.45[ ] or [ ]0.45	The estimate is significantly different from zero, and the CI covers a region of values that are larger in magnitude than the coverage value = 0.45. There is strong statistical evidence for a significant variable–factor relationship.

4.5 Disjoint Cluster Analysis

Cluster analysis is a multivariate statistical method to group cases or data points into clusters or groups suggested by the data, not defined a priori, so that the degree of association is strong between members of the same cluster and weak between members of different clusters. When exploring and describing large datasets, it is sometimes useful to summarize the information by assigning each observation to a cluster with similar characteristics. Clustering can be used to reduce the size of the data and to induce groupings. As a result, cluster analysis can reveal similarities in multivariate data, which may have been otherwise impossible to find. It tries to identify a set of groups that both minimize within-group variation and maximize between-group variation. Each group or cluster is homogeneous with respect to the multiattributes. The statistical theory, methods, and the computation aspects of cluster analysis are presented in detail elsewhere.[12,13]

4.5.1 Types of Cluster Analysis

Hierarchical cluster analysis (HCA). It is a nested method of grouping observations. In HCA, one cluster may be entirely contained within another cluster, but no other kind of overlap between clusters is allowed. Therefore, it is not suitable for clustering large databases. Prior knowledge of the number of clusters is not required, and once a cluster is assigned, it cannot be reassigned.

Disjoint (nonhierarchical) cluster analysis (DCA). DCA assigns each observation in only one cluster. First, the observations are arbitrarily divided into clusters, and then observations are reassigned one by one to different clusters on the basis of their similarity to the other observations in the cluster. The process continues until no items need to be reassigned. Disjoint clustering is generally more efficient for large datasets, and a priori knowledge of the number of clusters is required to perform DCA.

k-means cluster analysis. It is the most common form of disjoint cluster analysis and is often used in data mining. In *k*-means clustering, the cluster centers are derived from the means of the observations assigned to each cluster when the algorithm is run to complete convergence. In the *k*-means model, each iteration reduces the variation within the clusters and maximizes the differences among the distinct clusters until convergence is achieved. A set of points called *cluster seeds* is selected as a first guess of the means of the clusters. Each observation is assigned to the nearest seed to form temporary clusters. The seeds are then replaced by the means of the temporary clusters, and the process is repeated until no further changes occur in the clusters.

Optimum number of population clusters. While there is no perfect way to determine the number of clusters, we could use some statistics to help in the process. Cubic clustering (CCC), pseudo F-statistic (PSF), and pseudo

$t2$ statistic (PST2) are useful in determining the number of clusters in the data. An overlay plot of PST2 and PSF (Y-axis) versus the number of clusters (X-axis) plot can be used to select the number of potential clusters in the data. Starting from large values in the X-axis, move left until you find a big jump up in the PST2 value. Select the cluster number when the PST2 value has a relatively big drop. Similarly, a relatively large pseudo F-statistic (PSF) value indicates an optimum cluster number when you check the PSF value from left to right in the X-axis of the overlay plot. Values of the cubic clustering criterion greater than 2 indicate good clusters, and values between 0 and 2 indicate potential clusters. They should be considered with caution, however, because large negative values can indicate outliers. However, the scale of variables used (original versus standardized scale) can affect these guidelines reported by Sharma. It may be advisable to look for consensus among the three statistics; that is, the local peaks of the CCC and PSF combined with a big drop in the value of the PST2 statistic. It must be emphasized that these criteria are appropriate only for compact or slightly elongated clusters that are roughly multivariate normal.[19]

4.5.2 FASTCLUS: SAS Procedure to Perform Disjoint Cluster Analysis

PROC FASTCLUS performs a DCA on the basis of distances computed from one or more quantitative variables.[21] The observations are divided into clusters such that every observation belongs to only one cluster. The clusters do not form a tree structure as they do in the hierarchical clustering procedure. FASTCLUS finds disjoint clusters of observations using a k-means method applied to coordinate data. By default, the FASTCLUS procedure uses Euclidean distances, so the cluster centers are based on least-squares estimation. The initialization method used by the FASTCLUS procedure makes it sensitive to outliers. PROC FASTCLUS can be an effective procedure for detecting outliers because they often appear as clusters with only one member.

The FASTCLUS procedure is intended for use with large datasets having 300 or more observations. With small datasets, the results may be highly sensitive to the order of the observations in the dataset.[21] Another potential problem is that the choice of the number of clusters (k) may be critical. Quite different kinds of clusters may emerge when k is changed. Good initialization of the cluster centroids is also important since some clusters may even be left empty if their centroids lie initially far from the distribution of data.

Before using disjoint clustering, decide whether the multiattributes should be standardized in some way, since variables with large variances tend to have more effect on the resulting clusters than those with small variances. If all variables are measured in the same units, standardization may not be necessary. Otherwise, some form of standardization is strongly recommended. Removing all clusters with small frequencies improves cluster separation and provides visually sharper cluster outlines in scatter plots.

4.6 Biplot Display of PCA, EFA, and DCA Results

Biplot[15] display is a visualization technique for investigating the interrelationships between the observations and variables in multivariate data. To display a biplot, the data should be considered as a matrix, in which the column represents the variable space, while the row represents the observational space. The term *biplot* implies it is a plot of two dimensions with the observation and variable spaces plotted simultaneously. In PCA, relationships between PC scores and PCA loadings associated with any two PCs can be illustrated in a biplot display. Similarly, in FA, relationships between the factor scores and factor loadings associated with any two factors can be displayed simultaneously. In DCA, the success of cluster groupings can be verified by performing a canonical discriminant analysis and plotting cluster grouping and the first two canonical discriminate function scores.[14]

4.7 PCA and EFA Using SAS Macro FACTOR2

The FACTOR2 macro is a powerful user-friendly SAS application for performing principal component and factor analysis on multivariate attributes. The SAS procedure, FACTOR, is the main tool used in this macro since both PCA and EFA can be performed using PROC FACTOR.[16,17] In addition to using the PROC FACTOR as the main tool, other SAS procedures, CORR, GPLOT, SGPLOT, BOXPLOT, and IML modules, are also incorporated in the FACTOR2 macro. The following are the enhanced features of the FACTOR2 macro:

- Data description table, including all numerical and categorical variable names and labels and the macro input values specified by the user in the last run of FACTOR2 macro submission are generated.
- The scatter plot matrix and simple descriptive statistics of all multivariate attributes and the significance of their correlations are reported.
- *New:* Estimation of Cronbach coefficient alpha and their 95% confidence intervals when performing latent factor analysis are produced.
- Test statistics and *p*-values for testing multivariate skewness and kurtosis (SAS/IML module is required) are reported.
- Q-Q plot for detecting deviation from multivariate normality and outlier detection plot for detecting multivariate outliers are generated.
- Parallel analysis plots are overlaid with the scree plot to estimate the optimum number of PCs or factors.

- *New:* Factor pattern plots are generated using the New 9.2: statistical graphics feature before and after factor rotation.
- *New:* Assessing the significance and the nature of factor loadings are generated using the New 9.2: statistical graphics feature.
- *New:* Confidence interval estimates for factor loading when ML factor analysis is used.
- Biplot display showing the interrelationship between the principal component or factor scores and the correlations among the multiattributes are produced for all combinations of selected principal components or factors.
- Options for saving the output tables and graphics in WORD, HTML, PDF, and TXT formats are available.

Software requirements for using the FACTOR2 macro are the following:

- SAS/BASE, SAS/STAT, SAS/GRAPH, and SAS/IML must be licensed and installed at your site.
- SAS version 9.13 and above is required for full utilization.

4.7.1 Steps Involved in Running the FACTOR2 Macro

1. Create or open a temporary SAS dataset from $n \times p$ coordinate data containing p correlated continuous variables and n observations. If a coordinate $(n \times p)$ dataset is not available and only a correlation matrix is available, then create a special correlation SAS dataset (see Figure 4.15).
2. Open the FACTOR2.SAS macro-call file into the SAS EDITOR window. Instructions are given in Appendix 1 regarding downloading the macro-call and sample data files from this book's Web site. Click the RUN icon to submit the macro-call file FACTOR2.SAS to open the MACRO–CALL window called FACTOR2 (Figure 4.1).
   Special note to SAS Enterprise Guide (EG) CODE window users: Because the user-friendly SAS macro application included in this book uses SAS WINDOW/DISPLAY commands, and these commands are not compatible with SAS EG, open the traditional FACTOR2 macro-call file included in the\dmsas2e\maccal\nodisplay\ into the SAS editor. Read the instructions given in Appendix 3 regarding using the traditional macro-call files in the SAS EG/SAS Learning Edition (LE) code window.
3. Input the appropriate parameters in the macro-call window by following the instructions provided in the FACTOR2 macro help file in Appendix 2. Users can choose either the scatter plot analysis option or the PCA/EFA analysis option. Options for checking for multivariate normality assumptions and detecting for presence of outliers are also available. After inputting all the required macro parameters, check whether the cursor is in the last input field, and then hit the ENTER key (not the RUN icon) to submit the macro.

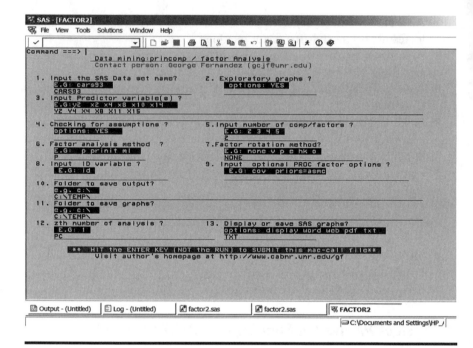

Figure 4.1 Screen copy of FACTOR2 macro-call window showing the macro-call parameters required for performing PCA.

4. Examine the LOG window (only in DISPLAY mode) for any macro execution errors. If you see any errors in the LOG window, activate the EDITOR window, resubmit the FACTOR2.SAS macro-call file, check your macro input values and correct if you see any input errors.

5. Save the output files. If no errors are found in the LOG window, activate the EDITOR window, resubmit the FACTOR2.SAS macro-call file, and change the macro input value from DISPLAY to any other desirable format. PCA or EFA SAS output files and exploratory graphs could be saved in user-specified format in the user-specified folder.

4.7.2 Case Study 1: Principal Component Analysis of 1993 Car Attribute Data

4.7.2.1 Study Objectives

1. Variable reduction: Reduce the dimension of 6 of highly correlated, multiattribute, coordinate data into fewer dimensions (2 or 3) without losing much of the variation in the dataset.

2. Scoring observations: Group or rank the observations in the dataset based on composite scores generated by an optimally weighted linear combination of the original variables.
3. Interrelationships: Investigate the interrelationship between the observations and the multiattributes and group similar observations and similar variables.

4.7.2.2 Data Descriptions

Data Name	SAS Dataset CARS93
Multiattributes	Y2: Midprice
	Y4: City gas mileage/gallon
	X4: HP
	X8: Passenger capacity
	X11: Width of the vehicle
	X15: Physical weight
Number of observations	92
Car93: Data source[18]	Lock, R. H. (1993)

Open the FACTOR2.SAS macro-call file in the SAS EDITOR window, and click RUN to open the FACTOR2 macro-call window (Figure. 4.1). Input the appropriate macro-input values by following the suggestions given in the help file (Appendix 2).

Exploratory analysis: Input Y2, Y4, X4, X8, X11, and X15 as the multiattributes in (#3). Input YES in the macro-call field #2 to perform data exploration and create a scatter plot matrix. (PCA will not be performed when you choose to run data exploration.) Submit the FACTOR2 macro, and SAS will output descriptive statistics, correlation matrices, and scatter plot matrices. Only selected output and graphics generated by the FACTOR2 macro are interpreted in the following text.

The descriptive simple statistics of all multiattributes generated by the SAS PROC CORR are presented in Table 4.2. The number of observations (N) per variable is useful in checking for missing values for any given attribute and providing information on the size of the $n \times p$ coordinate data. The estimates of central tendency (mean) and the variability (standard deviation) provide information on the nature of multiattributes that can be used to decide whether to use standardized or unstandardized data in the PCA analysis. The minimum and the maximum values describe the range of variation in each attribute and help to check for any extreme outliers.

Table 4.2 PROC CORR Output and Simple Statistics—FACTOR2 Macro

Variable	N	Mean	Standard Deviation	Sum	Minimum	Maximum	Label
Y2	93	19.50	9.65	1814	7.40	61.90	Midrange price (in $1000)
Y4	93	22.36	5.61	2080	15.00	46.00	City MPG (miles per gallon by EPA rating)
X4	93	143.82	52.37	13376	55.00	300.00	HP (maximum)
X8	93	5.08	1.038	473.00	2.00	8.00	Passenger capacity (persons)
X11	93	69.37	3.77	6452	60.00	78.00	Car width (inches)
X15	93	3073	589.89	285780	1695	4105	Weight (pounds)

The degree of linear association among the variables measured by the Pearson correlation coefficient (r), and their statistical significance are presented in Table 4.3. The value of r ranged from 0 to 0.87. The statistical significance of r varied from no correlation (p-value: 0.967) to a highly significant correlation (p-value < 0.0001). Among the 15 possible pairs of correlations, 13 pairs of correlations were highly significant, indicating that this data is suitable for performing PCA analysis. The scatter plot matrix among the six attributes presented in Figure 4.2 reveals the strength of correlation, presence of any outliers, and the nature of bidirectional variation. In addition, each scatter plot shows the linear regression line, 95% mean confidence interval band, and a horizontal line (*Y*-bar line), which passes through the mean of the *Y*-variable. If this *Y*-bar line intersects the confidence band lines, that is, the confidence band region does not enclose the *Y*-bar line, then the correlation between the *X* and *Y* variable is statistically significant. For example, among the 15 scatter plots present in Figure 4.2, only in two scatter plots (*Y*2 versus *X*8; *X*4 versus *X*8) did the *Y*-bar lines not intersect the confidence band. Only these two correlations are statistically not significant (Table 4.3).

To perform PCA, input *Y*2, *Y*4, *X*4, *X*8, *X*11, and *X*15 as the multiattributes in (#3). Leave the macro-call field #2 blank to perform PCA. (PCA will not be performed when you input YES to data exploration.) Input the appropriate macro-input values by following the suggestions given in the help file (Appendix 2).

Table 4.3 Pearson Correlation Coefficients and Their Statistical Significance Levels (*p*-values)—PROC CORR Output from FACTOR2—Macro

	Y2	Y4	X4	X8	X11	X15
Y2	1	−0.59[a]	0.78	0.05	0.45	0.64
Midrange price (in $1000)		<0.0001[b]	<0.0001	0.5817	<0.0001	<0.0001
Y4	−0.59	1	−0.67	−0.41	−0.72	−0.84
City MPG (miles per gallon by EPA rating)	<0.0001		<0.0001	<0.0001	<0.0001	<0.0001
X4	0.78	−0.67	1	0.009	0.64	0.73
HP (maximum)	<0.0001	<0.0001		0.9298	<0.0001	<0.0001
X8	0.05	−0.41	0.009	1	0.48	0.55
Passenger capacity (persons)	0.5817	<0.0001	0.9298		<0.0001	<0.0001
X11	0.45	−0.72	0.64	0.48	1	0.87
Car width (inches)	<0.0001	<0.0001	<0.0001	<0.0001		<0.0001
X15	0.64	−0.84	0.73	0.55	0.87	1
Weight (pounds)	<0.0001	<0.0001	<0.0001	<0.0001	<0.0001	

[a] Correlation coefficient.
[b] Statistically significant (*p*-value).

In PCA analysis, the dimensions of standardized multiattributes define the number of eigenvalues. An eigenvalue greater than 1 indicates that PC accounts for more of the variance than one of the original variables in standardized data. This can be confirmed by visually examining the improved scree plot (Figure 4.3) of eigenvalues and the parallel analysis of eigenvalues. This enhanced scree plot shows the rate of change in the magnitude of the eigenvalues for an increasing number of PCs. The rate of decline levels off at a given point in the scree plot that indicates the optimum number of PC to extract. Also, the intersection point between the scree plot and the parallel analysis plot reveals that the first two eigenvalues that account for 86.2% of the total variation could be retained as the significant PC (Figure 4.3).

If the data is standardized, that is, normalized to zero mean and 1 standard deviation, the sum of the eigenvalues is equal to the number of variables used. The magnitude of the eigenvalue is usually expressed as a percentage of the total variance. The information in Table 4.4 indicates that the first eigenvalue accounts for about 66% of the variation, the second for 20%, and the proportions drop off gradually for the rest of the eigenvalues. Cumulatively, the first two eigenvalues together account for 86% of the variation in the dataset. A two-dimensional view (of the six-dimensional dataset) can be created by projecting all data points onto the

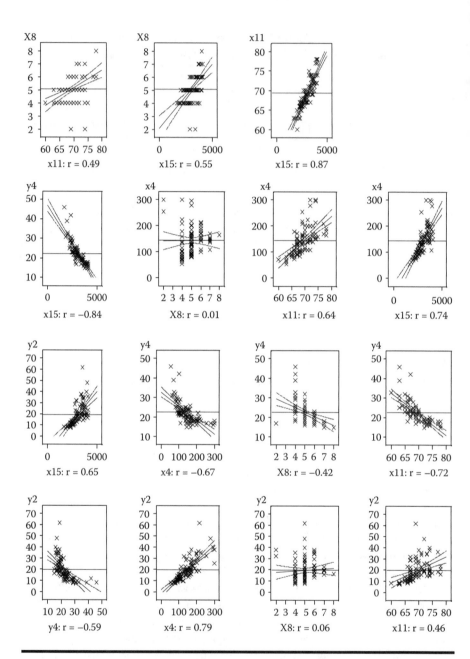

Figure 4.2 Scatter plot matrix illustrating the degree of linear correlation among the five attributes derived using the SAS macro FACTOR2.

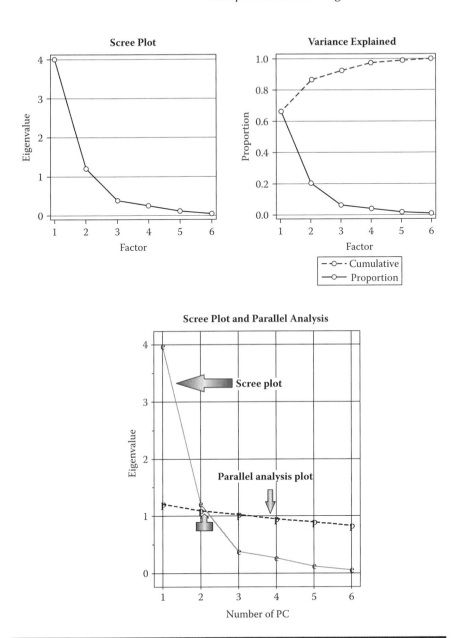

Figure 4.3   PCA scree plot (New: ODS Graphics feature) illustrating the relationship between number of PCs and the rate of decline of eigenvalue and the parallel analysis plot derived using the SAS macro FACTOR2.

Table 4.4   Eigenvalues in Principal Component Analysis—PROC FACTOR Output from FACTOR2 Macro

	Eigenvalue[a]	Difference	Proportion	Cumulative
1	3.97215807	2.76389648	0.6620	0.6620
2	1.20826159	0.83065407	0.2014	0.8634
3	0.37760751	0.11365635	0.0629	0.9263
4	0.26395117	0.14171050	0.0440	0.9703
5	0.12224066	0.06645966	0.0204	0.9907
6	0.05578100	—	0.0093	1.0000

[a] Eigenvalues of the correlation matrix: Total = 6, average = 1.

plane defined by the axes of the first two PC. This two-dimensional view will retain 86% of the information from the six-dimensional plot.

The new variables PC_1 and PC_2 are the linear combinations of the six standardized variables, and the magnitude of the eigenvalues accounts for the variation in the new PC scores. The eigenvectors presented in Table 4.5 provide the weights for transforming the six standardized variables into PCs. For example, PC_1 is derived by performing the following linear transformation using these eigenvectors.

$$PC_1 = 0.37781Y_1 - 0.44702Y_2 + 0.41786X_4 + 23403X_8 + 0.43847X_{11} + 0.48559X_{15}$$

The sum of the squared of eigenvectors for a given PC is equals to one.

PC loadings presented in Table 4.6 are the correlation coefficient between the first two PC scores and the original variables. They measure the importance of each variable

Table 4.5   Eigenvectors in PCA Analysis: PROC FACTOR Output from FACTOR2 Macro

	Variables	Eigenvectors	
		1	2
Y2	Midrange price (in $1000)	0.37781	−0.44215
Y4	City MPG (miles per gallon by EPA rating)	−0.44702	−0.05055
X4	HP (maximum)	0.41786	−0.42666
X8	Passenger capacity (persons)	0.23403	0.75256
X11	Car width (inches)	0.43847	0.19308
X15	Weight (pounds)	0.48559	0.12758

Table 4.6 Principal Component (PC) Loadings for the First Two PC: PROC FACTOR Output from FACTOR2 Macro

	Variables	FACTOR (PC) 1	FACTOR (PC) 2
Y2	Midrange price (in $1000)	0.75298	−0.48602
Y4	City MPG (miles per gallon by EPA rating)	−0.89092	−0.05557
X4	HP (maximum)	0.83281	−0.46899
X8	Passenger capacity (persons)	0.46643	0.82722
X11	Car width (inches)	0.87388	0.21224
X15	Weight (pounds)	0.96780	0.14024

in accounting for the variability in the PC. That is, the larger the loadings in absolute terms, the more influential the variables are in forming the new PC and vice versa. A high correlation between PC_1 and midrange price, city MPG, HP, car width, and weight indicate that these variables are associated with the direction of the maximum amount of variation in this dataset. The first PC loading patterns suggest that heavy, big, very powerful, and highly priced cars are less fuel efficient. A strong correlation between passenger capacity and PC_2 indicates that this variable is mainly attributed to the passenger capacity of the vehicle responsible for the next largest variation in the data perpendicular to PC_1. A visual display of the degree and the direction of PC loadings is presented in Figure 4.4. The regression plot between PC scores and the original variables derived using the SAS macro FACTOR2 displays the statistical significance of the linear association between the original variable and the derived PC scores (Figure 4.5).

A partial list of the first two PC scores presented in Table 4.7 is the scores computed by the linear combination of the standardized variables using the eigenvectors as the weights. The cars that have small negative scores for the PC_1 are less expensive, small, and less powerful, but they are highly fuel efficient. Similarly, expensive, large, and powerful cars with low fuel efficiency are listed at the end of the table with the large positive PC_1 scores.

A biplot display of both PC (PC_1 and PC_2) scores and PC loadings (Figure 4.6) is very effective in studying the relationships within observations, between variables, and the interrelationship between observations and the variables. The X-Y axis of the biplot of PCA analysis represents the standardized PC_1 and PC_2 scores, respectively. In order to display the relationships among the variables, the PC loading values for each PC are overlaid on the same plot after being multiplied by the corresponding maximum value of PC. For example, PC_1 loading values are multiplied by the maximum value of the PC_1 score, and the PC_2 loadings are multiplied by the maximum value of the PC_2 scores. This transformation places both the variables and the observations on the same scale in the biplot display since the range of PC loadings is usually shorter than the PC scores.

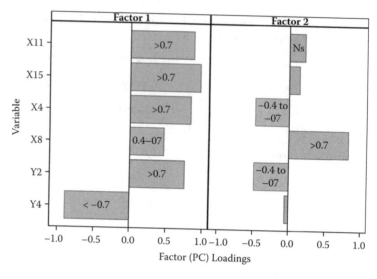

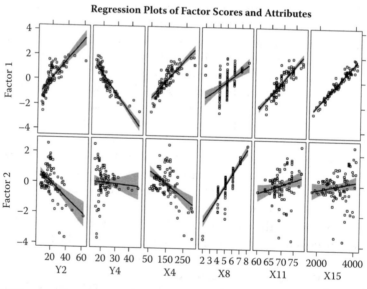

Figure 4.4 Factor loadings plot. (New: Statistical graphics feature and the regression plot between PC scores and the original variables derived using the SAS macro FACTOR2.)

Cars having larger (>75% percentile) or smaller (<25% percentile) PC scores are only identified by their ID numbers on the biplot to avoid crowding of too many ID values. Cars with similar characteristics are displayed together in the biplot observational space since they have similar PC$_1$ and PC$_2$ scores. For example, small compact cars with relatively higher gas mileage such as "Geo Metro (ID 12)" and

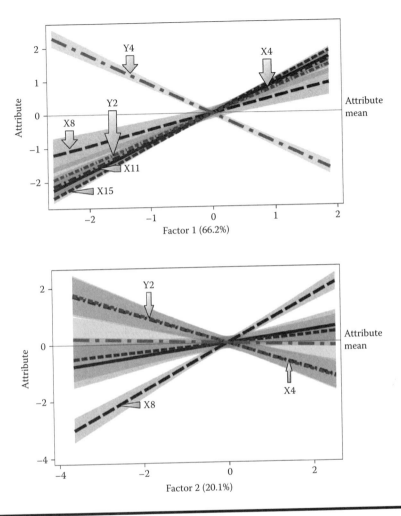

Figure 4.5 Assessing the significance and the nature of factor loadings. (New: Statistical graphics feature) derived using the SAS macro FACTOR2.

"Ford Fiesta (ID 7)" are grouped closer. Similarly, cars with different attributes are displayed far apart since their PC_1, PC_2, or both PC scores are different. For example, small compact cars with relatively higher gas mileage such as "Geo Metro (ID 12)" and large expensive cars with relatively lower gas mileage such as "Lincoln Town car (ID 42)" are separated far apart.

The correlations among the multivariate attributes used in the PCA analysis are revealed by the angles between any two PC loading vectors. For each variable, a PC load vector is created by connecting the *X-Y* origin (0,0) and the multiplied value of PC_1 and PC_2 loadings in the biplot. The angles between any two variable vectors will be:

Table 4.7 Standardized and Sorted PC Scores for the First Two Components: PROC FACTOR Output from FACTOR2 Macro

Top 10% of the Cases Based on PC_1 Scores

ID	Midrange Price (in $1000)	City MPG (Miles Per Gallon by EPA Rating)	HP (Maximum)	Passenger Capacity (Persons)	Car Width (Inches)	Weight (Pounds)	PC_1	PC_2
77	40.1	16	295	5	74	3935	1.87883	-1.59784
78	47.9	17	278	5	72	4000	1.83435	-1.88502
50	36.1	18	210	6	77	4055	1.71747	0.00432
52	16.6	15	165	8	78	4025	1.54625	2.53289
27	23.7	16	180	6	78	4105	1.51272	0.81571
25	61.9	19	217	5	69	3525	1.41424	-2.26522
31	20.9	18	190	6	78	3950	1.35396	0.81134
33	34.7	16	200	6	73	3620	1.31725	-0.11842
87	25.8	18	300	4	72	3805	1.25524	-1.83327

Bottom 10% of the Cases Based on PC_1 Scores

ID	Midrange Price (in $1000)	City MPG (Miles per Gallon by EPA Rating)	HP (Maximum)	Passenger Capacity (Persons)	Car Width (Inches)	Weight (Pounds)	PC_1	PC_2
66	8.4	46	55	4	63	1695	-2.57996	-0.35560
84	8.6	39	70	4	63	1965	-2.12509	-0.36468
45	8.4	33	73	4	60	2045	-2.01915	-0.45320
54	7.4	31	63	4	63	1845	-1.90694	-0.22098
4	12.1	42	102	4	67	2350	-1.65615	-0.51046
72	9.8	32	82	5	65	2055	-1.50750	0.32332
74	9	31	74	4	66	2350	-1.44827	-0.13032
17	8	29	81	5	63	2345	-1.42375	0.39433
3	9.1	25	81	4	63	2240	-1.39891	-0.29835

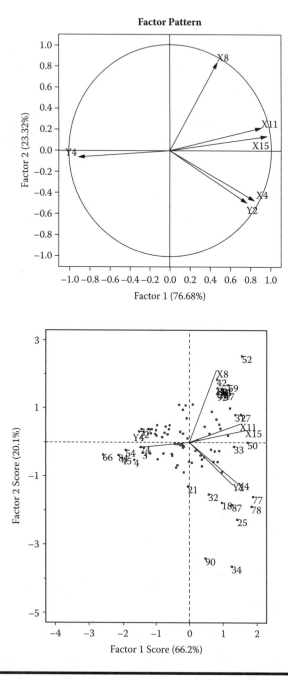

Figure 4.6 Factor pattern plot (New: ODS Graphics feature) and the biplot display of interrelationship between the first two PC scores and PC loadings derived using the SAS macro FACTOR2.

1. Narrower (<45°) if the correlations between these two attributes are positive and larger (e.g., Y_2 (midprice) and X_4(HP)).
2. Wider (around 90°) if the correlation is not significant (e.g., Y_2 (midprice) and X_8 (passenger capacity)).
3. Closer to 180° (>135°) if the correlations between these two attributes are negative and stronger (e.g., Y_4 (city gas mileage) and X_{15} (physical weight)).

Similarly, a stronger association between some cars (observations) and specific multiattributes (variables) (e.g., gas mileage (Y_4) and "Geo Metro" (ID:12); X_8 (passenger capacity) and "Chevrolet Astro Van" (ID 48)) are illustrated by the closer positioning near the tip of the variable vector.

4.7.3 Case Study 2: Maximum Likelihood FACTOR Analysis with VARIMAX Rotation of 1993 Car Attribute Data

4.7.3.1 Study Objectives

1. Suitability of multiattribute data for factor analysis: Check the multiattribute data for multivariate outliers and normality and for sampling adequacy for factor analysis.
2. Latent factor: Identify the latent common factors that is responsible for the observed significant correlations among the five car attributes.
3. Factor Scores: Group or rank the cars in the dataset based on the extracted common factor scores generated by an optimally weighted linear combination of the observed attributes.
4. Interrelationships: Investigate the interrelationship between the cars and the multiattributes, and group similar cars and similar attributes together based on factor scores and factor loadings.

4.7.3.2 Data Descriptions

Data Name	SAS Dataset CARS93
Multiattributes	Y2: Midprice
	Y4: City gas mileage/gallon
	X4: HP
	X11: Width of the vehicle
	X15: Physical weight
Number of observations	92
Data source[18]	Lock, R. H. (1993)

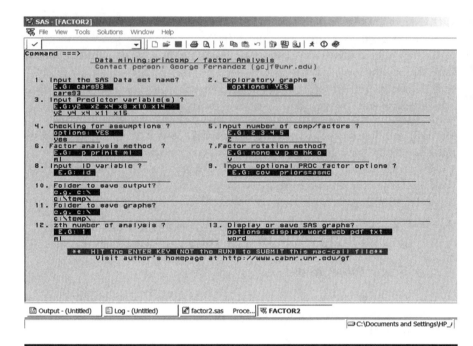

Figure 4.7   Screen copy of FACTOR2 macro-call window showing the macro-call parameters required for performing EFA using the ML method.

Open the FACTOR2.SAS macro-call file in the SAS EDITOR window, and click RUN to open the FACTOR2 macro-call window (Figure 4.7). Refer to the FACTOR2 macro help file in Appendix 2 for detail information about the features and options available.

Input $Y2$, $Y4$, $X4$, $X11$, and $X15$ as the multiattributes in macro input field (#3). Leave the #2 field blank to skip EDA and to perform ML factor analysis.

To check for both multivariate normality and the presence of influential outliers, input YES in the field #4.

Also, input ML in macro field #6 for factor analysis method and V in field #7 for VARIMAX factor rotation method.

Submit the FACTOR2 macro and SAS will output multivariate normality graphs and outlier detection statistics, and complete results from ML factor analysis. Only selected output and graphics generated by the FACTOR2 macro are presented and interpreted in Figure 4.7.

Exploratory analysis. Descriptive statistics, correlation matrix, and the scatter plot analysis of these five multiattributes are covered in PCA analysis Case Study 1. Measuring the internal consistency of multiattributes used in EFA using Cronbach's alpha is a new feature added to FACTOR2 macro. Evaluating the internal consistency of multiattributes is useful if the multiattributes are subscales of a latent factor.

The raw and standardized overall Cronbach's alpha, change in the Cronbach's alpha when a variable is deleted from the analysis, and 95% confidence interval estimates for the standardized Cronbach's alpha by various methods are presented in Table 4.8. However, assessing the internal consistency of the five attributes used in this case study is not meaningful, because these attributes are not subscale measurements of a common index variable. Therefore, the overall standardized Cronbach's alpha is 0.43 and is not expected to exceed 0.7. The change in Cronbach coefficient alpha after deleting Y_4 showed a significant increase in the standardized alpha coefficients to above 0.899, indicating that Y_4, city gas mileage/gallon, is not related to other variables, and deleting this variable improves the overall internal consistency.

Checking for multivariate normality assumptions and performing ML FACTOR analysis: Multivariate normality is a requirement in ML factor analysis, especially for testing the hypothesis on number of factors. Checking for multivariate normality and influential observations in multivariate data provides a valuable insight on the distribution aspects and influential effects on correlation among the multiattributes as well.

In the FACTOR2 macro, checking for multivariate normality is performed by computing Mardia's multivariate skewness, kurtosis measures, and testing by chi-square tests.[19] The estimates of multivariate normality test statistics and the corresponding *p*-values are presented in Table 4.9. Moderately large multivariate skewness and kurtosis values and highly significant *p*-values clearly indicated that the distribution of five multiattributes used in the EFA does not satisfy multivariate normality. This is confirmed by assessing visually the chi-squared quantile-quantile plot of the squared Mahalanobis distances (Figure 4.8). A strong departure from the 450 reference line clearly indicated that the distribution is not multivariate normal. However, performing EFA and interpreting factor scores are not affected by a violation in multivariate normality assumptions. However, the hypothesis tests on a number of factors extracted in the ML factor method were affected by the severe departure from multivariate normality.

Checking for the presence of multivariate influential observations is performed by computing robust distance square statistic (RDSQ) and the difference between RDSQ and the quantiles of the expected chi-square value (DIFF). Multivariate influential observations are identified when the DIFF values exceed 2.5. The estimates of RDSQ and DIFF values for the eight multivariate influential observations are presented in Table 4.9. Very small, compact, inexpensive cars, some sports cars, and very expensive luxury cars are identified as most influential observations. The presence of multivariate influential observations is also visually assessed by a graphical display of DIFF values versus the quantile of chi-square values (Figure 4.9). The impact of these extreme influential observations on EFA analysis outcomes can be verified by excluding these extreme observations one at a time and examining the effects on hypothesis tests and factor loadings.

Table 4.8 Overall Standardized Cronbach Coefficient Alpha after Deleted Variable and Various 95% CI Estimates PROC CORR and PROC IML Output from FACTOR2 Macro

Deleted Variable	Correlation with Total	Raw Cronbach Alpha	Standardized Correlation with Total	Standardized Cronbach Alpha	Label
Y2	0.670	0.147	0.644	0.014	Midrange price (in $1000)
Y4	–0.8469	0.189	–.807	0.899	City MPG (miles per gallon by EPA rating)
X4	0.743	0.021	0.785	–0.131	Horsepower (maximum)
X11	0.871	0.159	0.617	0.041	Car width (inches)
X15	0.737	0.228	0.727	–0.069	Weight (pounds)
Cronbach alpha	Raw: –0.161	Standardized = 0.43			
Confidence intervals method for standardized Cronbach alpha	95% Lower	95% Upper			
Bonett	0.214	0.589			
Feldt	0.226	0.596			
Fisher	0.250	0.584			

Hakstain and Whalen	0.231	0.599
Iacobucci and Duchachek	0.314	0.549
Koning and Frances Exact	0.227	0.595
Koning and Frances Asymptotic	0.216	0.588
Asymptotic distribution free	0.285	0.579

Note: Number of observations: 93.

Number of items: 5.

Confidence interval measurements are estimated based on Kromrey et al.[22]

Table 4.9 Multivariate Influential/Outlier Observations Obtained Using PROC IML from FACTOR2 Macro

ID	Model	RDSQ	CHISQ	Diff = (RDSQ – CHISQ) > 2.5
25	Mercedes-Benz 300E	50.6050	16.5511	34.0539
66	Geo Metro	33.6640	13.8929	19.7711
4	Honda Civic	32.0837	12.6235	19.4601
87	Dodge Stealth	28.8156	11.7724	17.0432
34	Chevrolet Corvette	22.5587	11.1273	11.4314
90	Mazda RX-7	18.8527	10.6057	8.2470
78	Infiniti Q45	12.9521	10.1665	2.7856
42	Volkswagen Euro van	12.4927	9.7863	2.7064

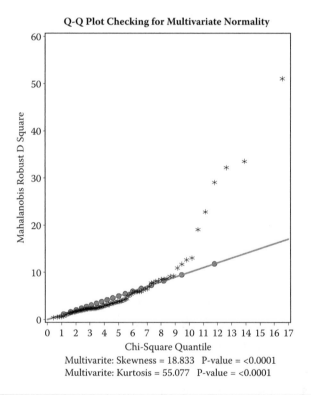

Figure 4.8 Quantile-quantile plot for assessing multivariate normality ML FACTOR analysis derived using the SAS macro FACTOR2.

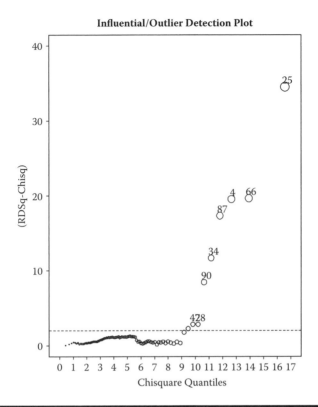

Figure 4.9 Multivariate outlier detection plot based on robust squared distances derived using the SAS macro FACTOR2.

Assessing the appropriateness of common factor analysis: A common (C) factor is an unobservable latent factor that contributes to the variance of at least two of the observed variables. A unique (U) factor is an unobservable hypothetical variable that contributes to the variance of only one of the observed variables. The model ($X1 = C + U$) for common factor analysis has one unique factor for each observed variable. If the multivariate data are appropriate for common factor analysis, the partial correlations between any two attributes controlling all the other variables should be small compared to the original correlations presented in Table 4.2. The partial correlation between the variables $Y2$ and $X4$, for example, is 0.63 (Table 4.10), slightly less than the original correlation of −0.78 in absolute values. The partial correlation between $Y2$ and $Y4$ is −0.02, which is much smaller in absolute value than the original correlation −0.58; this is a very good indication of appropriateness for common factor analysis. In general, all pairs of partial correlations are smaller than the original simple correlation. Only two out of the ten all-possible partial correlations are significant based on the default criteria (Table 4.10).

In addition to examining the partial correlation coefficients, Kaiser's measure of sampling adequacy (MSA) and the residual correlations also provide very effective

Table 4.10 Partial Correlations Coefficients Matrix (Factor Method: ML; Priors = SMC) PROC FACTOR Output from FACTOR2 Macro

	Partial Correlations Controlling All Other Variables					
	Label	Y2	Y4	X4	X11	X15
Y2	Midrange price (in $1000)	100*	−2	63*	−37	32
Y4	City MPG (miles per gallon by EPA rating)	−2	100*	−9	5	−52
X4	HP (maximum)	63*	−9	100*	23	3
X11	Car width (inches)	−37	5	23	100*	72*
X15	Weight (pounds)	32	−52	3	72*	100*

Note: Prior communality estimates computed based on squared multiple correlations (SMC). Printed correlation values are multiplied by 100 and rounded to the nearest integer. Values greater than 56.36 are flagged by an asterisk.

measures of assessing the suitability of performing common factor analysis. The Kaiser's MSA value ranges from 0 to 1. For each variable and for all variables together, it provides a composite index quantifying how much smaller the partial correlations are in relation to the original simple correlations. Values of 0.8 or 0.9 are considered good, while MSAs below 0.5 are unacceptable.[20] All five attributes used in this study have acceptable MSA values, and the overall MSA value (0.77) (Table 4.11) also falls in the acceptable region, confirming the suitability of the data for performing common factor analysis.

Table 4.12 shows the prior communality estimates for five attributes used in this analysis. The squared multiple correlations (SMCs), which are given in Table 4.12, represent the proportion of variance of each of the five attributes shared by all remaining attributes. A small communality estimate might indicate that the attribute may need to be modified or even dropped from the analysis. The Y_2 attribute has the prior communality estimate of 0.681, which means about 68% of the variance of the Y_2 midprice is shared by all other attributes included in the analysis, indicating that this attribute shares a common variance among the other attributes. Similarly, the prior communality estimates of the other four attributes are also high (>0.65), indicating that all five attributes contribute to a common variance in the data. The sum of all prior communality estimates, 3.81, is the estimate of the common variance among all attributes. This initial estimate of the common variance constitutes about 76% of the total variance present among all five attributes.

The amount of variances accounted for by each factor is presented in Table 4.13 as eigenvalues. The number of eigenvalues reported should equal the number of variables included in the study. Following the column of eigenvalues are three measures of each eigenvalue's relative size and importance. The first of these displays the difference between the eigenvalue and its successor. The last two columns display

Table 4.11   Kaiser's Measure of Sampling Adequacy (Factor Method: ML; Priors = SMC) and Computed by the PROC FACTOR Output from FACTOR2 Macro

Variable				
Y2	Y4	X4	X11	X15
0.718	0.879	0.817	0.730	0.734

Note: Kaiser's measure of sampling adequacy: Overall MSA = 0.771.

Table 4.12   Prior Communality Estimates (Factor Method: ML; Priors = SMC) and Computed by the PROC FACTOR Output from FACTOR2 Macro

Variables				
Y2	Y4	X4	X11	X15
0.681	0.717	0.728	0.798	0.886

Table 4.13   Preliminary Eigenvalues (Factor Method: ML) Computed by the PROC FACTOR in FACTOR2 Macro

	Eigenvalue	Difference	Proportion	Cumulative
1	18.3024528	16.7416537	0.9578	0.9578
2	1.5607991	1.4718931	0.0817	1.0395
3	0.0889060	0.3824374	0.0047	1.0442
4	−0.2935315	0.2567970	−0.0154	1.0288
5	−0.5503285	—	−0.0288	1.0000

Note: Prior communality estimates derived by squared multiple correlations (SMC). Preliminary Eigenvalues: total 19.1082979; Average = 3.82165958.

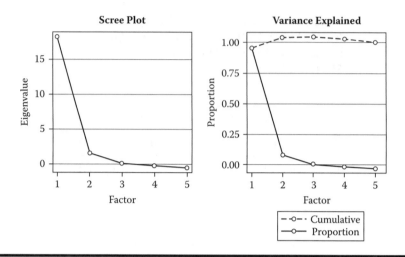

Figure 4.10   Scree plot illustrating the relationship between number of factors and the rate of decline of eigenvalue in ML FACTOR analysis (New: Statistical graphics feature) derived using the SAS macro FACTOR2.

the individual and cumulative proportions that the corresponding factor contributes to the total variation. The first two eigenvalues account for almost all the variability (95% and 5%) in the five-dimensional dataset.

Note: Prior communality estimates derived by squared multiple correlations (SMC). Preliminary Eigenvalues: total 19.1082979; average = 3.82165958

Determining the number of latent factors: In common factor analysis, the dimensions of standardized multiattributes define the number of extractable eigenvalues. An eigenvalue greater than one indicates that the factor accounts for more variance than one of the original variables in standardized data. This can be confirmed by visually examining the scree plot (Figure 4.10) of eigenvalues. A scree plot of the eigenvalues (Figure 4.10) could be used to determine the number of meaningful factors. The scree plot shows the rate of change in the magnitude of the eigenvalues for an increasing number of factors. The rate of decline levels off where the scree plot bends, indicating two as the optimum number of factors to extract.

According to the Kaiser and Guttman rule, only two factors can be extracted because only the first two factors have an eigenvalue greater than one (Table 4.13). These two large positive eigenvalues together account for 104% of the common variance, which is close to 100%. Negative eigenvalues occur only due to the restriction that the sum of eigenvalues be set equal to the estimated common variance (communality estimates), and not the total variance in common factor analysis. A large first eigenvalue (18.30) and a much smaller second eigenvalue (1.55) suggest the presence of a dominant global factor.

Assuming multivariate normality and larger samples, chi-square tests can be performed in the ML FACTOR method to test the number of meaningful factors.

Table 4.14 Significant Tests on the Number of Factors Based on Maximum Likelihood FACTOR Analysis Computed by the PROC FACTOR in FACTOR2 Macro

Test	DF	Chi-Square	Pr > ChiSq
H0: No common factors	10	415.9534	<0.0001
HA: At least one common factor			
H0: Two factors are sufficient	1	4.8530	0.0276
HA: More factors are needed			
Chi-square without Bartlett's correction		5.0640493	
Akaike's information criterion		3.0640493	
Schwarz's bayesian criterion		0.5314498	
Tucker and Lewis's reliability coefficient		0.9050865	

The probability levels for the chi-square test are <0.001 for the hypothesis of no common factors, and 0.0276 for two common factors (Table 4.14). Therefore, the two-factor model seems to be an adequate representation and reconfirms the results from scree plot and the Kaiser and Guttman rule. The Tucker and Lewis reliability coefficient, a measure of the goodness of fit of a ML FACTOR analysis, was 0.905. Tucker and Lewis's reliability coefficient indicates good reliability. Reliability is a value between 0 and 1, with a larger value indicating better reliability. The AIC and SBC criteria can be used to compare the significance of different FA solutions for the same data.

Interpreting Common Factors: Table 4.15 shows the initial unrotated factor loadings, which consist of the correlations between the five attributes and the two

Table 4.15 Initial Factor Loadings Computed by the PROC FACTOR in FACTOR2 Macro

		FACTOR1	FACTOR2
Y2	Midrange price (in $1000)	100*	0
X4	HP (maximum)	79*	31
X11	Car width (inches)	46	78*
X15	Weight (pounds)	65	75*
Y4	City MPG (miles per gallon by EPA rating)	−59	−61

Note: Prior communality estimates derived by squared multiple correlations (SMC). Printed values are multiplied by 100 and rounded to the nearest integer. Values greater than 0.650 are flagged by an asterisk.

Table 4.16   VARIMAX-Rotated Factor Loadings Computed by the PROC FACTOR in FACTOR2 Macro

		FACTOR1	FACTOR2
X15	Weight (pounds)	89*	42
X11	Car width (inches)	87*	23
Y4	City MPG (miles per gallon by EPA rating)	−75*	−41
Y2	Midrange price (in $1000)	27	96*
X4	HP (maximum)	52	67*

Note: Prior communality estimates derived by squared multiple correlations (SMCs). Printed values are multiplied by 100 and rounded to the nearest integer. Values greater than 0.650586 are flagged by an asterisk.

retained factors. The correlations greater than 0.649 are flagged by an asterisk. There are some split loadings where X_{15} and Y_4 significantly (>0.59) loaded on more than one factor. A commonly used rule is that there should be at least three variables per common factor. In this case study, the five variables we used are not adequate to extract two common factors following the rule of three variables per factor.

Table 4.16 shows the factor loadings of the two extracted factors after the VARIMAX rotation. The VARIMAX rotation is an orthogonal rotation in which the angle between the reference axes of factors is maintained at 90°. The rotated loading is usually somewhat simpler to interpret; that is, the rotated factor$_1$ can now be interpreted as *"size"* factor. The size variables, X_{15}, X_{11}, and Y_4 load higher on FACTOR$_1$. The large and heavy cars show a negative correlation with the gas mileage as well. The rotated factor seems to measure price performance. The midprice and HP variables load heavily on FACTOR$_2$. Even though the variance explained by the rotated FACTOR$_1$ is less than FACTOR$_2$ (Table 4.17), the cumulative variance explained by both common factors remains the same after the VARIMAX, an orthogonal rotation. Also, note that the VARIMAX rotation, as with any orthogonal rotation, has not changed the final communalities. Additionally, visual display of the degree and the direction of factor loadings before and after the rotation are presented in Figure 4.12.

Table 4.17   Percentage Variance Explained by Each Factor after VARIMAX Rotation Computed by the PROC FACTOR (Method: ML; Prior Communality Estimates: SMC) in FACTOR2 Macro

FACTOR	Weighted	Unweighted
FACTOR1	41.542	2.461
FACTOR2	10.0733	1.77

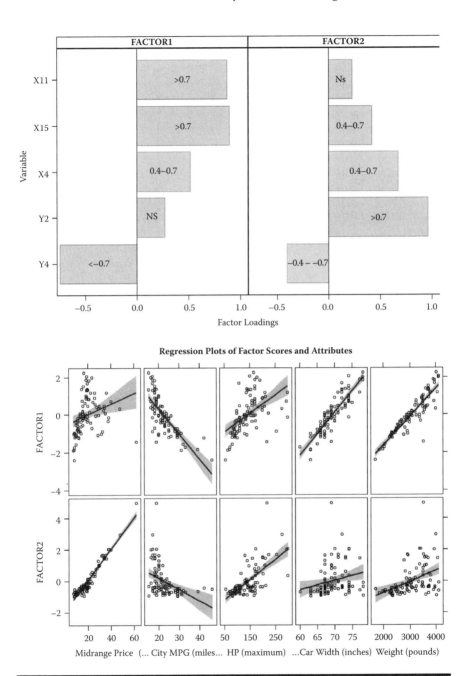

Figure 4.11 ML factor loadings plot (New: Statistical graphics feature) and the regression plot between factor scores and the original variables derived using the SAS macro FACTOR2.

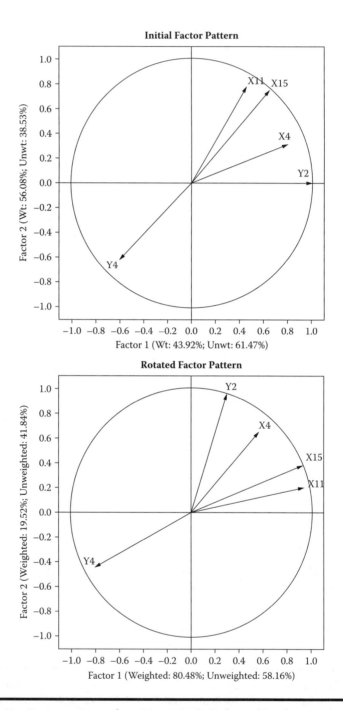

Figure 4.12 Factor pattern plots (New: Statistical graphics feature) before and after rotation derived using the SAS macro FACTOR2.

Checking the validity of common factor analysis: The final communality estimates are the proportion of variance of the variables accounted for by the extracted
common factors. When the factors are orthogonal, the final communalities are
calculated by taking the sum of the squares of each row of the factor loadings.
The final communality estimates show that all the variables are well accounted for
by the two factors, with final communality estimates ranging from 0.72 for X_4 to
1 for Y_2. The variable Y_2 has a communality of 1.0, and therefore has an infinite
weight that is displayed next to the final communality estimate infinite value.
The first eigenvalue is also infinite. Infinite values are ignored in computing the
total of the eigenvalues and the total final communality. The final communality estimates are all fairly close to the prior communality estimates reported in
Table 4.12. Only the communality for the variables Y_2 and X_{15} increased appre-
ciably. The inspection of the partial correlation matrix yields similar results: the
correlations among the five attributes after removing the factor contribution are
very low since the largest is 0.18 (Table 4.19). The root-mean-square off-diagonal
partials are also very low at 0.071876 (Table 4.20). The residual matrix provides
an indication of how well the factor model fit the data. The off-diagonal elements
of the residual correlation matrix (Table 4.21) are all close to 0.02, indicating
that the correlations among the five attributes can be reproduced fairly accurately
from the retained factors. The overall root-mean-square off-diagonal residual is
0.015 (Table 4.22), which is the average of the sum of off-diagonal values in the
residual matrix. These results confirm that the SMCs provide good and optimal
communality estimates.

It is possible to estimate the factor scores or a subject's relative standing on
each of the factors, if the original subject, x variable coordinate data, is available.
Standardized rotated factor score values for part of the observations in the dataset

Table 4.18 Final Communality Estimates and Variable Weights after VARIMAX Rotation Computed by the PROC FACTOR (Method: ML; Prior Communality Estimates: SMC) in FACTOR2 Macro

Variable	Communality	Weight
Y2	1.000	Infty
Y4	0.725	3.64
X4	0.720	3.572
X11	0.810	5.263
X15	0.976	43.134

Note: Total Communality: Weighted = 51.615638; Unweighted = 4.232617.

Table 4.19 Final Partial Correlations Controlling Factors Computed by the PROC FACTOR (Method: ML; Prior Communality Estimates: SMC) in FACTOR2 Macro

Variable	Label	Y2	Y4	X4	X11	X15
Y2	Midrange price (in $1000)	0	0	0	0	0
Y4	City MPG (miles per gallon by EPA rating)	0	100*	−4	11	−3
X4	HP (maximum)	0	−4	100*	18	−8
X11	Car width (inches)	0	11	18	100*	0
X15	Weight (pounds)	0	−3	−8	0	100*

Note: Printed values are multiplied by 100 and rounded to the nearest integer. Values greater than 40.5 are flagged by an asterisk.

Table 4.20 Root-Mean-Square of Off-Diagonal Partials Computed by the PROC FACTOR (Method: ML) in FACTOR2 Macro

Y2	Y4	X4	X11	X15
0.000	0.059	0.099	0.103	0.041

Note: Prior communality estimates: SMC. Overall RMS = 0.071.

Table 4.21 Residual Correlations with Uniqueness on the Diagonal Computed by the PROC FACTOR (Method: ML) in FACTOR2 Macro

	Label	Y2	Y4	X4	X11	X15
Y2	Midrange price (in $1000)	0	0	0	0	0
Y4	City MPG (miles per gallon by EPA rating)	0	27*	−1	2	0
X4	HP (maximum)	0	−1	28*	4	−1
X11	Car width (inches)	0	2	4	19*	0
X15	Weight (pounds)	0	0	−1	0	2

Note: Prior communality estimates: SMC. Printed values are multiplied by 100 and rounded to the nearest integer. Values greater than 8.8 are flagged by an asterisk.

Table 4.22 Root-Mean-Square (RMS) Off-Diagonal Residuals Computed by the PROC FACTOR (Method: ML) in FACTOR2 Macro

Y2	Y4	X4	X11	X15
0.000	0.013	0.021	0.023	0.003

Note: Prior Communality Estimates: SMC. Overall = 0.015.

are presented in Table 4.23. These factor scores can be used to rank the cars or develop scorecards based on size and price factors.

Investigating interrelationship between multiattributes and observations: Biplot display of both factor (FACTOR1 and FACTOR2) scores and factor loadings (Figure 4.13) is very effective in studying the relationships within observations, between variables, and the interrelationship between observations and variables. The *X-Y* axis of the biplot of rotated factor analysis represents the standardized FACTOR1 and FACTOR2 scores, respectively. In order to display the relationships among the variables, the factor loading for each factor is overlaid on the same plot after being multiplied by the corresponding maximum value of factor score. For example, FACTOR1 loading values are multiplied by the maximum value of FACTOR1 score, and FACTOR2 loadings are multiplied by the maximum value of FACTOR2 scores. This transformation places both the variables and the observations on the same scale in the biplot display since the range of factor loadings are usually shorter (−1 to +1) than the factor scores.

Cars having larger (>75% percentile) or smaller (<25% percentile) factor scores are only identified by their ID numbers on the biplot to avoid overcrowding of labeling. Cars with similar characteristics are displayed together in the biplot observational space since they have similar FACTOR1 and FACTOR2 scores. For example, small compacts cars with a relatively higher gas mileage, such as Geo Metro (ID 12) and Ford Fiesta (ID 7), are grouped closer together. Similarly, cars with different attributes are displayed far apart since their FACTOR1, FACTOR2, or both factor scores are different. For example, small compact cars with a relatively higher gas mileage, such as Geo Metro (ID 12), and large expensive cars with a relatively lower gas mileage, such as Lincoln Town car (ID 42), are separated far apart in the biplot display.

The correlations among the multivariate attributes used in the factor analysis are revealed by the angles between any two factor-loading vectors. For each variable, a factor-loading vector is created by connecting the origin (0,0) and the multiplied value of FACTOR1 and FACTOR2 loadings on the biplot. The angles between any two variable vectors will be:

■ Narrower (<45°) if the correlations between these two attributes are positive and larger—for example, *Y*2 (midprice) and *X*4 (HP).

Table 4.23 VARIMAX-Rotated Factor Scores Computed by the PROC FACTOR (Method: ML; Prior Communality Estimates: SMC) in FACTOR2 Macro

Top 10% of the Cases Based on FACTOR1 Scores

ID	Midrange Price (in $1000)	City MPG (Miles per Gallon by EPA Rating)	HP (Maximum)	Car Width (Inches)	Weight (Pounds)	FACTOR1	FACTOR2
52	16.6	15	165	78	4025	2.23778	-0.94645
59	19.1	17	151	74	4100	2.05867	-0.62677
27	23.7	16	180	78	4105	1.98739	-0.11167
22	18.8	17	170	77	3910	1.85890	-0.60250
31	20.9	18	190	78	3950	1.85000	-0.37404
42	19.7	17	109	72	3960	1.68177	-0.45553
81	16.3	18	170	74	3715	1.53655	-0.78025
47	19.5	18	170	74	3715	1.36149	-0.38640
36	14.9	19	140	73	3610	1.36076	-0.88112

66		8.4		46	55	63	1695	-2.39356	-0.51784
54		7.4		31	63	63	1845	-1.87937	-0.77097
84		8.6		39	70	63	1965	-1.83132	-0.65546
45		8.4		33	73	60	2045	-1.70580	-0.71251
72		9.8		32	82	65	2055	-1.57230	-0.59967
25		61.9		19	217	69	3525	-1.44541	4.97000
4		12.1		42	102	67	2350	-1.22432	-0.45069
12		12.2		29	92	66	2295	-1.20263	-0.44607
3		9.1		25	81	63	2240	-1.19046	-0.78305

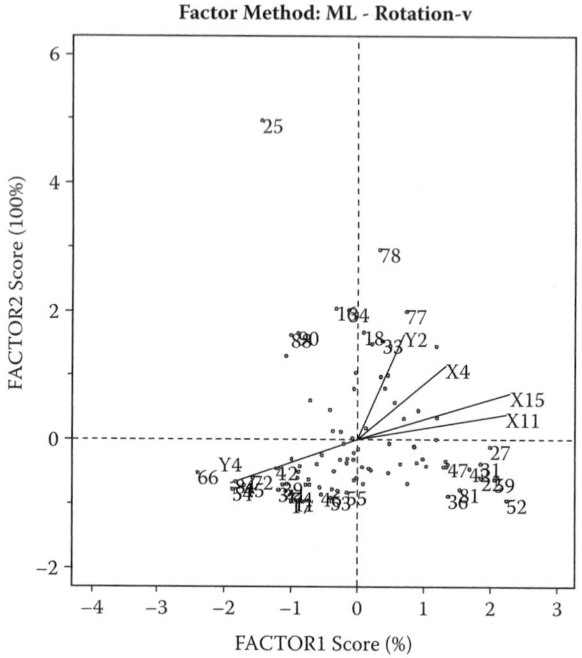

Factor Method: ML - Rotation-v

Figure 4.13 Biplot display of interrelationship between the first two rotated ML FACTOR scores and ML-FACTOR loadings derived using the SAS macro FACTOR2.

■ Wider (around 90°) if the correlation is not significant—for example, *Y*2 (midprice) and *X*11 (car width).

■ Closer to 180° (>135°) if the correlations between these two attributes are negative and stronger—for example, *Y*4 (city gas mileage) and *X*15 (physical weight).

Similarly, a stronger association between some cars (observations) and specific multiattributes (variables)—for example, gas mileage (*Y*4) and Geo Metro (ID 12), *Y*2 (Midprice) and Mercedes-Benz 300E (ID 44)—are illustrated by the closer positioning near the tip of the variable vector.

4.7.3 CASE Study 3: Maximum Likelihood FACTOR Analysis with VARIMAX Rotation Using a Multivariate Data in the Form of Correlation Matrix

4.7.3.1 Study Objectives

1. *Suitability of multiattribute data for factor analysis*: Check the multiattribute data for multivariate outliers and normality and for sampling adequacy for factor analysis. This is not feasible since the original data is not available.

2. *Latent factor*: Identify the latent common factors that are responsible for the observed significant correlations among the five car attributes.
3. *Factor scores*: Group or rank the cars in the dataset based on the extracted common factor scores generated by an optimally weighted linear combination of the observed attributes. (Because only the correlation matrix is used as the input, scoring is not feasible.)
4. *Interrelationships*: Investigate the interrelationship between the cars and the multiattributes, and group similar cars and similar attributes together based on factor scores and factor loadings. This option is not feasible since factor scores are not generated. Only the correlation structures between the multiattributes are generated.

4.7.3.2 Data Descriptions

Data name	Special correlation SAS data dataset CARS93 (see Figure 4.15 for a sample correlation data matrix)
Correlation matrix of the five multiattributes	Y2: Midprice
	X2: Number of cylinders
	X4: HP
	X13: Rear seat room length in inches
	X8: Passenger capacity
	X11: Car width in inches
Number of observations	N = 92 is included in the special correlation data matrix
Data source[18]	Lock, R. H. (1993)

■ Open the FACTOR2.SAS macro-call file in the SAS EDITOR window, and click RUN to open the FACTOR2 macro-call window (Figure 4.14). Refer to the FACTOR2 macro help file in Appendix 2 for detailed information about the features and options available.
■ Input the name of the special correlation data matrix (CORRMATRIX) in the macro-call input #1 (see Figure 4.15).

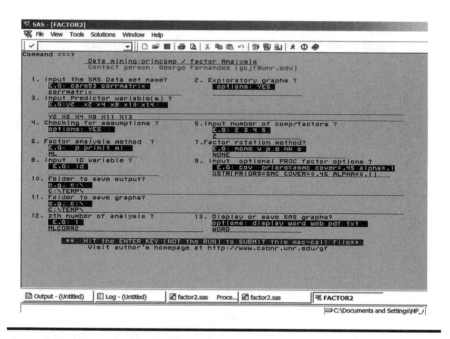

Figure 4.14   Screen copy of FACTOR2 macro-call window showing the macro-call parameters required for performing EFA using the correlation matrix as data input.

	A	B	C	D	E	F	G	H	I	J	K	L	M	N	O	P	Q	R	S	T	U	V
	TYPE	_NAME_	Y1	Y2	Y3	Y4	Y5	X1	X2	X3	X4	X5	X6	X7	X8	X9	X10	X11	X12	X13	X14	X15
1																						
2	MEAN		16.83	19.17	21.52	23.09	29.98	0.854	4.854	2.589	140	5328	2368	16.11	4.939	183.2	103.2	68.9	38.62	27.54	13.89	2988
3	STD		8.922	9.959	11.45	5.594	5.011	0.722	1.297	1.004	51.06	583.7	497.3	3.01	0.709	15.27	6.467	3.694	3.165	2.841	2.998	565.9
4	N		82	82	82	82	82	82	82	82	82	82	82	82	82	82	82	82	82	82	82	82
5	CORR	Y1	1	0.971	0.91	-0.66	-0.66	0.585	0.738	0.705	0.795	-0.09	-0.49	0.736	0.327	0.611	0.676	0.544	0.478	0.418	0.413	0.777
6	CORR	Y2		1	0.983	-0.62	-0.63	0.595	0.675	0.642	0.787	-0.03	-0.44	0.707	0.274	0.552	0.628	0.495	0.425	0.342	0.367	0.737
7	CORR	Y3			1	-0.57	-0.57	0.578	0.599	0.567	0.75	0.011	-0.38	0.656	0.222	0.485	0.566	0.438	0.366	0.269	0.315	0.677
8	CORR	Y4				1	0.945	-0.44	-0.68	-0.75	-0.7	0.327	0.712	-0.8	-0.46	-0.72	-0.67	-0.69	-0.65	-0.34	-0.49	-0.84
9	CORR	Y5					1	-0.41	-0.64	-0.67	-0.69	0.234	0.608	-0.74	-0.41	-0.62	-0.57	-0.59	-0.56	-0.28	-0.37	-0.78
10	CORR	X1						1	0.478	0.519	0.549	-0.15	-0.41	0.497	0.248	0.504	0.535	0.568	0.537	0.228	0.289	0.598
11	CORR	X2							1	0.888	0.79	-0.37	-0.74	0.725	0.42	0.74	0.737	0.798	0.645	0.435	0.586	0.832
12	CORR	X3								1	0.772	-0.52	-0.82	0.821	0.526	0.852	0.859	0.89	0.787	0.517	0.681	0.917
13	CORR	X4									1	0.024	-0.59	0.796	0.241	0.643	0.663	0.689	0.598	0.279	0.359	0.86
14	CORR	X5										1	0.507	-0.24	-0.38	-0.44	-0.4	-0.49	-0.48	-0.27	-0.52	-0.35
15	CORR	X6											1	-0.63	-0.48	-0.75	-0.73	-0.79	-0.72	-0.42	-0.59	-0.78
16	CORR	X7												1	0.451	0.794	0.793	0.772	0.674	0.445	0.613	0.9
17	CORR	X8													1	0.646	0.649	0.516	0.49	0.74	0.653	0.51
18	CORR	X9														1	0.911	0.885	0.811	0.585	0.713	0.88
19	CORR	X10															1	0.85	0.746	0.66	0.734	0.879
20	CORR	X11																1	0.821	0.429	0.673	0.869
21	CORR	X12																	1	0.462	0.585	0.787
22	CORR	X13																		1	0.652	0.486
23	CORR	X14																			1	0.637
24	CORR	X15																				1

Figure 4.15   Screen copy of the suitable correlation data format required for performing EFA using correlation matrix data.

- Input $Y2$, $X2$, $X4$, $X8$, $X11$, and $X13$ as the multiattributes in macro input field (#3) in the second line (leave the input line 1 blank—see Figure 4.14).
- Leave the #2 field blank to skip EDA and to perform ML factor analysis.
- No multivariate normality check: leave field #4 blank.
- Also, input ML in macro field #6 for factor analysis method and NONE or V in field #7 for VARIMAX factor rotation method.
- Input the following string in the macro call #9: %str (PRIORS=SMC COVER=0.45 ALPHA=0.1).
- PRIORS=SMC: Use SMC as the prior method.
- COVERS=0.45. Request estimating confidence interval for factor loading and the cutoff value for detecting significance $= 0.45$.
- Use 90% as the level of confidence in estimating CI for the factor loadings.

Because the output generated when performing ML factor analysis is discussed in detail in Case Study 3, only the selected new output is discussed in the following text.

1. The significance of factor loadings based on 90% coverage display estimation is presented in Table 4.24. This new option is available in SAS 9.2 when the ML factor analysis method is chosen.
   Further explanation of the 90% CI coverage display of the factor loadings is reported in Table 4.24.

Table 4.24 Factor Pattern with 90% Confidence Limits Cover |*| = 0.45 Computed by the PROC FACTOR Using FACTOR2 Macro

Variables	FACTOR1	FACTOR2
X2	0.87086[a]	−0.11854
	0.05882[b]	0.39272
	0.73346[c]	−0.64941
	0.93988[d]	0.49001
	0*[][e]	[*0]
X4	0.86965	−0.38729
	0.18116	0.36739
	0.10853	−0.80741
	0.98798	0.29344
	0[*]	[*0]

(continued)

Table 4.24 Factor Pattern with 90% Confidence Limits Cover |*| = 0.45 Computed by the PROC FACTOR Using FACTOR2 Macro (Continued)

Variables	FACTOR1	FACTOR2
X11	0.81438	0.03686
	0.04371	0.37462
	0.72891	−0.52278
	0.87485	0.57430
	0*[]	[0*]
Y2	0.76059	−0.25744
	0.12070	−0.34245
	0.48277	−0.69965
	0.89932	0.32738
	0*[]	[*0]
X13	0.56750	0.54525
	0.24361	0.30082
	0.05272	−0.09230
	0.84399	0.86572
	0[*]	[0*]
X8	0.59719	0.72536
	0.35073	0.21338
	−0.20499	0.17633
	0.91945	0.93016
	[0*]	0[*]

Note: (New SAS 9.2 option; Method: ML; Prior Communality Estimates: SMC).
[a] Estimate.
[b] Standard error.
[c] Lower confidence level.
[d] Upper confidence level.
[e] Coverage display.

[0*]	[*0]	The estimate is not significantly different from zero or the
X8-F1	Y2-F2	COVER = 0.45 value. The population value might have been
X13-F2	X4-F2	larger or smaller in magnitude than the COVER = 0.45
X11-F2	X2-F2	value. There is no statistical evidence for the salience of the variable–FACTOR relationship.
0[*]		
X4-F1		The estimate is significantly different from zero but not
X13-F1		from the COVER = 0.45. This is marginal statistical evidence
X8-F2		for the salience of the variable–FACTOR relationship.
0*[]		
X2-F1		The estimate is significantly different from zero, and the
X11=F1		CI covers a region of values that are larger in magnitude than the COVER = 0.45. This is strong statistical evidence
Y2-F1		for the salience of the variable–FACTOR relationship.

Note: $F1$ = FACTOR1; $F2$ = FACTOR2.

2. Scree plot, variance-explained plot, and initial factor loadings and the association between loadings and the FACTOR1 and FACTOR2 are presented in Figure 4.16. These graphics are produced by the SAS ODS graphics option in SAS version 9.2. Based on the scree plot, two factor models are selected.

4.8 Disjoint Cluster Analysis Using SAS Macro DISJCLS2

The DISJCLS2 macro is a powerful SAS application for exploring and visualizing multivariate data and for performing disjoint cluster analysis using the k-means algorithms. The SAS procedure FASTCLUS[20,21] is the main tool used in the DISJCLS2 macro. Also, the SAS/STAT procedures CLUSTER, STEPDISC, CANDISC, GPLOT, BOXPLOT, and IML modules are also utilized in the DISJCLS2 macro to perform enhanced disjoint cluster analysis.

The FASTCLUS procedure is used to extract a user-specified number of clusters based on k-means cluster analysis. To verify the user-specified optimum cluster number, cubic clustering criterion (CCC), pseudo F-statistic (PSF), and pseudo T^2 (PST2) for a number of clusters ranging from 1 to 20 are generated using Ward's method of cluster analysis in PROC CLUSTER. To perform variable selection that significantly discriminates the clusters, the backward selection method in stepwise discriminant analysis using the STEPDISC procedure is used. To test the hypothesis

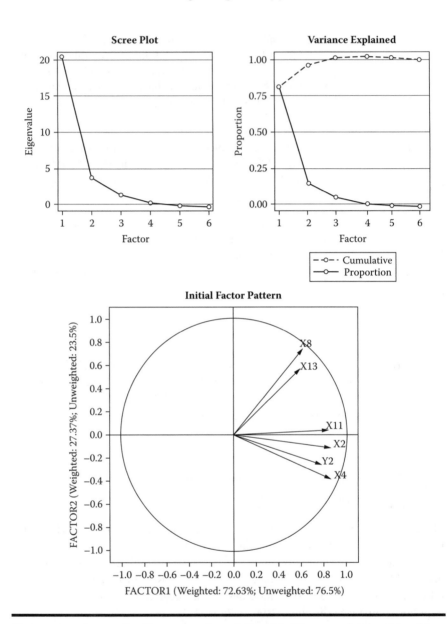

Figure 4.16 Scree plot illustrating the relationship between the number of factors and the rate of decline of eigenvalue and factor pattern plot when performing EFA using correlation matrix as data (New: Statistical graphics features) derived using the SAS macro FACTOR2.

that significant differences exist among the clusters for multiattributes, canonical discriminant analysis based on the CANDISC is used. SAS IML is also used to test the hypothesis of multivariate normality, which is a requirement for canonical discriminant analysis hypothesis testing. PROC GPLOT is used to generate scatter plots by cluster, quantile-quantile plots for testing of multivariate normality, diagnostic plots based on CCC, PSF, and PST2 for selecting the optimum cluster numbers, and biplot display of canonical discriminant analysis. The new SGPLOT procedure is used to generate the box plot matrix of cluster group by multiattribute display. The BOXPLOT procedure is used to show the between-cluster differences for intracluster distances and canonical discriminant functions. For more details of these SAS procedures and the available options, readers are encouraged to see the online references.[21,22] The enhanced features of the DISJCLS2 macro for performing complete cluster analysis are:

- *New*: Data description table, including all numerical and categorical variable names and labels and the macro input values specified by the user in the last run of DISJCLS2 macro submission are generated.
- Scatter plot matrix of all multivariate attributes by cluster groups are displayed.
- Test statistics and *p*-values for testing multivariate skewness and kurtosis after accounting for the variation among the cluster groups are reported.
- Q-Q for detecting deviation from multivariate normality and plots for detecting multivariate outliers after accounting for the variation among the cluster groups are produced.
- Graphical displays of CCC, PSF, and PST2 by cluster numbers ranging from 1 to 20 are generated to verify that the user-specified number of clusters in the DCA is the optimum cluster solution.
- To verify that the variables used in DCA are significant in discriminating the clusters by performing a stepwise discriminant analysis.
- Options for performing a disjoint cluster analysis on standardized multiattributes are included since variables with large variances tend to have more influence on the resulting clusters than those with small variances.
- DCA based on principal components of highly correlated multiattributes is also available in the DISJCLS2 macro.
- Options for detecting statistical significance among cluster groups by performing canonical discriminant analysis are included.
- Options for displaying interrelationship between the canonical disciminant scores for cluster groups and the correlations among the multiattributes in a biplot graphs are included.
- Options for displaying cluster group differences for each multivariate attribute by box plot display are included.
- Options for saving the output tables and graphics in WORD, HTML, PDF, and TXT formats are available.

Software requirements for using the DISJCLS2 macro are the following:

■ SAS/BASE, SAS/STAT, SAS/GRAPH, and SAS/IML must be licensed and installed at your site to perform the complete analysis presented here.
■ SAS version 9.13 and above is required for full utilization.

4.8.1 Steps Involved in Running the DISJCLS2 Macro

■ Step 1. Create or open a SAS dataset containing continuous variables.
■ Step 2. Open the DISJCLS2.SAS macro-call file into the SAS EDITOR window. Instructions are given in Appendix 1 regarding downloading the macro-call and sample data files from this book's Web site.
   Click the RUN icon to submit the macro-call file DISJCLS2.SAS to open the MACRO–CALL window called DISJCLS2 (Figure 4.17).
■ *Special note to SAS Enterprise Guide (EG) CODE window users:* Because the user-friendly SAS macro application included in this book uses SAS WINDOW/DISPLAY commands, and these commands are not compatible with SAS EG, open the traditional DISJCLUS2 macro-call file included in

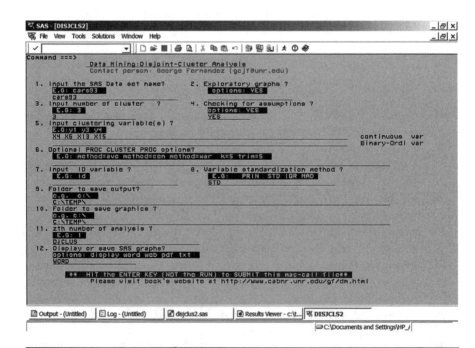

Figure 4.17 Screen copy of DISJCLS2 macro-call window showing the macro-call parameters required for performing complete disjoint cluster analysis.

the \dmsas2e\maccal\nodisplay\ into the SAS editor. Read the instructions given in Appendix 3 regarding using the traditional macro-call files in the SAS EG/SAS Learning Edition (LE) code window.

▪ Step 3: Input the appropriate parameters in the macro-call window by following the instructions provided in the DISJCLS2 macro-help file (Appendix 2). Users can choose both the cluster exploration option, disjoint cluster analysis, and the checking for multivariate normality assumptions option. After inputting all the required macro parameters, check to see that the cursor is in the last input field and that the RESULTS VIEWER window is closed; then hit the ENTER key (not the RUN icon) to submit the macro.

▪ Step 4: Examine the LOG window (only in DISPLAY mode) for any macro execution errors. If you see any errors in the LOG window, activate the EDITOR window, resubmit the DISJCLS2.SAS macro-call file, check your macro input values, and correct if you see any input errors.

▪ Step 5: Save the output files. If no errors are found in the LOG window, activate the EDITOR window, resubmit the DISJCLS2.SAS macro-call file, and change the macro input value from DISPLAY to any other desirable format (see Section 4.8.2). The printout of disjoint cluster analysis and exploratory graphs can be saved as a user-specified format file in the user-specified folder.

4.8.2 Case Study 4: Disjoint Cluster Analysis of 1993 Car Attribute Data

4.8.2.1 Study Objectives

1. Checking for suitability of car93 data for performing k-means cluster analysis: Perform the preliminary k-means cluster analysis with the user-specified cluster number, and check the multiattribute data for multivariate outliers and normality after adjusting for between-cluster variations. This is a requirement for hypothesis tests on cluster differences based on canonical discriminant analysis. Perform stepwise discriminant analysis using the backward elimination routine to confirm that the multiattributes used in the cluster analysis are statistically significant.

2. k-means clustering: Extract disjoint clusters from the multivariate dataset, and characterize the cluster differences.

3. Canonical discriminant analysis: Verify the results of disjoint cluster groupings using MANOVA test statistics and canonical discriminant analysis.

4. Interrelationships: Investigate the interrelationship between the cluster groupings and the multiattributes based on canonical discriminant function scores and structure loadings.

4.8.2.2 Data Descriptions

Data Name	SAS Dataset CARS93
Multiattributes	*X4:* HP Horsepower (maximum)
	X5: RPM (Revolutions per minute at maximum horsepower)
	X13: RSEATRM (Rear seat room in inches)
	X15: Weight (pounds)
Number of observations	92
Data source[18]	Lock, R. H. (1993)

Open the DISJCLS2 macro-call window in SAS (Figure 4.17), and input the appropriate macro-input values by following the suggestions given in the help file (Appendix 2). To generate scatter plot matrix of cluster separation, variable selection, and optimum cluster number estimation: Input X_4, X_5, X_{13}, and X_{15} as the multiattributes in (#2). Input YES in field #3 to perform scatter plot matrix analysis of cluster groupings and stepwise variable selections. Submit the DISJCLS2 macro to output the disjoint cluster analysis results and scatter plot matrix plot. Only selected output and graphics generated by the DISJCLS2 macro are interpreted in Figure 4.17.

Scatter plot matrix of cluster separation: Characteristics of standardized multiattributes included in the disjoint cluster analysis are presented in Table 4.25. Total standard deviation (STD) for all attributes are approximately equal to one since all the variables are standardized to a zero mean and unit standard deviation prior to

Table 4.25 Characteristics of Standardized Multiattributes Used in Disjoint Cluster Analysis Computed by the PROC FASTCLUS in DISJCLS2 Macro

Variable	Total STD	Within STD	R-Square	RSQ/(1 – RSQ)
X4	0.93284	0.58144	0.620128	1.632467
X5	0.98644	0.65352	0.570839	1.330127
X13	1.00000	0.84822	0.296511	0.421486
X15	1.00906	0.57061	0.687326	2.198219
OVERALL	0.98253	0.67273	0.541619	1.181591

Note: Overall cluster differences based on pseudo F-statistic = 51.99.

Table 4.26 Cluster Mean Values of Standardized Multiattributes Used in Disjoint Cluster Analysis Computed by PROC FASTCLUS in DISJCLS2 Macro

	Cluster Means			
Cluster	X4	X5	X13	X15
1	1.556	0.931	−0.049	0.927
2	−0.652	0.435	−0.476	−0.806
3	0.256	−0.978	0.695	0.833

clustering. The pooled within-cluster standard deviation (within STD) describes the average within-cluster variability. A relatively homogenous cluster should have a smaller within-STD. The percentage of variability in each standardized variable attributed to the cluster differences is given by the R-squared statistic. For example, 69% of the variation in $X15$ (physical weight) could be attributed to between-cluster differences. The ratio of between-cluster variance to within-cluster variance is provided in the last column under ($R^2/(1-R^2)$). Variable $X15$ accounts for most of the variation among the clusters, and $X13$ accounts for the least amount of between-cluster variations. The overall row provides the average estimates of between- and within-cluster estimates pooled across all variables. The estimated pseudo F-statistic for disjoint clustering is estimated by the following formula:

$$((R^2)/(c-1))/((1-R^2)/(n-c))$$

where R^2 is the observed overall R^2, c is the number of clusters, and n is the number of observations in the dataset. The pseudo F-statistic is an overall indicator of the measure of fit. The general goal is to maximize the pseudo F-statistic when selecting the optimum number of clusters or significant attributes used in clustering.

Cluster means and standard deviations for the three extracted clusters are presented in Tables 4.26 and 4.27, respectively. Based on the cluster mean values, we

Table 4.27 Cluster Standard Deviation of Standardized Multiattributes Used in Disjoint Cluster Analysis Computed by PROC FASTCLUS in DISJCLS2 Macro

	Cluster Standard Deviations			
Cluster	X4	X5	X13	X15
1	0.875	0.4191	1.037	0.390
2	0.533	0.682	0.747	0.598
3	0.5300	0.670	0.915	0.576

can come to the following conclusions: Clusters 1 and 2 are mainly differentiated by X_4 and X_{15}. Variables X_4 and X_5 separate clusters 1 and 3. All four attributes equally separate clusters 2 and 3. Cluster standard deviation provides measures of within-cluster homogeneity. Overall, clusters 2 and 3 are more homogeneous than cluster 1.

The scatter plot matrix among the four attributes presented in Figure 4.18 reveals the strength of correlation, presence of any outlying observations, and the nature of bidirectional variation in separating the cluster groupings. Variables X15 and X4 appear to have strong correlations in the scatter plot. The best bivariable cluster separation is observed in the scatter plot between X4 and X5. Even though clusters 2 and 3 are more homogeneous than cluster 1 based on within-cluster standard deviation, a few extreme observations belonging to cluster 2 show up in the scatter plot based on attribute X4.

Significant variable selection: The backward elimination method in the stepwise discriminant (PROC STEPDISC) analysis is used to select the significant variables for effective clustering. After trying out many combinations of variables, the final four attributes presented in Table 4.28 are selected as the best subset of variables for effective clustering of the car93 data. The backward elimination variable selection results presented in Table 4.28 show that based on the STEPDISC default cutoff values, none of the four variables can be dropped from the analysis. However, X5 and X4 can be identified as the most significant variables with largest partial R2 and F-statistic responsible for effective clustering.

Determining the optimum number of clusters: To determine the optimum cluster solutions, Ward's method of cluster analysis available in the PROC CLUSTER is used to generate clustering criteria such as cubic clustering (CCC), pseudo F-statistic, (PSF) and pseudo t2 statistic (PST2). A trend plot of CCC (Y-axis) versus the number of clusters (X-axis) plot presented in Figure 4.19 can be used to select the optimum number of clusters in the data. According to Khattree and Naik,[19] values of the CCC derived from unstandardized data greater than 2 or 3 indicate good clusters. Values between 0 and 2 indicate potential clusters, but should be evaluated with caution; large negative values may indicate the presence of outliers. However, in the case of car93 data, all CCC values are negative, which indicates the presence of multivariate outliers. However, there is a big jump for the CCC values at the three-cluster solution (Figure 4.19). Thus, the three clusters are selected tentatively.

An overlay plot of PST2 and PSF (Y-axis) versus the number of clusters (X-axis) plot (presented in Figure 4.19) can also be used to select the optimum number of clusters in the data. Starting from the large values in the X-axis, when we move right, a big jump in the PST2 value occurs for cluster number 2. Similarly, a relatively large PSF occurs at cluster number 2 when we move from left to right in the X-axis of the overlay plot. Both PST2 and PSF statistics indicate two clusters

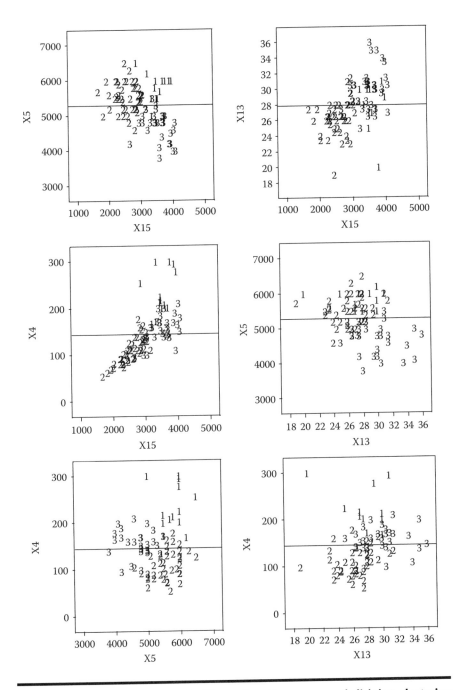

Figure 4.18 Scatter plot matrix illustrating the success of disjoint clustering in a two-dimensional display of multiattributes derived using the SAS macro DISJCLS2.

Table 4.28   Summary Statistics of Backward Elimination Method in Testing for Cluster Separations Using Stepwise Discriminant Analysis Employing the PROC STEPDISC in DISJCLS2 Macro

Variable	Label	Partial R-Square	F-Value	Pr > F
*X*4	HP (maximum)	0.2999	18.64	<0.0001
*X*5	RPM (revolutions per minute at maximum HP)	0.2800	16.91	<0.0001
*X*13	Rear seat room (inches)	0.0571	2.63	0.0774
*X*15	Weight (pounds)	0.1252	6.22	0.0030

Note: Statistics for removal, DF = 2, 84. No variables can be removed based on minimum partial R² value 0.02.

as the optimum solution. Based on these results, two is selected as the number of potential clusters. These graphical methods suggest that between 2 and 3 clusters is the optimal number. Thus, a three-cluster solution is selected for *k*-means cluster analysis and verified subsequently by canonical discriminant analysis. The results of canonical discriminant analysis are presented in the next section.

In the second phase of analysis, the results of checking for multivariate normality, cluster solution verification by canonical discriminant analysis and the biplot analysis results are discussed. To perform the second phase of the analysis, open the DISJCLS2 macro-call window in SAS (Figure 4.17), and input the appropriate macro-input values by following the suggestions given in the help file (Appendix 2). Input *X*4, *X*5, *X*13, and *X*15 as the multiattributes in (#2). Leave the macro field #3 blank to verify the cluster solution by canonical discriminant analysis (CDA) and to skip scatter plot matrix analysis of cluster groupings and stepwise variable selections. Also, to check for both multivariate normality and the presence of influential observations, input YES in field #4. Submit the DISJCLS2 macro to output the CDA results and multivariate normality check.

Checking for multivariate normality and the presence of multivariate influential observations: After adjusting for between-cluster differences, multivariate normally distributed assumption of individual clusters are verified graphically (Figure 4.20). Also, detection for influential observations is performed by computing robust distance square statistic (RDSQ) and the difference between RDSQ and the quantiles of the expected chi-square value (DIFF). Multivariate influential observations are identified when the DIFF values exceed 2.5. The estimates of RDSQ and DIFF values for the several multivariate influential observation are presented in Table 4.29. Observation number 90 belongs to cluster 1 and is identified as the most influential

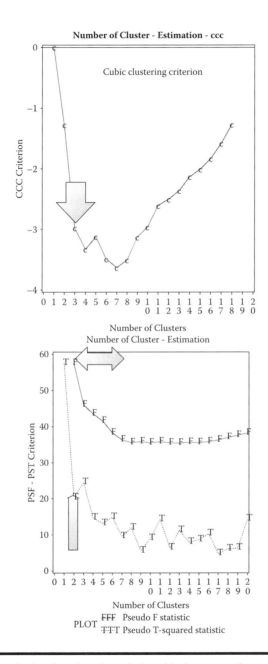

Figure 4.19 Trend plot showing the relationship between the number of cluster formed and the cubic clustering criterion (CCC), and an overlaid plot showing the relationship between the number of cluster formed and the pseudo F- and T-squared statistics used in deciding the number of optimal clusters derived using the SAS macro DISJCLS2.

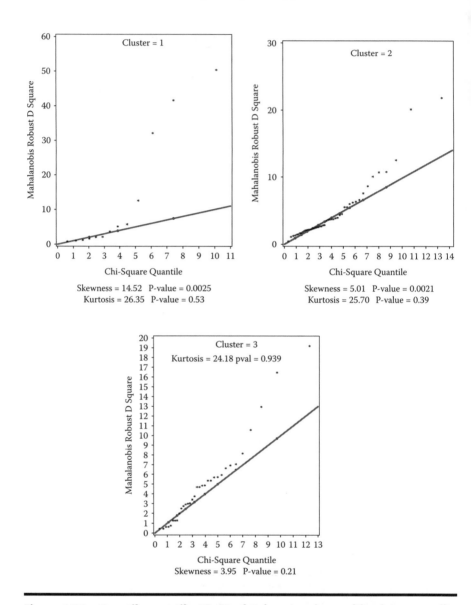

Figure 4.20 Quantile-quantile (Q-Q) plot for assessing multivariate normality within each clusters derived from the SAS macro DISJCLS2.

observation based on large DIFF value. The presence of several multivariate influential observations are also visually assessed by a graphical display of DIFF values versus the quantile of chi-square values (Figure 4.21). The impact of these extreme influential observations on DCA analysis outcomes can be verified by excluding this extreme observation and examining the effects on hypothesis tests in CDA.

Table 4.29 Checking for the Presence of Multivariate Influential Observations by Cluster Using PROC IML in DISJCLS2 Macro

CLUSTER	ID	Robust Distance Square (RDSQ)	Chi-Square (Chi)	Diff = (RDSQ – Chi)
1	90	50.3686	10.1195	40.2491
1	34	41.6322	7.4179	34.2143
1	87	32.1024	6.0930	26.0093
1	77	12.5418	5.1809	7.3609
2	21	20.1297	10.5649	9.56479
2	6	21.8219	13.1342	8.68771
2	54	12.5205	9.3377	3.18282
2	73	10.7100	7.8886	2.82137
2	63	10.0606	7.3821	2.67851
3	93	19.2061	12.3157	6.89036
3	33	16.4678	9.7181	6.74978
3	42	12.9472	8.4714	4.47574
3	59	10.5879	7.6312	2.95673

Validating cluster differences using CDA: A matrix of square distance values between the clusters estimated using the CDA is presented in Table 4.30. The differences between cluster distances show that clusters 2 and 3 are separated far apart relative to the distances between clusters 1 and 2 and 1 and 3. In addition, highly significant *p*-values for testing the Mahalanobis distances reveal that the distances between the clusters are statistically significant (Table 4.31).

Checking for significant cluster groupings by CDA: CDA extracts canonical discriminant functions, which are linear combinations of the original variables. Assuming that multivariate normality assumption is not seriously violated, cluster differences for combined canonical functions can be examined by multivariate ANOVA (MANOVA).

CDA analysis produced an output of four multivariate ANOVA test statistics and the *p*-values for testing the significance of between-cluster differences for canonical functions. All four MANOVA tests clearly indicate that large significant differences exist for at least one of the clusters (Table 4.32). The characteristics of each canonical function and their significance can be performed using the canonical correlations.

This maximal multiple correlations between the first canonical function and the cluster differences are called the *first canonical correlation*. The second canonical

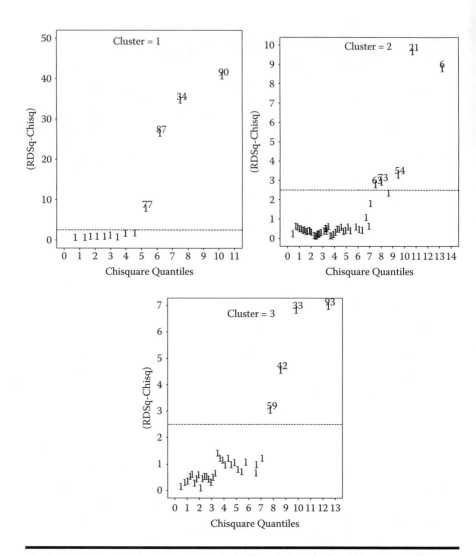

Figure 4.21 Multivariate outlier detection plot based on robust squared distances within each clusters derived using the SAS macro DISJCLS2.

correlation is obtained by finding the linear combination uncorrelated with the first canonical variable that has the highest possible multiple correlations with the groups. In CDA, the process of extracting canonical functions is repeated until you extract the maximum number of canonical functions, which is equal to the number of groups minus one, or the number of variables in the analysis, whichever is smaller. Approximately 71% of the cluster variations can be accounted for by the first canonical function. The second independent functions account for the rest of the variation between clusters (Table 4.33). Both functions are statistically significant (Table 4.34).

Table 4.30 A Table of Square Distance Matrix between the Extracted Clusters Estimated Using the PROC CANDISC in DISJCLS2 Macro

Squared Distance to CLUSTER			
From CLUSTER	1	2	3
1	0	16.221	14.230
2	16.221	0	11.800
3	14.230	11.800	0

Table 4.31 A Table of *p*-values Indicating the Statistical Significance of Square Distance Matrix between the Extracted Clusters Using the CDA Employing the PROC CANDISC in DISJCLS2 Macro

Prob > Mahalanobis Distance for Squared Distance to CLUSTER			
From CLUSTER	1	2	3
1	1.0000	<0.0001	<0.0001
2	<0.0001	1.0000	<0.0001
3	<0.0001	<0.0001	1.0000

Table 4.32 MANOVA Results Testing the Statistical Significance of Extracted Clusters Estimated Using the Canonical Discriminant Analysis Employing the PROC CANDISC in DISJCLS2 Macro

Statistic	Value	F-Value	Num DF	Den DF	Pr > F
Wilks' lambda	0.1117	43.32	8	174	<0.0001
Pillai's trace	1.3185	42.56	8	176	<0.0001
Hotelling–Lawley trace	4.1002	44.29	8	121.98	<0.0001
Roy's greatest root	2.6435	58.16	4	88	<0.0001

Table 4.33 Estimates of Canonical Correlations between the Extracted Clusters and the Canonical Functions in the Canonical Discriminant Analysis Using PROC CANDISC in DISJCLS2 Macro

	Canonical Correlation	Adjusted Canonical Correlation	Approximate Standard Error	Squared Canonical Correlation	Eigenvalues of $Inv(E)*H = CanRsq/(1-CanRsq)$			
					Eigenvalue	Difference	Proportion	Cumulative
1	0.851	0.841	0.028	0.725	2.643	1.186	0.644	0.6447
2	0.770	.	0.042	0.592	1.456		0.355	1.0000

Table 4.34   Testing for the Significance of Canonical Correlations between the Extracted Clusters and the Canonical Functions in the Canonical Discriminant Analysis Using PROC CANDISC in DISJCLS2 Macro

	Likelihood ratio	Approximate F Value	Numerator Degrees of Freedom	Denominator Degrees of Freedom	Pr > F
1	0.1117	43.32	8	174	<0.0001
2	0.4070	42.73	3	88	<0.0001

Note: Test of H0: The canonical correlations in the current row and all that follow are zero.

The degree of cluster separation by each canonical function and the distributional properties of canonical functions by clusters can be visually examined in box plots (Figure 4.22). The first canonical discriminant function discriminates cluster 2 from the other two clusters more effectively (Figure 4.22), while the second canonical discriminant function separates cluster 1 from the other two clusters (Figure 4.22).

The structure coefficients presented in Table 4.35 indicate the simple correlations between the variables and the canonical discriminant functions. These structure

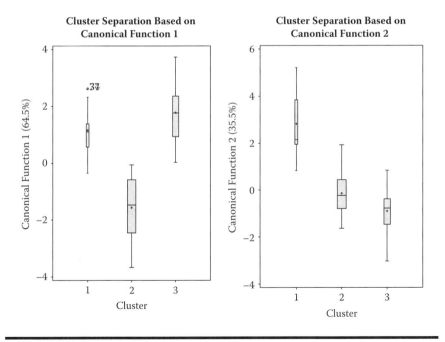

Figure 4.22   Box plot display of canonical discriminant analysis functions 1 and 2 by clusters assessing the success of cluster separation derived from the SAS macro DISJCLS2.

Table 4.35   Total Canonical Structure (Correlation Coefficients between the Multiattributes and the Canonical Functions) in the Canonical Discriminant Analysis Using the PROC CANDISC in DISJCLS2 Macro

Variable	Label	CAN1	CAN2
X4	HP (maximum)	0.706	0.702
X5	RPM (revolutions per minute at maximum HP)	−0.602	0.703
X13	Rear seat room (inches)	0.604	−0.231
X15	Weight (pounds)	0.956	0.117

loadings are commonly used when interpreting the meaning of the canonical variable because the structure loadings appear to be more stable, and they allow for the interpretation of canonical variables in a manner that is analogous to factor analysis. The first canonical function has larger positive correlations with HP, WEIGHT, and RSEATRM and negative loadings for RPM. The second canonical function has significant positive loadings for HP and RPM.

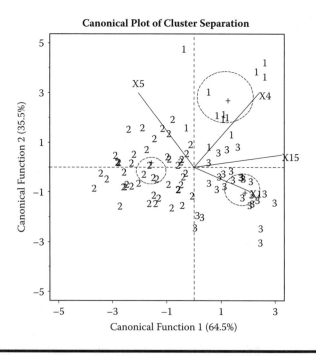

Figure 4.23   Biplot display of interrelationship between the first two canonical discriminant functions and cluster groupings derived using the SAS macro DISJCLS2.

Biplot display of canonical discriminant function scores and the cluster groupings: The extracted canonical function scores can be used to plot pairs of canonical functions in two-dimensional biplots to aid visual interpretation of cluster differences. Interrelationships among the four multiattributes and the discriminations of the three clusters are presented in Figure 4.23. The first canonical function, which has the largest loadings on all four attributes, discriminated the small less powerful cars in cluster 2 from the other two cluster groups, 1 and 3, successfully. The second canonical function, which has larger loadings on X_4 and X_5, discriminated the cluster 1 group containing very powerful sports cars from clusters 3 and 2 successfully. The cluster mean value for each cluster is shown by "+" symbol in the biplot. The circle line around the "+" symbol in each cluster shows the 95% confidence circle for the cluster mean values. The larger circle for the cluster group 1 indicates that the within-cluster variability is relatively higher for cluster 1. The smaller cluster size might be one of the reasons for this large variation in cluster 1.

These canonical function scores can be used as the basis for group member scoring and ranking the observations within each cluster. In addition, Figure 4.24 shows

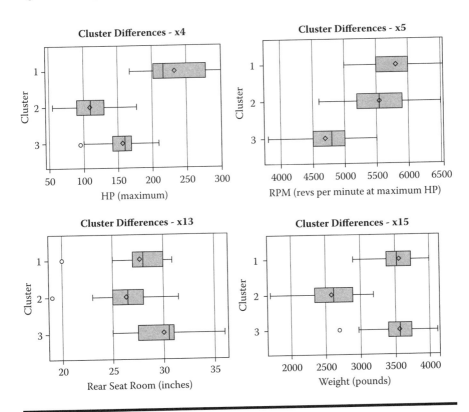

Figure 4.24 Box plot display showing cluster differences for each multivariate attribute derived in disjoint cluster analysis using the SAS macro DISJCLS2.

the box plot matrix of cluster differences for each multivariate attribute included in the disjoint cluster analysis, including the 5-number summary statistics, the shape of distribution, and presence of outliers. This information can be used to evaluate the usefulness of the cluster groups derived.

4.9 Summary

The methods of performing unsupervised learning methods to reduce dimensionality, latent factor extraction, and cluster segmentation of continuous multiattribute data using the user-friendly SAS macro applications FACTOR2 and DISJCLS2 are covered in this chapter. Both descriptive summary statistics and graphical analysis are used to summarize total variation, reduce dimensionality of multiattributes, check for multivariate outliers and normality, segment observations, and explore interrelationships between attributes and observations in biplots. Steps involved in using the user-friendly SAS macro applications FACTOR2 and DISJCLS2 for performing unsupervised learning methods are presented using the cars93 dataset.

References

1. Sharma, S., *Applied Multivariate Techniques*, John Wiley & Sons, New York, 1996, chap. 4,5,7.
2. Johnson, R. A. and Wichern, D. W., *Applied Multivariate Statistical Analysis*, 5th ed., Prentice Hall, NJ, 2002, chap 8,9,12.
3. Khattree, R. and Naik, D. N., *Multivariate Data Reduction and Discrimination with SAS Software*, 1st ed., SAS Institute, Inc., Cary, NC, 2000, chap. 2.
4. Khattree, R. and Naik, D. N., *Applied Multivariate Statistics with SAS Software*, 1st ed., SAS Institute, Inc., Cary, NC, 1995, chap. 2.
5. Neter, J., Kutner, M. H., Nachtsheim, C. J., and Wasserman W., *Applied Linear Regression Models*. Richard D. Irwin, 1996, chap. 6.
6. Hatcher, L., *A Step-By-Step Approach to Using the SAS System for FACTOR Analysis and Structural Equation Modeling*, 1st ed., SAS Institute, Cary, NC, 1994, chap. 1.
7. Sharma, S., *Applied Multivariate Techniques*, John Wiley & Sons, New York, 1996, chap. 4.
8. Sharma, S., *Applied Multivariate Techniques*, John Wiley & Sons, New York, 1996, chap. 5.
9. Johnson, R. A. and Wichern, D. W., *Applied Multivariate Statistical Analysis*, 5th ed., Prentice Hall, NJ, 2002, chap 9.
10. SAS Institute, Inc., The Factor procedure: Confidence Intervals and the Salience of Factor Loadings in SAS ONLINE Documentation at http://support.sas.com/onlinedoc/913/getDoc/en/statug.hlp/factor_sect16.htm (last accessed 05-21-09).
11. Khattree, R. and Naik, D. N., *Multivariate Data Reduction and Discrimination with SAS Software*, 1st ed., SAS Institute, Inc., Cary, NC, 2000, chap. 4.

12. Sharma, S., *Applied Multivariate Techniques*, John Wiley & Sons, New York, 1996, chap. 7.

13. Johnson, R. A. and Wichern, D. W., *Applied Multivariate Statistical Analysis*. Prentice Hall, NJ, 5th ed., 2002, chap. 12.

14. Khattree, R. and Naik, D. N., *Multivariate Data Reduction and Discrimination with SAS Software*, 1st ed., SAS Institute, Inc., Cary, NC, 2000, chap. 6.

15. Gabriel, K. R., Biplot display of multivariate matrices for inspection of data and diagnosis, In V. Barnett (Ed.). *Interpreting Multivariate Data*, John Wiley & Sons, London, 1981.

16. SAS Institute, Inc., Comparison of the PRINCOMP and FACTOR Procedures in SAS ONLINE Documentation at http://support.sas.com/onlinedoc/913/getDoc/en/statug.hlp/intromult_sect2.htm (last accessed 05-21-09).

17. SAS Institute, Inc., The FACTOR procedure in SAS ONLINE Documentation at http://support.sas.com/onlinedoc/913/getDoc/en/statug.hlp/FACTOR_index.htm (last accessed 05-21-09).

18. Lock, R. H., New car data, *Journal of Statistics Education*, 1, No. 1 1993.

19. Khattree, R. and Naik, D. N., *Applied Multivariate Statistics with SAS Software*, 1st ed., SAS Institute, Inc., Cary, NC, 1995, chap 1.

20. SAS Institute, Inc., Introduction to Clustering Procedures in SAS ONLINE Documentation at http://support.sas.com/onlinedoc/913/getDoc/en/statug.hlp/intro-clus_index.htm (last accessed 05-21-09).

21. SAS Institute, Inc., The FASTCLUS procedure in SAS ONLINE Documentation at http://support.sas.com/onlinedoc/913/getDoc/en/statug.hlp/fastclus_index.htm (last accessed 05-21-09).

22. Kromrey, J. D., Romano, J., and Hibbard, S. T., 2008. ALPHA_CI: A SAS® Macro for Computing Confidence Intervals for Coefficient Alpha SAS Global Forum Paper 230-2008 at http://www2.sas.com/proceedings/forum2008/230-2008.pdf.

Chapter 5

Supervised Learning
Methods: Prediction

5.1 Introduction

The goal of supervised predictive models is to find a model that will correctly associate the input (independent, predictor) variables with the targets (dependent, response) variables. Automated data collection, data warehousing, and ever-faster computing combine to make it possible to fit many variations of predictive models. The combination of many predictors, large databases, and powerful software make it easy to build such models that hold the potential to reveal hidden trends. Thus, supervised predictive models play a key role in data mining and knowledge discovery.

The supervised predictive models include both regression and classification models. Classification models use categorical response, while regression models use continuous and binary or ordinal variables as targets. In regression we want to approximate the regression function, while in classification problems we want to approximate the probability of class membership as a function of the input variables.

In modeling, training data gives the model a chance to learn and find a solution that identifies essential patterns that are not overly specific to the sample data. One way to accomplish this is to fit the model to data. The variable to be predicted and its predictors are carefully monitored, and the observations are randomly selected in the training datasets that are used to build such models. This training data provides the predictive model a chance to identify essential patterns that are specific to the whole database. After training, the fitted model must be validated with data independent of the training set that gives a way to measure the model's ability to generalize what it has learned. As in all other data-mining techniques, these supervised predictive methods are not immune to badly chosen training data. Therefore, the observations for the training set *must* be carefully chosen; remember the golden rule is "garbage in, garbage out."

Three supervised predictive model techniques, multiple linear regression (MLR), binary logistic regression (BLR), and ordinal logistic regression are discussed in this chapter. The main objective of predictive modeling is to model the relationship between several predictor variables and a response variable. The association between these two sets of variables is described by a linear equation in the case of MLR, and by a nonlinear logistic function in the case of BLR that predicts the response variable from a function of predictor variables. In most situations, predictive models merely provide useful approximations of the true unknown model. However, even in cases where theory is lacking, a predictive model may provide an excellent predictive equation if the model is carefully formulated from a large representative database. In general, predictive models allow the analysts to

- Determine which predictor variables are associated with the response
- Determine the form of the relationship between the response and the predictor variables
- Estimate the best fitted predictive model
- Estimate the model parameters and their confidence intervals
- Test hypotheses about the model parameters
- Estimate the predicted scores for new cases

A brief account on nonmathematical description and application of these supervised predictive methods are discussed in this chapter. For a mathematical account of MLR and BLR, the readers are encouraged to refer to the indicated statistics books.[1,2]

5.2 Applications of Supervised Predictive Methods

Predictive modeling is a powerful tool that is being incorporated more and more into data mining. MLR could be used to build business intelligence solutions for problems such as predicting insurance claim losses, credit card holder balances, amount of online orders, and cell phone usage. MLR is also ideal for solving predictive modeling questions involving continuous outcomes such as

- How much will a customer spend on his or her next purchase?
- How large a balance will a credit card holder carry?
- How many minutes will someone use on long distance this month?

Given binary (yes/no) outcomes, BLR can estimate the probability that a treatment for a serious illness will succeed, or the probability that a policyholder will file an insurance claim. In addition, BLR models could be also used to answer the following:

- Will a software customer upgrade current software?
- Will a homeowner refinance their mortgage in the next quarter?
- Will a customer respond to a direct mail offer?

It is very clear from these applications that predictive modeling can be one of the most powerful tools for decision-making business endeavors.

5.3 Multiple Linear Regression Modeling

In MLR, the association between the two sets of variables is described by a linear equation that predicts the response variable from a function of predictor variables. The estimated MLR model contains regression parameters that are estimated by using the least-squares criterion in such a way that prediction is optimized. In most situations, MLR models merely provide useful approximations of the true unknown model. However, even in cases where theory is lacking, a MLR model may provide an excellent predictive equation if the model is carefully formulated from a large representative database. The major conceptual limitation of MLR modeling based on observational studies is that one can only ascertain relationships but never be sure about the underlying causal mechanism. Significant regression relationship does not imply cause-and-effect relationships in uncontrolled observational studies. However, MLR modeling is considered to be the most widely used technique by all disciplines. The statistical theory, methods, and computation aspects of MLR are presented in detail elsewhere.[1,2]

5.3.1 Multiple Linear Regressions: Key Concepts and Terminology

- *Overall model fit*: In MLR, the statistical significance of the overall fit is determined by an *F-test* by comparing the regression model variance to the error variance. The R^2 estimate is an indicator of how well the model fits the data (e.g., an R^2 close to 1.0 indicates that the model has accounted for almost all of the variability in the response variable with the predictor variables specified in the model). The concept of R^2 can be visually examined in an overlay plot of ordered and centered response variables (describing the total variation) and the corresponding residuals (describing the residual variation) versus the ascending observation sequence. The area of total variation not covered by the residual variation illustrates the model-explained variation. Other model and data violations also show up in the explained variation plot (see Figures 5.16

and 5.34 for examples of explained variation plots). Whether a given R^2 value is considered to be large or small depends on the context of the particular study. R^2 is not recommended for selecting the best model since it does not account for the presence of redundant predictor variables. Instead, $R^2_{(adjusted)}$ is recommended because the sample size and number of predictors are used in adjusting the R^2 estimate. Caution must be exercised with the interpretation of R^2 for models with no intercept term. As a general rule, the no-intercept model should be fit when only theoretical justification exists and the data appears to fit a no-intercept framework.

■ *Regression parameter estimates*: In MLR, the regression model is estimated using the *least-squares* criterion by finding the best-fitted line, which minimizes the error in the regression. The regression model contains a Y-intercept (β_0) and regression coefficients (β_i) for each predictor variable. Coefficient β_i measures the partial contributions of each predictor variable to the prediction of the response. Thus, β_i estimates the amount by which the mean response changes when the predictor is changed by one unit while all the other predictors are unchanged. However, if the model includes interactions or higher-order terms, it may not be possible to interpret individual regression coefficients. For example, if the equation includes both linear and quadratic terms for a given variable, you cannot physically change the value of the linear term without also changing the value of the quadratic term. To interpret the direction of the relationship between the predictor variable and the response, look at the signs (plus or minus) of the partial regression (β) coefficients. If a β coefficient is positive, then the relationship of this variable with the response is positive, and if the β coefficient is negative, then the relationship is negative. In an observational study where the true model form is unknown, interpretation of parameter estimates becomes even more complicated. A parameter estimate can be interpreted as the expected difference in response between two observations that differ by one unit on the predictor in question and have the same values for all other predictors. We cannot make inferences about changes in an observational study since we have not actually changed anything. It may not even be possible, in principle, to change one predictor independently of all the others. Nor can you draw conclusions about causality without experimental manipulation.

■ *Standardized regression coefficients*: Two regression coefficients in the same model can be directly compared only if the predictors are measured in the same units. Standardized regression coefficients are sometimes used to compare the effects of predictors measured in different units. Standardizing the variables (zero mean with unit standard deviation) effectively makes standard deviation the unit of measurement. This makes sense only if the standard deviation is a meaningful quantity, which is usually the case only if the observations are sampled from well-defined databases.

■ *Significance of regression parameters*: The statistical significance of regression parameters is determined based on the partial sums of squares *Type 2 (SS_2)*

and the *t-statistics* derived by dividing the parameter estimates (β) by its standard error (*se*). If higher-order model terms such as quadratic and cross products are included in the regression model, the *p-values* based on *SS2* are incorrect for the linear and main effects of the parameters. Under these circumstances, the right significance tests for the linear and main effects could be determined using the *SS1* (*Type I*) sequential sums of squares. Although a *p*-value based on the *t-test* provides the statistical significance of a given variable in predicting the response in that sample, it does not necessarily measure the importance of a predictor. An important predictor can have a large (non-significant) *p*-value if the sample is small, if the predictor is measured over a narrow range, if there are large measurement errors, or if another closely related predictor is included in the equation. An unimportant predictor can have a very small *p*-value in a large sample. Computing a confidence interval for a parameter estimate gives you more useful information than just looking at the *p*-value, but confidence intervals do not solve problems of measurement errors in predictors or highly correlated predictors.

■ *Model estimation in MLR with categorical variables*: When categorical variables are used as predictors, separate regression models are estimated for each level or a combination of levels, within all categorical variables included in the model. One of the levels is treated as the baseline, and differences in the intercept and slope estimates for all other levels compared with the base level is estimated and tested for significance. The main effects of the categorical variables, and the interaction between the categorical variable and continuous predictors, must be specified in the model statement to estimate the differences in the intercepts and slopes, respectively.[3] MLR models with categorical variables can be modeled more efficiently in the SAS GLM procedure[3] where the GLM generates the suitable design matrix when the categorical variables are listed in the class statement. The influence of the categorical variables on the response could be examined graphically in scatter plots between the response and each predictor variable by each categorical variable. The significance of the categorical variable and the need for fitting the heterogeneous slope model could be checked visually by examining the interaction between the predictor and the categorical variable. See Figure 5.32 for examples of regression diagnostic plots suitable for testing the significance of the categorical variable. For additional details regarding the fitting of MLR with categorical variables, see References 3 and 4.

■ *Predicted and residual scores*: After the model has been fit, predicted and residual values are usually estimated. The regression line expresses the best prediction of the response for a given set of the predictor variable. The deviation of a particular observed value from the regression line (its predicted value) is called the *residual value*. The smaller the variability of the residual values around the regression line relative to the overall variability, the better the prediction.

■ *Miscellaneous terms*: *Standardized* residual is the ratio of the residual to its standard error. *Studentized residual* is useful in detecting outliers since an observation with greater than 2.5 in absolute terms could be treated as an outlier. The *predicted residual* for *i*th observation is defined as the residual for the *i*th observation based on the regression model that results from dropping the *i*th observation from the parameter estimates. The sum of squares of predicted residual errors is called the *PRESS* statistic. Another R^2 statistic, called $R^2_{prediction}$, is useful in estimating the predictive power of the regression based on the press statistic. A big drop in the $R^2_{prediction}$ value from the R^2, or a negative $R^2_{prediction}$ value, is an indication of a very unstable regression model with low predictive potential. There are two kinds of interval estimates for the predicted value. For a given level of confidence, the *confidence interval* provides an interval estimate for the mean value of the response, whereas the *prediction interval* is an interval estimate for an individual value of a response. These interval estimates are useful in developing scorecards for observations in the database.

5.3.2 Model Selection in Multiple Linear Regression[9]

MLR modeling from large databases containing many predictor variables presents big challenges to the data analyst in selecting the best model. The regression model assumes that we have specified the correct model but, often, theory or intuitive reasoning does not suggest such a model. It is customary to use an automated procedure that employs information on data to select a suitable subset of variables. SAS software offers nine model-selection methods in the regression procedure to help the analyst in selecting the best model.[9] In this section, one of the selection methods, *Maximum R^2 Improvement* (*MAXR*), which is implemented within the SAS macro *REGDIAG2*, is discussed.

The MAXR improvement technique does not settle for a single model. Instead, it compares all possible combinations and tries to find the best variable subsets for one-variable models, two-variable models, and so on. The MAXR method may require much more computer time than the STEPWISE methods. In addition, R^2, $R^2_{(adjusted)}$, *Akakike Information Criterion* (*AIC*), *Schwarz's Bayesian Criterion* (*SBC*) root-mean-square error (*RMSE*), and Mallows Cp statistics are generated for each model generated in the model-selection methods. Because the model selection option is not implemented within PROC GLM, when categorical predictors are included for model selection, the REGDIAG2 macro uses PROC REG for model selection after creating dummy variables for all the categorical predictors using PROC GLMMOD.

R^2 is defined as the proportion of variance of the response that is predictable from the predictor variables. The $R^2_{(adjusted)}$ statistic is an alternative measure to R^2 that is adjusted for the number of parameters and the sample size. *RMSE* is the measure of the MLR model error standard deviation. *AIC* and *SBC* are the MLR model error variance statistic adjusted for the sample size and number of parameters. Minimum

RMSE and *AIC*, and maximum R^2 and $R^2_{(adjusted)}$, are characteristics of an optimum subset for a given number of variables. For a subset with p parameters including the *intercept*, the Cp statistic is a measure of the total squared error estimated by adding the model error variance and the bias component introduced by not including important variables. If the $Cp:p$ ratio is plotted against p, Mallows recommends selecting the model where $Cp:p$ first approaches 1.[9,10] Parameter estimates are unbiased for the best model since $Cp:p$ is approximately equal to 1 (see Figure 5.5 for an example of the Cp model selection plot).

5.3.2.1 Best Candidate Models Selected Based on AICC and SBC

Best candidate model selection is a new feature implemented in the REGDIAG2 SAS macro. AICC is considered the corrected AIC and is computed by adding a penalty factor to the AIC statistic computed by the PROC REG.

$$AICC = AIC + ((2*P*(P + 1))/(N - P - 1))$$

where
$AIC = AIC$ value computed by the PROC REG
P = Total number of regression parameters in the mode (including the intercept)
N = Sample size

Delta *AICC* ($\Delta AICC = AICC_i - AICC_{min}$) and delta *SBC* ($\Delta SBC = SBC_i - SBC_{min}$) less than 2 are used as the selection criteria in selecting the best candidates. In addition to selecting the best candidate models, the standardized regression coefficients for all the selected models and the following additional statistics are also computed and reported in the best candidate models selected by the REGDIAG2 macro:

$AICC_{weight} = Exp(-0.5*\Delta AICC_i)/$Sum of $(Exp(-0.5*\Delta AICC_i))$, all best-candidate models

$SBC_{weight} = Exp(-0.5*\Delta SBC_i)/$Sum of $(Exp(-0.5*\Delta SBC_i))$, all best-candidate models

$$AICC_{weight\ ratio} = AICC_{weight}/Max (AICC_{weight})$$

$$SBC_{weight\ ratio} = SBC_{weight}/Max (SBC_{weight})$$

5.3.2.2 Model Selection Based on the New SAS PROC GLMSELECT

If the number of predictor variables exceeds 20, running all possible subsets models will take very long time to complete. Therefore, under these circumstances, prescreening could be recommended to discard the least-contributing predictors. To discard the least important predictors and to select the user-specified number of effects, the LASSO[11] method implemented in the GLMSELECT[12] the new SAS procedure in SAS version 9.2 is also utilized in this REGDIAG2 macro.

The main features of the GLMSELECT procedure are as follows:

■ Supports any degree of interaction (crossed effects) and nested effects
■ Enables selection from a very large number of effects (tens of thousands)
■ Offers selection of individual levels of classification effects
■ Provides effect selection based on a variety of selection criteria
■ Provides stopping rules based on a variety of model evaluation criteria
■ Produces graphical representation of selection process

For more information of the theory and selection features, refer SAS Institute.[13]

The LASSO model selection options—CHOOSE=NONE and SELECT= SBC—are used in REGDIAG2 macro. The FIT CRITERIA and the COEFFICIENT EVALUATION plots generated by the SAS ODS GRAPHICS features were utilized in the prescreening evaluation to identify the potential subset ranges and to discard potential insignificant predictors, and to select less than or equal to 15 potentially significant predictors. The LASSO selection method adds or drops an effect and computes several information criteria (IC) statistics in each step. The FIT CRITERIA plots display the trend of six IC statistics in each step, and the best subset is identified by a STAR symbol (see Figure 5.6 for an example). The new powerful model selection procedure can also handle categorical predictors.

5.3.3 Exploratory Analysis Using Diagnostic Plots

Simple scatter plots are very useful in exploring the relationship between a response and a single predictor variable in a simple linear regression. But these plots are not effective in revealing the complex relationships of predictor variables or data problems in multiple linear regressions. However, partial regression plots are considered useful in detecting influential observations and multiple outliers, nonlinearity and model specification errors, multicollinearity, and heteroscedasticity problem.[5] These partial plots illustrate the partial effects or the effects of a given predictor variable after adjusting for all other predictor variables in the regression model. Two types of partial regression plots are considered superior in detecting regression model problems. The mechanics of these two partial plots are described using two variable MLR models.

1. **Augmented partial residual plot**[6] between response (Y) and the predictor variable $X1$ (see Figure 5.1 for an example)

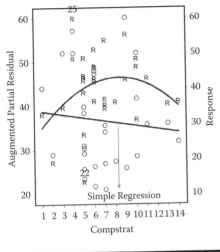

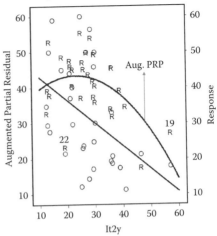

Figure 5.1   Regression diagnostic plot using SAS macro REGDIAG2: Overlay plot of simple linear regression and augmented partial residual plots.

Step 1: Fit a quadratic regression model.

$$Y_i = \beta_0 + \beta_1 X_1 + \beta_2 X_2 + \beta_3 X_1^2 + \varepsilon_i \tag{5.1}$$

Step 2: Add the X_1 linear $(\beta_1 X_1)$ and the X_1 quadratic $(\beta_3 X_1^2)$ components back to the residual (ε_i).

$$APR = \varepsilon_i + \beta_1 X_1 + \beta_3 X_1^2 \tag{5.2}$$

Step 3: Fit a simple linear regression between augmented partial residual (APR) and the predicated variable X_1.

This augmented partial residual plot is considered very effective in detecting outliers, nonlinearity, and heteroscedasticity.

2. **Partial Leverage plot (PL)**[7] (Figure 5.2) between response (Y) and the predictor variable X_1

Step 1: Fit two MLR models. Remove the effects of all other predictors from the response (Y_i) and the predictor (X_1) in question.

$$Y_i = \beta_0 + \beta_2 X_2 + \varepsilon_i \tag{5.3}$$

$$X_i = a_0 + b_2 X_2 + e_i \tag{5.4}$$

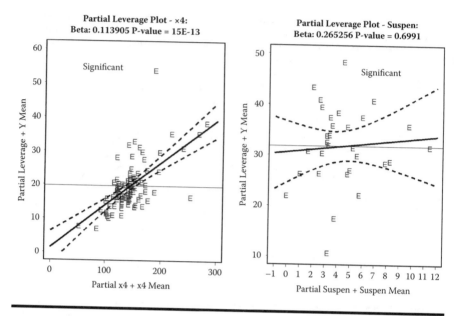

Figure 5.2 Regression diagnostic plot using SAS macro REGDIAG2: Partial leverage plot for detecting significant linear effect.

Step 2: Add the *Y-mean* to Y-residual (ε_i) and X_1-*mean* to X_1-residual (e_i) and compute partial regression estimate for *Y*, (*PRy*) and partial leverage estimate for X_1, (PL_{x1}).

$$PR_y = \varepsilon_i + Y_{mean} \tag{5.5}$$

$$PL_{x1} = e_i + X_{1mean} \tag{5.6}$$

Step 3: Fit a simple linear regression between **PR_y** and **PL_{x1}**. Include 95% confidence band around the regression line and draw a horizontal line through the Y_{mean}.

The slope of the regression line will be equal to the regression coefficient for *X*1 in the MLR.

The residual from the *PL* plot will be equal to the residual from the MLR. Also based on the position of the horizontal line through response mean and the confidence curves, the following conclusions can be made regarding the significance of the slope.

Confidence curve crosses the horizontal line = Significant slope

Confidence curve asymptotic to horizontal line = Boarder line significance

Confidence curve does not cross the horizontal line = Nonsignificant slope

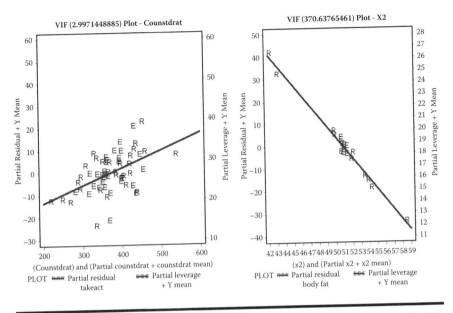

Figure 5.3   Regression diagnostic plot using SAS macro REGDIAG2: VIF plot for detecting multicollinearity—an overlay plot of augmented partial residual and partial leverage plots.

Thus the PL plot is effective in showing the statistical significance of the partial effects of predictor variable in a multiple linear regression.

3. **VIF plot**: Both augmented partial residual and partial leverage plots in the original format fail to detect the presence of multicollinearity. Stine[8] proposed overlaying the partial residual and partial leverage points on the same plot to detect multicollinearity. Thus, by overlaying the augmented partial residual and partial leverage points with the centered X_i values on the X-axis, the degree of multicollinearity can be detected clearly by the amount of shrinkage of partial regression residuals. Since the overlaid plot is mainly useful in detecting multicollinearity, I named this plot VIF plot[5] (Figure 5.3).

4. **Two-factor interaction plots between continuous variables**: A 3-D plot is generated, and the statistical significance of the interaction term is displayed on the title statement if the interaction term is significant at a ***P-value*** $\leqslant 0.15$. For example, in the case of a multiple linear regression with two continuous variables, the interaction effect is computed as follows and plotted on the z-axis.

Full model with the interaction $Y_{i\,hat} = \beta_0 + \beta_1 X_1 + \beta_2 X_2 + \beta_3 X_1{}^* X_2$

Reduced model without the interaction $Y_{r\,hat} = \beta_0 + \beta_1 X_1 + \beta_2 X_2$

Interaction term $= (Y_{i\,hat} - Y_{r\,hat}) + Y_{mean}$

Then, on the *X*-axis the *X*1 variable, and on the *Y* axis the *X*2 variable, are plotted. See Figure 5.15 for an example of the 3-D interaction plot.

5.3.4 Violations of Regression Model Assumptions[14]

If the sample regression data violate one or more of the MLR assumptions, the results of the analysis may be incorrect or misleading. The assumptions for a valid MLR are:

- Model parameters are correctly specified.
- Residuals from the regression are **independent** and have **zero mean, constant variance**, and **normal distribution**.
- Influential outliers are absent.
- There is lack of multicollinearity.

5.3.4.1 Model Specification Error

When important predictor variables or significant higher-order model terms (quadratic and cross-product) are omitted from the regression model, the residual error term no longer has the random error property. The augmented partial residual plot is very efficient in detecting the need for a nonlinear (quadratic) term. The need for an interaction between any two predictor variables could be evaluated in the interaction test plot. A simple scatter plot between a predictor and the response variable by an indicator variable could indicate the need for an interaction term between a predictor and the indicator variable. The significance of any omitted predictor variable can only be evaluated by including it in the model and following the usual diagnostic routine.

5.3.4.2 Serial Correlation among the Residual

In a time series or spatially correlated data, the residuals are usually not independent and positively correlated among the adjacent observations. This condition is known as serial correlation or first-order autocorrelation. When the first-order autocorrelation is severe (> 0.3), the *se* for the parameter estimates are underestimated. The significance of the first-order autocorrelation could be evaluated by the Durbin–Watson test and an approximate test based on the $2/n$ critical value criteria. The cyclic pattern of residual associated with the significant the positive autocorrelation can be evaluated by examining the trend plot between residuals by the observation sequence (see Figure 5.20 for an example of the autocorrelation detection plot). The SAS AUTOREG procedure available in the SAS/ETS[15] module provides an effective method of adjusting for autocorrelation.

5.3.4.3 Influential Outliers

The presence of significant outliers produces biased regression estimates and reduces the predictive power of the regression model. An influential outlier may act as a high-leverage point, distorting the fitted equation and perhaps fitting the model poorly. The SAS/REG procedure has many powerful influential diagnostic statistics.[16] If the absolute value of the student residual for a given observation is greater than 2.5, then it could be treated as a significant outlier. High-leverage data points are highly influential and have significant hat-matrix values. The DFFITS statistics shows the impact of each data point by estimating the change in the predicted value in standardized units when the *i*th observation is excluded from the model, DIFFITS statistic greater than 1.5 could be used as a cutoff value in influential observation detection. An outlier detection bubble plot between *student* and *hat-value* identifies the outliers if they fall outside the 2.5 boundary line and detect influential points if the diameter of the bubble plot, which is proportional to DFFITS, is relatively big.

5.3.4.4 Multicollinearity

When a predictor variable is nearly a linear combination of other predictors in the model, the affected estimates are unstable and have high standard errors. If multicollinearity among the predictors is strong, the partial regression estimates may have the wrong sign and size, and are unstable. If you remove a predictor involved in a collinear relationship from the model, the sign and size of the remaining predictor can change dramatically. The fitting of higher-order polynomials of a predictor variable with a mean not equal to zero can create difficult multicollinearity problems. PROC REG provides VIF, COLLINOINT options for detecting multicollinearity. Condition indices >30 indicate the presence of severe multicollinearity. The VIF option provides the Variance Inflation Factors (VIF), which measure the inflation in the variances of the parameter estimates due to multicollinearity that exists among the predictor variables. A VIF value greater than 10 is usually considered significant. The presence of severe multicollinearity could be detected graphically in the VIF plot when the partial leverage points shrink and form a cluster near the mean of the predictor variable relative to the partial residual. One of the remedial measures for multicollinearity is to redefine highly correlated variables. For example, if *X* and *Y* are highly correlated, they could be replaced in a linear regression by *X* + *Y* and *X* – *Y* without changing the fit of the model or statistics for other predictors.

5.3.4.5 Heteroscedasticity in Residual Variance

Nonconsistency of error variance occurs in MLR if the residual variance is not constant and shows a trend with the change in the predicted value. The *se* of the parameters become incorrect, resulting in incorrect significance tests and confidence interval estimates. A fan pattern like the profile of a megaphone, with a noticeable flare either to the right or to the left in the residual plot against the predicted value, is the indication of significant heteroscedasticity. The Breusch–Pagan test,[17] based on the significance of the linear model using the squared absolute residual as the response and all combination of variables as predictors is recommended for detecting heteroscedasticity. However, the presence of significant outliers and nonnormality may confound heteroscedasticity and interfere with the detection. If both nonlinearity and unequal variances are present, employing a transformation on response may have the effect of simultaneously improving the linearity and promoting equality of the variances. SAS PROC MIXED[18] provides additional capabilities to adjust for unequal variances using the repeated statement.

5.3.4.6 Nonnormality of Residuals

MLR models are fairly robust against violation of nonnormality, especially in large samples. Signs of nonnormality are significant skewness (lack of symmetry) and/or kurtosis light-tailedness or heavy-tailedness. The normal probability plot (normal Q-Q plot), along with the normality test,[19] can provide information on the normality of the residual distribution. In the case of nonnormality, fitting generalized linear models based on the SAS GENMOD[20] procedure or employing a transformation on response and/or one or more predictor variables may result in a more powerful test. However, if there are only a small number of data points (<32), nonnormality can be hard to detect. If the sample size is large (>300), the normality test may detect statistically significant but trivial departures from normality that will have no real effect on the multiple linear regressions tests. See Figure 5.21 for an example of model violation detection plots.

5.3.5 Regression Model Validation

The regression model estimated using the training dataset could be validated by applying the model to independent validation data and by comparing the model fit. If both models produce similar R^2 and show comparable predictive models, then the estimated regression model could be used for prediction with reasonable accuracy. Model validation could be further strengthened if both training and the validation residual plots show a similar pattern. See Figure 5.37 for examples of comparing prediction and residual pattern between the training and validation datasets.

5.3.6 Robust Regression

Robust regression is an important regression tool for modeling data containing influential outliers. If the data contain influential outliers, then the application of some form of robust regression that downweights the influence of the troublesome outliers is beneficial. It can be used to detect outliers and to provide resistant (stable) coefficients in the presence of outliers. Robust regression is a form of weighted least-squares regression and is done iteratively. At each step a new set of weights are estimated based on the residuals from the previous step. In general, the larger the residuals, the smaller the weights that are applied. This generates an iteration process, and it goes on until the change in the parameter estimates is small enough. In robust regression, the method most commonly used today is Huber M estimation, and this is implemented within the REGDIAG2 macro as an additional estimation method when you choose to remove extreme outliers. For additional information, refer to the SAS Help files (version 9.2) on PROC ROBUSTREG.

5.3.7 Survey Regression

Due to the variability of attributes of items in the population, survey researchers apply scientific sample designs in the sample selection process to reduce the risk of a distorted view of the population, and they make inferences about the population based on the information from the sample survey data. In order to make statistically valid inferences for the population, they must incorporate the sample design in the data analysis. The regression analysis for sample survey data can be performed by using the SAS SURVEYREG procedure. This PROC can be used in modeling finite survey data coming from complex survey sample designs such as designs with stratification, clustering, and unequal weighting. The SURVEYREG procedure

- Fits linear models for survey data and computes regression coefficients and their variance–covariance matrix.
- Performs significance tests for the model effects and for any specified estimable linear functions of the model parameters.
- Can compute predicted values for the sample survey data.
- Computes the regression coefficient estimators by generalized least-squares estimation using elementwise regression.
- Assumes that the regression coefficients are the same across strata and primary sampling units (PSUs).
- Uses either the Taylor series (linearization) method or replication (resampling) methods to estimate sampling errors of estimators based on complex sample designs.
- Can specify classification effects using the same syntax as in the GLM procedure.
- Can do hypothesis tests for the model effects for any specified estimable linear functions of model parameters and for custom hypothesis tests of linear combinations of the regression parameters.

This option is implemented in the REGDIAG2 macro. See Appendix 2 for the instructions in the REGDIAG2 help file to implement SURVEYREG estimates. For more information on SURVEYREG, refer to the SAS version 9.1 software help file.

5.4 Binary Logistic Regression Modeling

Logistic regression is a powerful modeling technique used extensively in data mining applications. It allows more advanced analyses than the *Chi-square* test, which tests for the independence of two categorical variables. It allows analyst to use binary responses (yes–no, true–false) and both continuous and categorical predictors in modeling. Logistic regression does allow an ordinal variable, for example, a rank order of the severity of injury from 0 to 4, as the response variable. Binary Logistic Regression (BLR) allows construction of more complex models than the straight linear models, so interactions among the continuous and categorical predictors can also be explored.

BLR uses maximum likelihood estimation (MLE) after converting the binary response into a logit value (the natural log of the odds of the response occurring or not) and estimates the probability of a given event occurring. MLE relies on large-sample asymptotic normality, which means that the reliability of estimates decline when there are few cases for each observed combination of X variables. In BLR, changes in the log odds of the response, not changes in the response itself, are modeled. BLR does not assume linearity of relationship between the predictors, and the residuals do not require normality and homoscedasticity. The success of BLR could be evaluated by investigating the classification table showing correct and incorrect classifications of the binary response. Also, goodness-of-fit tests such as *model chi-square* are available as indicators of model appropriateness as is the *Wald* statistic to test the significance of individual parameters. The BLR model assumes the following:

- All relevant variables in the regression model are included.
- No multicollinearity exists among the continuous predictor variables.
- Error terms are assumed to be independent.
- Predictor variables are measured without errors.
- No overdispersion occurs.

The statistical theory, methods, and computation aspects of BLR are presented in detail elsewhere.[21,22,23]

5.4.1 Terminology and Key Concepts

Probability: It is the chance of an occurrence of an event. The probabilities are one category divided by the total. Note that probability values always sum to 1, and the odds are the ratio of the probabilities of the binary event. Thus, if there is a 25% chance of raining, then the odds of raining will be 0.25/0.75 = 1/3.

Odds ratio[24,25]: Odds and probability describe how often something happens relative to its opposite happening. Odds can range from zero to plus infinity, with the odds of 1 indicating neutrality, or no difference. Briefly, an odds ratio is the ratio of two odds, and relative risk is the ratio of two probabilities. The odds ratio is the ratio of two *odds* and is a comparative measure (effect size) between two levels of a categorical variable or a unit change in the continuous variable. An odds ratio of 1.0 indicates that the two variables are statistically independent. The odds ratio of summer to winter in predicting raining means the odds for summer is the denominator, and the odds for winter is the numerator, and the odds ratio describes the change in the odds of raining from summer months to winter months. In this case the odds ratio is a measure of the strength and direction of relationship between probability of raining and two raining seasons. If the 95% confidence interval of the odds ratio includes the value of 1.0, the predictor is not considered significant. Odds ratios can be computed for both categorical and continuous data; also, they can vary only from 0 to .999, while in the case of an increase it can vary from 1.001 to infinity. With these measures, one must be very careful when coding the values for the response and the predictor variables. Reversing the coding for the binary response also inverts the interpretation. The interpretation of the odds ratio is valid only when a unit change in the predictor variable is relevant. If a predictor variable is involved in a significant quadratic relationship or interacting with other predictors, then the interpretation of the odds ratio is not valid.

Logits: These are used in the BLR equation to estimate (predict) the log odds that the response equals 1 and contain exactly the same information as the odds ratios. Logit values range from minus infinity to plus infinity, but because they are logarithms, the numbers usually range from +5 to −5 even when dealing with very rare occurrences. Unlike the odds ratio, logit is symmetrical and, therefore, it can be compared more easily. A positive logit means that, when that independent variable increases, the odds that the dependent variable equals 1 increase. A negative logit means that, when the independent variable decreases, the odds that the dependent variable equals 1 decrease. The logit can be converted easily into an odds ratio of the response simply by using the exponential function (raising the natural $\log e$ to the β_1 power). For instance, if the *logit* $\beta_1 = 2.303$, then its log odds ratio (the exponential function, e_1^{β}) is 10, and we may say that, when the response variable increases one unit, the odds that the response event = 1 increase by a factor of 10 when other variables are controlled.

Percent increase in odds: Once the logit has been transformed back into an odds ratio, it may be expressed as a percent increase in odds. Let the logit coefficient for *current asset/net sales* is 1.52, with the response variable being bankruptcy. The odds ratio, which corresponds to a logit of +1.52, is approximately 4.57 ($e^{1.52}$). Therefore, we can conclude that, as each additional unit increases in the *current asset/net sales,* the odds of bankruptcy increases by 352% (4.52−1)*100% while controlling for other inputs in the model. However, saying that the probability of bankruptcy increases by 352% is incorrect.

Standardized logit coefficients: These correspond to beta (standardized regression) coefficients. These coefficients may be used to compare the relative importance of the predictor variables. Odds ratios are preferred for this purpose; however, when using standardized logit coefficients, we are measuring the relative importance of the predictor in terms of effect on the response-variables-logged odds, which is less intuitive than the actual odds of the response variable measured when odds ratios are used.

Testing the model fit [27,28]: This involves the following concepts.

- *Wald statistic.* This is used to test the significance of individual logistic regression coefficients (to test the null hypothesis that a particular logit (effect) coefficient is zero). It is the ratio of the unstandardized logit coefficient to its standard error.

- *Log likelihood ratio tests.* These are an alternative to the Wald statistic. If the log-likelihood test statistic is significant, ignore the Wald statistic. Log-likelihood tests are also useful in model selection. Models are run with and without the variable in question, for instance, and the difference in *−2log likelihood* (*−2LL*) between the two models is assessed by the chi-square statistic with the *df* being equal to the difference in the number of parameters between the two models.

- *Deviance.* Because *−2LL* has approximately a chi-square distribution, it can be used for assessing the significance of logistic regression, analogous to the use of the sum of squared errors in OLS regression. The *−2LL* statistic is the scaled deviance statistic for logistic regression. Deviance measures the error associated with the model after all the predictors are included in the model. It thus describes the unexplained variance in the response. The deviance for the null model describes the error associated with the model when only the intercept is included in the model, that is, *−2LL* for the model, which accepts the null hypothesis that all the β coefficients are 0.

Assessing the model fit: Terminology and key concepts are as follows.

- *Hosmer and Lemeshows goodness-of-fit test* [28,29]: The test divides subjects into deciles based on predicted probabilities, and then computes a chi square from observed and expected frequencies. Then a *p*-value value is computed from the chi-square distribution with 8 degrees of freedom to test the fit of the logistic model. If the *p*-value of Hosmer and Lemeshow goodness-of-fit test statistic (H–L) is 0.05 or less, we reject the null hypothesis that there is no difference between the observed and model-predicted values of the dependent. If the *p*-values H–L goodness-of-fit test statistic is greater than 0.05, we do not reject the null hypothesis that there is no difference, implying that the model estimates fit the data at an acceptable level. This does not mean that the model necessarily explains much of the variance in the dependent; only that, however much or little it does explain is significant. As with other tests, as the sample size gets larger, the H–L test's power to detect differences from the null hypothesis improves.

- *Brier score*[29]: This is a unitless measure of predictive accuracy computed from the classification table based on a cutpoint probability of 0.5. It ranges from 0 to 1. The smaller the score, the better the predictive ability of the model. The Brier score is useful in model selection and assessing model validity based on an independent validation dataset.

- *Adjusted generalized coefficient of determination* (R^2): This is a model assessment statistic similar to the R^2 in OLS regression that can reach a maximum value of 1. The statistic is computed using the ratio between the *–2LL* statistic for the null-model- and full-model-adjusted sample size.[29,30]

- *The c statistic and the receiver-operating characteristic* (*ROC*) *curve*: This is a measure of the classification power of the logistic equation. It varies from 0.5 (the models predictions are no better than chance) to 1.0 (the model always assigns higher probabilities to correct cases than to incorrect cases). Thus, the *c statistic* is the percentage of all possible pairs of cases in which the model assigns a higher probability to a correct case than to an incorrect case.[29] The ROC is a graphical display illustrating the predictive accuracy of the logistic curve. The ROC curve is constructed by plotting the *sensitivity* (measure of accuracy of predicting events) versus *1-specificity* (measure of error in predicting nonevents).[31,32] The area under the ROC curve is equal to the *c-statistic*. The ROC curve rises quickly, and the area under the ROC is larger for a model with high predictive accuracy. For an example of the ROC curve, see Figure 5.70. An overlay plot between the percentages of false positive and false negative versus the cutpoint probability could reveal the optimum cutpoint probability when both the false positive and false negative could be minimized. See examples of the overlay plot of the false-positive and false-negative curves in Figure 5.71.

5.4.2 Model Selection in Logistic Regression

The SAS software offers four model selection methods in the LOGISTIC procedure to help the analyst in selecting the best model. However, these selection methods are not guaranteed to find the best model in most cases. Therefore, in this second edition, a new approach called the *best candidate models selected based on AICC* and SBC is implemented.

AICC is considered the corrected AIC and is computed by adding a penalty factor to the AIC statistic computed by the PROC LOGISTIC. The SBC statistic is computed by the PROC LOGISTIC.

$$AICC = AIC + ((2*P*(P + 1))/(N - P - 1))$$

where

AIC = AIC value computed by the PROC LOGISTIC

P = Total number of regression parameters in the mode (including the intercept)

N = Sample size

Delta $_{AICC}$ **($\Delta AICC = AICC_i - AICC_{min}$) and delta** SBC **($\Delta SBC = SBC_i - SBC_{min}$)** less than 2 are used as the criteria in selecting the best candidates. In addition to selecting the best candidate models, the standardized regression coefficients for all the selected models and the following additional statistics are also computed and reported in the best-candidate models selected by the LOGIST2 macro:

$AICC_{weight} = Exp(-0.5*\Delta AICC_i)/Sum\ of\ (Exp(-0.5*\Delta AICC_i))$
all best-candidate models

$SBC_{weight} = Exp(-0.5*\Delta SBC_i)/Sum\ of\ (Exp(-0.5*\Delta SBC_i))$
all best-candidate models

$AICC_{weight\ ratio} = AICC_{weight}/Max\ (AICC_{weight})$

$SBC_{weight\ ratio} = SBC_{weight}/Max\ (SBC_{weight})$

The first step in best-candidate model selection is to perform sequential stepwise selection. In sequential stepwise selection, stepwise variable selection is carried forward until all the predictor variables are entered by changing the default SLENTER and SLSTAY p-values from 0.15 to 1. However, in each sequential stepwise selection step, both AICC and SBC statistics are recorded, and the best subsets with the minimum AICC and minimum SBC are identified (see Figure 5.42 for an example).

In the second step, all possible combination of multiple logistic regression models are performed between (SBC: 2 − 1) a one-variable subset and (AIC: 3 + 1) four-variable subsets, and the best-candidate models based on delta AICC and delta SBC are identified (see Figure 5.43 for an example). The variables identified by the best-candidate models are selected and carried forward to data exploration and diagnostics.

5.4.3 Exploratory Analysis Using Diagnostic Plots

Simple logit plots are very useful in exploring the relationship between a binary response and a single continuous predictor variable in a BLR with a single predictor variable. However, these plots are not effective in revealing the complex relationships among the predictor variables or data problems in BLR with many predictors. The partial delta logit plots proposed here are useful in detecting significant predictors, nonlinearity, and multicollinearity. The partial delta logit plot illustrates the effects of a given continuous predictor variable, after adjusting for all other predictor variables, on the change in the logit estimate when the variable in question is dropped from the BLR. By overlaying the simple logit and partial delta logit plots, many features of the BLR could be revealed. The mechanics of these two logit plots are described using the two-variable BLR model.

1. Simple logit model for the binary response of the predictor variable X_1
   Step 1: Fit a simple BLR model.

$$Logit_{(Pi)} = \beta_0 + \beta_1 X_1 \tag{5.7}$$

2. Obtain the delta logit estimate for a given predictor
   Step 1: Fit the full BLR model with a quadratic term for X_1.

$$Logit_{(full)} (P_i) = \beta_0 + \beta_1 X_1 + \beta_2 X_2 + \beta_3 X_1^2 \tag{5.8}$$

Step 2: Fit the reduced BLR model.

$$Logit_{(reduced)} (P_i) = \beta_0 + \beta_2 X_2 \tag{5.9}$$

Step 3: Estimate the delta logit: the difference in logit between the full and the reduced model.

$$\Delta logit = Logit_{(full)} - Logit_{(reduced)} \tag{5.10}$$

Step 4: Compute the partial residual for X_1, and add X_1-mean.

$$X_i = a_0 + b_2 X_2 + e_i \tag{5.11}$$

$$PR_{x1} = e_i + X1 \, mean \tag{5.12}$$

Step 5: Overlay simple logit and partial delta logit plots.

Simple logit plot: **$Logit (P_i)$ vs. X_1**

Partial delta logit plot: **$\Delta logit$ vs. PR_{x1}**

5.4.3.1 Interpretation

The positive or negative slope in the partial delta logit plot shows the significance of the predictor variable in question. The quadratic trend in the partial delta logit plot confirms the need for quadratic term for X_i in BLR. The clustering of delta logit points near the mean of X_i in the partial delta logit plot confirms the presence of multicollinearity among the predictors. Large differences between the simple logit and the partial delta logit line illustrate the difference between the simple and the partial effects for a given variable X_i. See Figures 5.62 and 5.63 for some examples of these diagnostic plots.

5.4.3.2 Two-Factor Interaction Plots between Continuous Variables

A 3-D plot is generated, and the statistical significance of the interaction term is displayed on the title statement if the interaction term is significant at a p-value $\leq$ 0.15. For example, in a logistic linear regression with two continuous variables, the interaction effect is computed as follows and plotted on the z-axis.

Full model with the interaction: $log\ (p/1{-}p)_{full} = \beta_0 + \beta_1 X_1 + \beta_2 X_2 + \beta_3 X_1 X_2$

Reduced model without the interaction: $log\ (p/1{-}p)_{red} = \beta_0 + \beta_1 X_1 + \beta_2 X_2$

Interaction term = $(log\ (p/1{-}p)_{full} - log\ (p/1{-}p)_{red})$

Then, on the x-axis the $X1$ variable and on the y-axis the $X2$ variable are plotted. See Figure 5.49 for an example of the 3-D interaction plot.

5.4.4 Checking for Violations of Regression Model Assumptions

If logistic regression data violate one or more of the BLR assumptions, the results of the analysis may be incorrect or misleading. The assumptions for a valid MLR are:

■ Model parameters are incorrectly specified.
■ Influential outliers are absent.
■ Multicollinearity is lacking.
■ There is no overdispersion.

5.4.4.1 Model Specification Error

When important predictor variables or significant higher-order model terms (quadratic and cross-product) are omitted from the BLR, the predicted probability will be biased. The partial delta logit plot is effective for detecting the need for a quadratic term.

5.4.4.2 Influential Outlier

The presence of significant outliers produces biased logit estimates and reduces the predictive power of the BLR. An influential outlier may act as a high-leverage point, distorting the fitted equation and perhaps fitting the model poorly. The SAS/ LOGISTIC procedure has many powerful influential diagnostic statistics.[33] The DIFDEV statistic detects the ill-fitted observation that is responsible for the differences between the data and the predicted probabilities. It shows the change in the deviance due to deleting given observations. Observations with DIFDEV greater than four could be examined for the outlier. High-leverage data points are highly influential and have significant *hat-values*. The C-bar statistic measures the impact on change in the confidence intervals as a result of deleting a given observation. The leverage statistic, *h*, is useful in identifying cases with high-leverage effects. The leverage statistic varies from 0 (no influence on the model) to 1 (completely determines the model). The leverage of any given case may be compared to the average leverage, which equals p/n, where $p = (k + 1)/n$, k = the number of predictors, and n = the sample size. Displaying *DIFDEV, Hat, and C-bar* statistics on the same plot is an effective way of identifying both outliers and influential observations. An example of the influential outlier detection plot is given in Figure 5.72.

5.4.4.3 Multicollinearity

When a predictor variable is nearly a linear combination of other predictors in the model, the affected estimates are unstable and have high standard errors. If multicollinearity among the predictors is strong, the partial logit estimates may have wrong sign and size, and are unstable. If you remove a predictor involved in a collinear relationship from the model, the sign and size of the remaining predictor can change dramatically. PROC LOGISTIC does not have options for detecting multicollinearity. However, the partial delta logit plot proposed in this chapter provides graphical methods for detecting multicollinearity.

5.4.4.4 Overdispersion[34]

The expected variance of the binary response in BLR is $np(1 - p)$, where n is the sample size and p is the probability of the binary event. When the observed variance exceeds the expected variance, we have overdispersion. Model- and data-specific errors can contribute to overdispersion. Pearson *chi-square* and *deviance chi-square* tests are available for detecting significant overdispersion and adjusting the *se* of the parameters by the degree of overdispersion. If there is moderate discrepancy, standard errors will be underestimated, and we should use adjusted standard error. Adjusted standard error will make the confidence intervals wider. However, if there are large discrepancies, this indicates a need to respecify the model, or that the sample was not random, or other serious design problems.

5.5 Ordinal Logistic Regression[37]

Many ordinal response variables are difficult, if not impossible, to measure on an interval scale. However, numeric values are assigned arbitrary to ordinal levels and parametric modeling commonly performed, which may lead to erroneous conclusions when analyzed. One option is to transform the ordinal scale to the binary and then run a binary logistic regression. However, the loss of information and decrease in statistical power are often too high a price to pay. Moreover, the resulting odds ratios may depend on the cut point chosen to dichotomize the outcome, and this choice is often arbitrary.

The proportional odds or cumulative odds ordinal logistic regression is one of the recommended models for the analysis of ordinal data and comes from the class of generalized linear models. It is a generalization of a binary logistic regression model when the response variable has more than two ordinal categories. The unique feature of the proportional odds model is that the odds ratio for each predictor is considered as constant across all possible collapsing of the response variable, The proportional odds model is used to estimate the odds of being at or below a particular level of the response variable. If the ordinal outcome has four levels (a, b, c, and d), three logits will be modeled, one for each of the following cut points: a versus b,c,d; a,b vs. c,d; and a,b,c vs. d.

This model can estimate the odds of being at or beyond a particular level of the response variable as well because, below and beyond a particular category are just two complementary directions. However, ignoring the discrete ordinal nature of the response variable would make the analysis lose some useful information and lead to misleading. This option is implemented in the LOGIST2 macro. For an example of performing ordinal logistic regression, see Case Study 5.

5.6 Survey Logistic Regression

When a complex sample design is used to draw a sample from a finite population, the sample design should be incorporated in the analysis of the survey data in order to make statistically valid inferences for the finite population.The logistic regression analysis for the sample survey data can be performed by the SAS SURVEYLOGISTIC procedure, which fits linear logistic regression models for discrete response survey data by the method of maximum likelihood. This PROC can be used in the logistic regression modeling of finite survey data coming from complex survey sample designs, which incorporates complex survey sample designs, including designs with stratification, clustering, and unequal weighting.

The survey logistic regression procedure

- Uses the maximum-likelihood estimation with either the Fisher scoring algorithm or the Newton–Ralphson algorithm.
- Can specify starting values for the parameter estimates.
- Logit link function in the ordinal logistic regression models can be replaced by the probit function or the complementary log–log function.
- Odds ratio estimates are displayed along with parameter estimates. You can also specify the change in the explanatory variables for which the odds ratio estimates are desired.
- Variances of the regression parameters and odds ratios are computed by using either the Taylor series (linearization) method or replication (resampling) methods to estimate sampling errors of estimators based on complex sample designs.

- Can specify categorical variables (also known as CLASS variables) as explanatory variables.
- Can specify interaction terms in the same way as in the LOGISTIC procedure.

This option is implemented in the LOGIST2 macro. See Appendix 2 for the instructions in the LOGIST2 help file to implement SURVEY LOGISTIC regression estimates. For more information about this PROC, refer to the SAS software version 9.2 help file.

5.7 Multiple Linear Regression Using SAS Macro REGDIAG2

The REGDIAG2 macro is a powerful SAS application that is useful for performing multiple linear regression analysis. Options are available for obtaining various regression diagnostic graphs and tests. SAS procedures, REG, and GLM, are the main tools used in the macro.[35,36] In addition to these SAS procedures, GLMSELECT, ROBUSTREG, RSREG, SURVEYREG, BOXPLOT, GPLOT, and SGPLOTS and procedures are also utilized in the REGDIAG2 macro. The enhanced features implemented in the REGDIAG2 macro are:

- *New*: Variable screening step using GLMSELECT, and best-candidate model selection using AICC and SBC, are implemented.
- Regression diagnostic plots such as augmented partial residual plot, partial leverage plot, and Vif plot for each predictor variable are automatically generated.
- *New*: Interaction diagnostic plots for detecting significant interaction between two continuous variables or between a categorical and continuous variable are generated.
- Plots for checking for violations of model assumptions (residual plots for detecting heteroscedasticity and autocorrelation, outlier detection plot, normal probability and distribution plots of residual) are also generated.
- Test statistics and P-values for testing normally distributed errors and model specification errors are automatically produced, as well as *Breusch–Pagan* tests for detecting heteroscedasticity.
- In the case of simple linear regression, plots of linear and quadratic regression plots with a 95% confidence band are generated automatically; if you fit MLR, the overall model fit plot is produced.
- Options are available for validating the MLR model obtained from a training dataset using an independent validation dataset by comparing fitted lines and the residual.
- *New*: Options are implemented to run the ROBUST regression using SAS ROBUSTREG when extreme outliers are present in the data.

- *New*: Options are implemented to run SURVEYREG regression using SAS SURVEYREG when the data are coming from survey data and the design weights are available.
- Options for saving the output tables and graphics in WORD, HTML, PDF, and TXT formats are available.

Software requirements for using the REGDIAG2 macro are:

- SAS/BASE, SAS/STAT, and SAS/GRAPH must be licensed and installed at your site.
- SAS version 9.0 and above is required for full utilization.

5.7.1 Steps Involved in Running the REGDIAG2 Macro

- Step 1: Create an SAS temporary dataset containing at least one continuous response (target) variable and many continuous and/or categorical predictor (input) variables.
- Step 2: Open the REGDIAG2.SAS macro-call file into the SAS EDITOR window. Instructions are given in Appendix 1 regarding downloading the macro-call and sample data files from this book's Web site. Click the RUN icon to submit the macro-call file REGDIAG2.SAS to open the MACRO-CALL window called REGDIAG2 (Figure 5.4).
- *Special note to SAS Enterprise Guide (EG) Code Window Users*: Because these user-friendly SAS macro applications included in this book use SAS WINDOW/DISPLAY commands, and these commands are not compatible with SAS EG, open the traditional REGDIAG2 macro-call file included in the \dmsas2e\maccall\nodisplay\ into the SAS editor. Read the instructions given in Appendix 3 regarding using the traditional macro-call files in the SAS EG/SAS Learning Edition (LE) code window.
- Step 3: Input the appropriate parameters in the macro-call window by following the instructions provided in the REGDIAG2 macro help file in Appendix 2. Users can choose whether to include regression diagnostic plots for each predictor variable and to exclude large extreme observations from model fitting. After inputting all the required macro parameters, check that the cursor is in the last input field, and then hit the ENTER key (not the RUN icon) to submit the macro.
- Step 4: Examine the LOG window (only in DISPLAY mode) for any macro execution errors. If you see any errors in the LOG window, activate the EDITOR window, resubmit the REGDIAG2.SAS macro-call file, check your macro input values, and correct any input errors.
- Step 5: If no errors are found in the LOG window, activate the EDITOR window, resubmit the REGDIAG2.SAS macro-call file, and change the macro

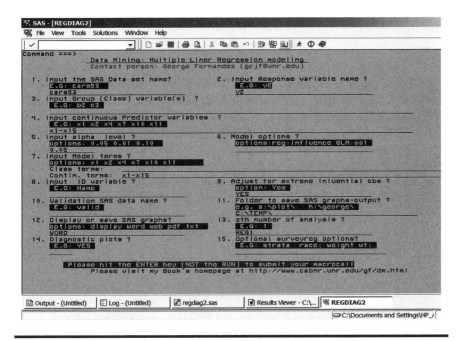

Figure 5.4   Screen copy of REGDIAG2 macro-call window showing the macro-call parameters required for performing model selection in MLR.

input value from DISPLAY to any other desirable format (see Appendix 2). The SAS output files from complete regression modeling and exploratory graphs can be saved as user-specified format files in the user-specified folder.

5.8 Lift Chart Using SAS Macro LIFT2

The LIFT2 macro is a powerful SAS application for producing a LIFT chart (IF–THEN ANALYSIS) in linear and logistic regression models. Options are available for graphically comparing the predicted response from the full model and the predicted response from the reduced model (keeping the variable of interest at a constant level). SAS procedures, REG, GLM, and LOGISTIC can be used in the macro. In addition to these SAS procedures, the GPLOT procedure is also utilized in the LIFT2 macro. There are currently no procedures or options available in the SAS system to produce LIFT charts automatically.

Software requirements for using the LIFT2 macro:

■ SAS/BASE, SAS/STAT, and SAS/GRAPH must be licensed and installed at your site.
■ SAS version 9.0 and above is required for full utilization.

5.8.1 Steps Involved in Running the LIFT2 Macro

■ Step 1: Create a temporary SAS dataset containing at least one response (dependent) variable and many continuous and/or categorical predictor (independent) variables.

■ Step 2: Open the LIFT2.SAS macro-call file into SAS EDITOR Window. Instructions are given in Appendix 1 regarding downloading the macro-call and sample data files from this book's Web site. Click the RUN icon to submit the macro-call file LIFT2.SAS to open the macro-call window called LIFT2 (Figure 5.23).

■ Step 3: Input the appropriate parameters in the macro-call window by following the instructions provided in the LIFT2 macro help file in Appendix 2. After inputting all the required macro parameters, check that the cursor is in the last input field, and then hit the ENTER key (not the RUN icon) to submit the macro.

■ *Special note to SAS Enterprise Guide (EG) Code window users*: Because these user-friendly SAS macro applications included in this book use SAS WINDOW/ DISPLAY commands, and these commands are not compatible with SAS EG, open the traditional LIFT2 macro-call file included in the \dmsas2e\maccal\nodisplay\ into the SAS editor. Read the instructions given in Appendix 3 regarding using the traditional macro-call files in the SAS EG/ SAS Learning Edition (LE) code window.

■ Step 4: Examine the LOG window (only in DISPLAY mode) for any macro execution errors. If you see any errors in the LOG window, activate the EDITOR window, resubmit the LIFT2 macro-call file, check your macro input values, and correct any input errors.

■ Step 5: If no errors are found in the LOG window, activate the EDITOR window, resubmit the LIFT2 macro-call file, and change the macro input value from DISPLAY to any other desirable output format. The SAS output files from LIFT2 analysis and LIFT charts can be saved as user-specified format files in the user-specified folder.

5.9 Scoring New Regression Data Using the SAS Macro RSCORE2

The RSCORE2 macro is a powerful SAS application for scoring new datasets using the established regression model estimates. Options are available for checking the residuals graphically if observed response variables are available in the new dataset. Otherwise, only predicted scores are produced and saved in a SAS dataset. In addition to the SAS REG procedure, GPLOT is also utilized in the RSCORE2 macro. There are currently no procedures or options available in the SAS systems to compare the residuals from the new data automatically.

Software requirements for using the RSCORE2 macro:

▪ SAS/BASE, SAS/STAT, and SAS/GRAPH must be licensed and installed at your site.
▪ SAS version 9.0 and above is required for full utilization.

5.9.1 Steps Involved in Running the RSCORE2 Macro

▪ Step 1: Create a new scoring temporary SAS dataset containing one response (optional) variable and continuous predictor (independent) variables. This new dataset should contain all the predictor variables that were used to develop the original regression model.
▪ Step 2: Verify that the regression parameter estimates are available in an SAS dataset. If you use the REGDIAG2 macro to fit the original regression model, it should have created an SAS dataset called REGEST. Check this REGEST data and create a temporary dataset so that you can use it for scoring your new dataset.
▪ Step 3: Open the RSCORE2.SAS macro-call file into the SAS EDITOR window. Instructions are given in Appendix 1 regarding downloading the macro-call and sample data files from this book's Web site. Click the RUN icon to submit the macro-call file RSCORE2 and open the MACRO–CALL window called RSCORE2 (Figure 5.25).
▪ *Special note to SAS Enterprise Guide (EG) Code window users*: Because these user-friendly SAS macro applications included in this book use SAS WINDOW/DISPLAY commands, and these commands are not compatible with SAS EG, open the traditional RSCORE2 macro-call file included in the \dmsas2e\maccal\nodisplay\ into the SAS editor. Read the instructions given in Appendix 3 regarding using the traditional macro-call files in the SAS EG/ SAS Learning Edition (LE) code window.
▪ Step 4: Input the appropriate parameters in the macro-call window by following the instructions provided in the RSCORE2 macro help file in Appendix 2. After inputting all the required macro parameters, check that the cursor is in the last input field, and then hit the ENTER key (not the RUN icon) to submit the macro.
▪ Step 5: Examine the LOG window (only in DISPLAY mode) for any macro execution errors. If you see any errors in the LOG window, activate the EDITOR window, resubmit the RSCORE2.SAS macro-call file, check your macro input values, and correct any input errors.
▪ Step 6: If no errors are found in the LOG window, activate the EDITOR window, resubmit the RSCORE2.SAS macro-call file, and change the macro input value from DISPLAY to any other desirable output format. The SAS output files from RSCORE analysis and RSCORE charts can be saved as user-specified format files in the user-specified folder.

5.10 Logistic Regression Using SAS Macro LOGIST2

The LOGIST2 macro is a powerful SAS application for performing complete and user-friendly logistic regressions with and without categorical predictor variables. Options are available for performing model selection and various logistic regression diagnostic graphs and tests. The SAS procedure, LOGISTIC, is the main tool used in the macro. In addition to these SAS procedures, SURVEYLOGISTIC, GPLOT, GCHART, and SGPLOT are also utilized in the LOGIST2 macro to obtain diagnostic graphs. The enhanced features implemented in the LOGIST2 macro are:

- *New*: Best-candidate model selection using AICC and SBC criteria by comparing all possible combination of models within the optimum number of subsets determined by the sequential stepwise selection using AICC is implemented.
- Logistic regression diagnostic plots such as overlaid partial delta logit and simple logit plots (PROC GPLOT) for each continuous predictor variables are produced.
- *New*: Interaction diagnostic plots for detecting significant interaction between two continuous variables or between a categorical and continuous variable are generated.
- Outlier detection plot, ROC, and false-positive and false-negative plots, and LIFT charts for assessing the overall model fit are automatically generated.
- Option for excluding extreme outliers (delta deviance >4) and then performing logistic regression using the remaining data points is given.
- Option for validating the fitted logistic model obtained from a training dataset using an independent validation dataset by comparing the Brier score is given.
- *New*: Options are implemented to run survey logistic regression using SAS PROC SURVEYLOGISTIC when the data are coming from survey data and the design weights are available.
- Options for saving the output tables and graphics in WORD, HTML, PDF, and TXT formats are available.

Software requirements for using the LOGIST2 macro:

- SAS/BASE, SAS/STAT, and SAS/GRAPH must be licensed and installed at your site.
- SAS version 9.0 and above is required for full utilization.
- Active Internet connection is required for downloading compiled LOGIST2 macro from the book's Web site.

5.11 Scoring New Logistic Regression Data Using the SAS Macro LSCORE2

The LSCORE2 macro is a powerful SAS application for scoring new datasets using the established logistic regression model estimates. Options are available for checking the residuals graphically if the observed binary response variable is available in the new dataset. Otherwise, only predicted probability scores are produced and saved in an SAS dataset. In addition to the SAS LOGISTIC procedure, GPLOT is also utilized in the LSCORE2 macro. There is, currently, no procedure or options available in SAS systems to compare the residuals from the scoring data automatically.

Software requirements for using the LSCORE2 macro:

- SAS/BASE, SAS/STAT, and SAS/GRAPH must be licensed and installed at your site.
- SAS version 9.0 and above is recommended for full utilization.
- Active Internet connection is required for downloading compiled LSCORE2 macro from the book's Web site.

5.12 Case Study 1: Modeling Multiple Linear Regressions

5.12.1 Study Objectives

1. **Model selection**: Using the LASSO method in PROC GLMSELECT, eliminate less-contributing predictor variables and run all possible subset selection using MAXR2 in PROC REG, and select the best-candidate models using AICC and SBC criteria.
2. **Data exploration using diagnostic plots**: Check for linear and nonlinear relationships and significant outliers (augmented partial residual plot), significant regression relationship (partial leverage plot), multicollinearity (VIF plot), and detection of significant interaction (interaction plot) for each predictor variable. Check for the significance of linear, quadratic, and cross products, and the overall significance of each predictor variable.
3. **Regression model fitting and prediction**: Perform hypothesis testing on overall regression and on each parameter estimates. Estimate confidence intervals for parameter estimates, predict scores, and estimate their confidence intervals.
4. **Checking for any violations of regression assumptions**: Perform statistical tests and graphical analysis to detect influential outliers, multicollinearity among predictor variables, heteroscedasticity in residuals, and departure form normally distributed residual.

5. **Save _score_ and regest datasets for future use**: These two datasets are created and saved as temporary SAS datasets in the work folder and also exported to Excel worksheets and saved in the user-specified output folder. The _score_ data contains the observed variables; the predicted scores, including observations with missing response value; residuals; and confidence-interval estimates. This dataset could be used as the base for developing the scorecards for each observation. Also, the second SAS data called regest contains the parameter estimates that could be used in the RSCORE macro for scoring different datasets containing the same variables.

6. **If–then analysis and lift charts**: Perform IF-THEN analysis and construct a lift chart to estimate the differences in the predicted response when one of the continuous predictor variables is fixed at a given value.

Multiple Linear Regression Analysis of 1993 Car Attribute Data

Data name	SAS dataset CARS93
Multiattributes	Y2: Midprice
	X1 Air bags (0 = none, 1 = driver only, 2 = driver and passenger)
	X2 Number of cylinders
	X3 Engine size (liters)
	X4 HP (maximum)
	X5 RPM (revs per minute at maximum HP)
	X6 Engine revolutions per mile (in highest gear)
	X7 Fuel tank capacity (gallons)
	X8 Passenger capacity (persons)
	X9 Car length (inches)
	X10 Wheelbase (inches)
	X11 Car width (inches)
	X12 U-turn space (feet)
	X13 Rear seat room (inches)
	X14 Luggage capacity (cubic feet)
	X15 Weight (pounds)
Number of observations	92
Car93 Data Source:	Lock[38]

5.12.1.1 Step 1: Preliminary Model Selection

Open the REGDIAG2.SAS macro-call file in the SAS EDITOR window and click RUN to open the REGDIAG2 macro-call window (Figure 5.4). Input the appropriate macro-input values by following the suggestions given in the help file (Appendix 2). Leave the group variable option blank since all the predictors used are continuous. Leave the macro field #14 BLANK to skip regression diagnostics and to run MLR.

■ *Special note to SAS Enterprise Guide (EG) Code Window Users*: Because these user-friendly SAS macro applications included in this book, use SAS WINDOW/DISPLAY commands, and these commands are not compatible with SAS EG, open the traditional REGDIAG macro-call file included in the \dmsas2e\maccal\nodisplay\ into the SAS editor. Read the instructions given in Appendix 3 regarding using the traditional macro-call files in the SAS EG/ SAS Learning Edition (LE) code window.

Model selection: *Variable selection using MAX R^2 selection method*: The REGDIAG2 macro utilizes all possible regression models using the MAXR^2 selection methods and output the best two models for all subsets (Table 5.1). Because 15 continuous predictor variables were used in the model selection, the full model had 15 predictors. Fifteen subsets are possible with 15 predictor variables. By comparing the R^2, $R^2(adj)$, RMSE, $C(p)$, and AIC values between the full model and all subsets, we can conclude that the 6-variable subset model is superior to all other subsets.

The *Mallows $C(p)$* measures the total squared error for a subset that equals total error variance plus the bias introduced by not including the important variables in the subset. The $C(p)$ plot (Figure 5.5) shows the $C(p)$ statistic against the number of predictor variables for the full model and the best two models for each subset. Additionally, the *RMSE* statistic for the full model and best two regression models in each subset is also shown in the $C(p)$ plot. Furthermore, the diameter of the bubbles in the $C(p)$ plot is proportional to the magnitude of *RMSE*. Consequently, dropping any variable from the six-variable model is not recommended because, the $C(p)$, *RMSE*, and *AIC* values jump up so high. These results clearly indicate that $C(p)$, *RMSE*, and *AIC* statistics are better indicators for variable selection than R^2 and $R^2(adj)$. Thus, the $C(p)$ plot and the summary table of model selection statistics produced by the REGDIAG2 macro can be used effectively in selecting the best subset in regression models with many (5 to 25) predictor variables.

LASSO, the new model selection method implemented in the new SAS procedure GLMSELECT, is also utilized in the REGDIAG2 macro for screening all listed predictor variables and examine and visualize the contribution of each predictor in the model selection. Two informative diagnostic plots (Figures 5.6 and 5.7) generated by the ODS graphics feature in the GLMSELECT can be used to visualize the importance of the predictor variables. The fit criteria plot (Figure 5.6) displays the trend plots of six model selection criteria versus the number of model parameters, and

Table 5.1 Macro REGDIAG2—Best Two Subsets in All Possible MAXR^2 Selection Method

Number in Model	R-Square	Adjusted R-Square	C(p)	AIC	Root MSE	SBC	Variables in Model
1	0.6670	0.6627	48.5856	266.1241	5.10669	270.91297	X4
1	0.6193	0.6145	66.5603	276.9593	5.45992	281.74819	X15
2	0.7006	0.6929	37.8970	259.4966	4.87278	266.67998	X4 X7
2	0.6996	0.6919	38.2671	259.7618	4.88076	266.94510	X1 X4
3	0.7364	0.7261	26.4105	251.1920	4.60207	260.76981	X4 X11 X15
3	0.7276	0.7169	29.7340	253.8557	4.67837	263.43350	X4 X10 X11
4	0.7710	0.7589	15.3699	241.8019	4.31775	253.77414	X1 X2 X11 X15
4	0.7666	0.7544	16.9950	243.3118	4.35818	255.28403	X1 X4 X11 X15
5	0.7960	0.7824	7.9336	234.4305	4.10214	248.79718	X1 X2 X4 X7 X11
5	0.7943	0.7805	8.5844	235.1128	4.11945	249.47949	X1 X2 X7 X11 X15
6	0.8162	0.8013	2.2959	227.9613	3.91941	244.72243	X1 X2 X4 X7 X10 X11
6	0.8079	0.7924	5.4282	231.5424	4.00701	248.30352	X1 X2 X4 X7 X11 X15
7	0.8188	0.8015	3.3136	228.8049	3.91809	247.96048	X1 X2 X4 X6 X7 X10 X11
7	0.8185	0.8011	3.4231	228.9346	3.92123	248.09024	X1 X2 X4 X7 X10 X11 X15
8	0.8245	0.8050	3.1778	228.2320	3.88305	249.78208	X1 X2 X4 X6 X7 X10 X11 X15

8	0.8208	0.8009	4.5708	229.9193	3.92370	251.46934	X1 X2 X4 X7 X10 X11 X12 X15
9	0.8259	0.8038	4.6640	229.6007	3.89509	253.54524	X1 X2 X4 X6 X7 X10 X11 X12 X15
9	0.8248	0.8026	5.0546	230.0812	3.90666	254.02565	X1 X2 X4 X6 X7 X8 X10 X11 X15
10	0.8261	0.8013	6.5653	231.4789	3.91986	257.81784	X1 X2 X4 X6 X7 X9 X10 X11 X12 X15
10	0.8261	0.8013	6.5721	231.4873	3.92007	257.82622	X1 X2 X4 X6 X7 X8 X10 X11 X12 X15
11	0.8266	0.7989	8.4032	233.2784	3.94328	262.01177	X1 X2 X4 X6 X7 X8 X10 X11 X12 X13 X15
11	0.8265	0.7988	8.4254	233.3058	3.94395	262.03921	X1 X2 X4 X6 X7 X8 X9 X10 X11 X12 X15
12	0.8270	0.7964	10.2462	235.0837	3.96740	266.21150	X1 X2 X4 X6 X7 X8 X9 X10 X11 X12 X13 X15
12	0.8268	0.7963	10.3043	235.1558	3.96917	266.28364	X1 X2 X4 X6 X7 X8 X10 X11 X12 X13 X14 X15
13	0.8273	0.7938	12.1050	236.9082	3.99257	270.43044	X1 X2 X4 X5 X6 X7 X8 X9 X10 X11 X12 X13 X15
13	0.8272	0.7937	12.1621	236.9792	3.99432	270.50152	X1 X2 X4 X6 X7 X8 X9 X10 X11 X12 X13 X14 X15
14	0.8276	0.7911	14.0000	238.7775	4.01946	274.69420	X1 X2 X4 X5 X6 X7 X8 X9 X10 X11 X12 X13 X14 X15
14	0.8273	0.7907	14.1049	238.9081	4.02270	274.82480	X1 X2 X3 X4 X5 X6 X7 X8 X9 X10 X11 X12 X13 X15
15	0.8276	0.7878	16.0000	240.7775	4.05026	279.08865	X1 X2 X3 X4 X5 X6 X7 X8 X9 X10 X11 X12 X13 X14 X15

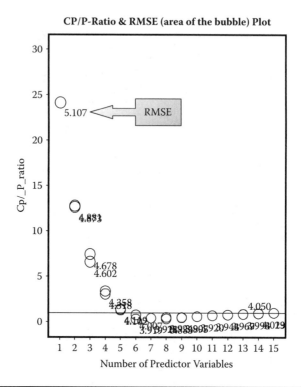

Figure 5.5 Model selection using SAS macro REGDIAG2: CP plot for selecting the best subset model.

in this example, all six criteria identify the 13-parameter model as the best model. However, beyond the six variables, no substantial gain was noted. The coefficient progression plot displayed in Figure 5.7 shows the stability of the standardized regression coefficients as a result of adding new variables in each model-selection step. The problem of multicollinearity among the predictor variables was not evident since all the standardized regression coefficients have values less than ±1. The following six variables, $X4$, $X7$, $X1$, $X2$, $X13$, and $X11$, were identified as the most contributing variables in the model selection sequence. Although $X15$ was included in the second step, it was later excluded from the model. Thus, these features enable the analysts to identify the most contributing variables and help them perform further investigations.

Because this model-selection step only includes the linear effects of the variables, it is recommended that this step be used as a preliminary model selection step rather than the final concluding step. Furthermore, the REGDIAG2 macro also has a feature for selecting the best-candidate models using AICC and SBC (Tables 5.2 and 5.3). Next we will examine the predictor variables selected in the best-candidate models.

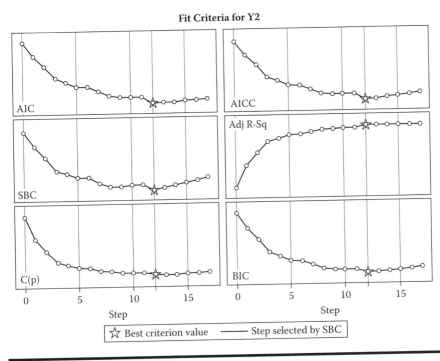

Figure 5.6 Model selection using SAS macro REGDIAG2: Fit criteria plots derived from using the ODS graphics feature in GLMSELECT procedure.

Both minimum AICC and SBC criteria identified the same six-variable model (X1, X2, X4, X7, X11, and X10) as the best model. The first five variables were also selected as the best contributing variables by the LASSO method (Figure 5.6), and the CP method picked the same six variables as the best model (Table 5.1). The ∆SBC criterion is very conservative and picked only one model as the best candidate where as ∆ AICC method identified five models as the best candidates. The standardized regression coefficients of the best candidate model's predictors were very stable, indicating the impact of multicollinearity is very minimal. Then, based on the preliminary model selection step, the following X1, X2, X4, X7, X11, and X10 variables were identified as the best linear predictors, and we can proceed with the second step of the analysis.

5.12.1.2 Step 2: Graphical Exploratory Analysis and Regression Diagnostic Plots

Open the REGDIAG2.SAS macro-call file in the SAS EDITOR window and click RUN to open the REGDIAG2 macro-call window (Figure. 5.8). Input the appropriate macro-input values by following the suggestions given in the help file

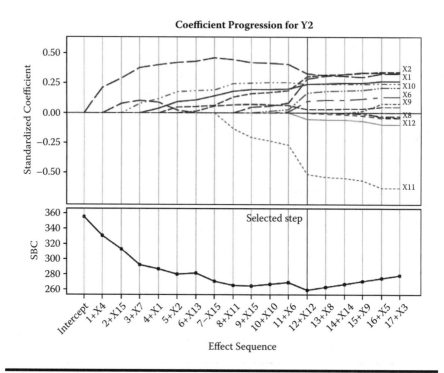

Figure 5.7   Model selection using SAS macro REGDIAG2: Standardized regression coefficient and SBC progression plots by model selection steps derived from using the ODS graphics feature in GLMSELECT procedure.

(Appendix 2). Leave the group variable option blank, because all the predictors used are continuous. Input YES in macro field #14 to request additional regression diagnostics plots using the selected predicted variables in step 1.

The three model selection plots—CP plot (Figure 5.9), fit criteria plot (Figure 5.10), and coefficient progression plot (Figure 5.11)—on the predicted variables selected in step 1 (6 variables: $X1$, $X2$, $X4$, $X7$, $X11$, and $X10$) further confirmed that these are the best linear predictors in all model selection criteria. Thus, in the second step, data exploration and diagnostic plot analysis were carried out using these six predictor variables.

Simple linear regression and augmented partial residual (APR) plots for all six predictor variables are presented in Figure 5.12. The linear/quadratic regression parameter estimates for the simple and multiple linear regressions and their significance levels are also displayed in the titles of the APR plots. The simple linear regression line describes the relationship between the response and a given predictor variable in a simple linear regression. The APR line shows the quadratic regression effect of the ith predictor on the response variable after accounting for

Table 5.2 Macro REGDIAG2—Standardized Regression Coefficient Estimates and the Several Model Selection Criteria for the Best-Candidate Models in All Possible MAXR^2 Selection Methods Using the Selection Criterion Delta AICC < 2

Dependent Variable	Intercept	Air Bags (0 = none, 1 = driver only, 2 = driver and passenger) X1	Number of Cylinders X2	Engine Size (liters) X3	HP (maximum) X4	RPM (revs per minute at maximum HP) X5	4 Engine Revolutions per Mile (in highest gear) X6	Fuel Tank Capacity (gallons) X7	Passenger Capacity (persons) X8	Car Length (inches) X9	Wheelbase (inches) X10	Car Width (inches) X11	U-Turn Space (feet) X12	Rear Seat Room (inches) X13	Luggage Capacity (cu ft X14	Weight (pounds) X15	Number of Parameters in Model
Y2	19.3219	2.44057	3.14933	.	3.53493	.	.	3.31439	.	.	2.80894	-6.1088	.	.	.	.	7
Y2	19.4307	2.21502	3.31102	.	2.66918	.	1.27614	2.33768	.	.	2.22848	-5.9390	.	.	.	3.10061	9
Y2	19.3455	2.38315	3.37800	.	3.52546	.	0.77161	3.25036	.	.	2.98053	-5.7254	.	.	.	.	8
Y2	19.3628	2.36426	3.02281	.	3.03817	.	.	2.80552	.	.	2.30391	-6.3801	.	.	1.81189	.	8
Y2	19.3221	2.49279	3.10444	.	3.56846	.	.	3.32973	.	.	2.90899	-5.7885	-0.523	.	.	.	8

(continued)

Table 5.2 Macro REGDIAG2—Standardized Regression Coefficient Estimates and the Several Model Selection Criteria for the Best-Candidate Models in All Possible MAXR² Selection Methods Using the Selection Criterion Delta AICC < 2 (Continued)

Adjusted r-Squared	Schwarz's Bayesian Criterion	AICC	DELTA_AICC	DELTA_SBC	W_AICC	W_AICCR
0.80133	244.722	229.279	0.00000	0.00000	0.33421	1.00000
0.80500	249.782	230.401	1.12177	5.05964	0.19074	0.57070
0.80147	247.960	230.519	1.24024	3.23805	0.17977	0.53788
0.80115	248.090	230.649	1.37000	3.36781	0.16847	0.50409
0.79975	248.658	231.217	1.93826	3.93607	0.12681	0.37941

Table 5.3 Macro REGDIAG2—Standardized Regression Coefficient Estimates and the Several Model-Selection Criteria for the Best-Candidate Models in All Possible MAXR² Selection Methods Using the Selection Criterion Delta SBC <2

Dependent Variable	Intercept	Air Bags (0 = none, 1 = driver only, 2 = driver and passenger) X1	Number of Cylinders X2	Engine Size (liters) X3	HP (maximum) X4	RPM (revs per minute at maximum HP) X5	4 Engine Revolutions per Mile (in highest gear) X6
Y2	19.3219	2.44057	3.14933	.	3.53493	.	.

Fuel Tank Capacity (gallons) X7	Passenger Capacity (persons) X8	Car Length (inches) X9	Wheelbase (inches) X10	Car Width (inches) X11	U-Turn Space (feet) X12	Rear Seat Room (inches) X13	Luggage Capacity (cu ft) X14	Weight (pounds) X15	Number of Parameters in Model
3.31439	.	.	2.80894	−6.1088	.	.	.	.	7

Adjusted r-squared	Schwarz's Bayesian criterion	AICC	DELTA_AICC	DELTA_SBC	W_SBC	W_SBCR
0.80133	244.722	229.279	0	0	1	1

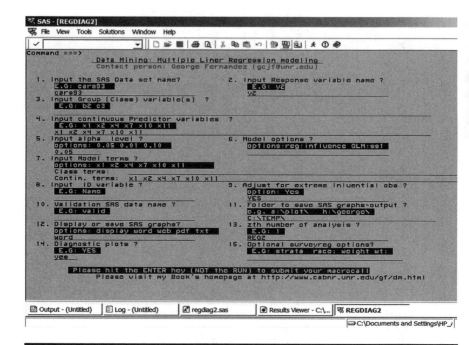

Figure 5.8 Screen copy of REGDIAG2 macro-call window showing the macro-call parameters required for performing regression diagnostic plots in MLR.

the linear effects of other predictors on the response. The APR plot is very effective in detecting significant outliers and nonlinear relationships. Significant outliers and/or influential observations are identified and marked on the APR plot if the absolute STUDENT value exceeds 2.5, or the DFFITS statistic exceeds 1.5. These influential statistics are derived from the MLR model involving all predictor variables. If the correlations among all predictor variables are negligible, the simple and the partial regression lines should have similar slopes.

The APR plots for the six-predictor variables showed significant linear relationships between the six predictors and median price. A big difference in the magnitude of the partial (adjusted) and the simple (unadjusted) regression effects for all six predictors on median price were clearly evident (Figure 5.12). The quadratic effects of all six predictor variables on the median price were not significant at the 5% level. Five significant outliers were also detected in these APR plots.

Partial leverage plots (PL) for all six predictor variables are presented in Figure 5.13. The PL display shows three curves: (a) the horizontal reference line that goes through the response variable mean, (b) the partial regression line, which quantifies the slope of the partial regression coefficient of the ith variable in the MLR, and (c) the 95% confidence band for partial regression line. The partial regression parameter estimates for the ith variable in the multiple linear regression

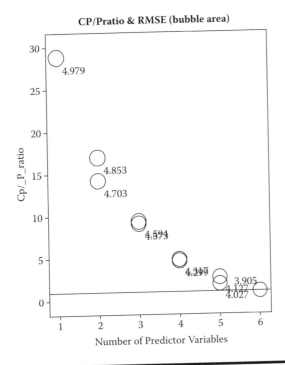

Figure 5.9 Confirming model selection using SAS macro REGDIAG2: CP plot for selecting the best-candidate model.

and their significance levels are also displayed in the titles. The slope of the partial regression coefficient is considered statistically significant at the 5% level if the response mean line (the horizontal reference line) intersects the 95% confidence band. If the response mean line lies within the 95% confidence band without intersecting it, then the partial regression coefficient is considered not significant. The PL plots for the six-predictor variables showed significant linear relationships between the six predictors and median price. Except for $X11$ (car width), all other predictor variables have positive partial effects on the median price.

The VIF plots for all six-predictor variables are presented in Figure 5.14. The VIF plot displays two overlaid curves: (a) The first curve shows the relation ship between partial residual + response mean and the ith predictor variable; (b) the second curve displays the relationship between the partial leverage + response mean and the partial ith predictor value + mean of ith predictor value. The slope of the both regression lines should be equal to the partial regression coefficient estimate for the ith predictor. When there is no high degree multicollinearity, both the partial residual (symbol R) and the partial leverage (Symbol E) values should be evenly distributed around the regression line. But, in the presence of severe multicollinearity, the partial leverage values, E shrinks and is distributed around the mean of the ith

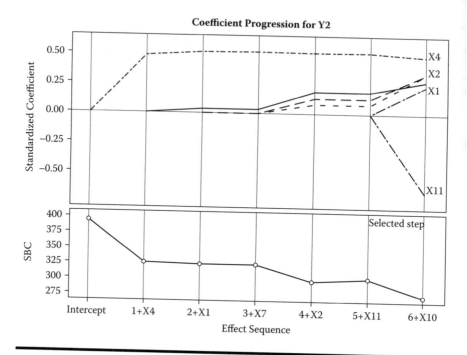

Figure 5.10   Model selection using SAS macro REGDIAG2: Standardized regression coefficient and SBC progression plots for the selected best-candidate model derived from using the ODS graphics feature in GLMSELECT procedure.

predictor variable. Also, the partial regression for the ith variable shows a nonsignificant relationship in the partial leverage plots, whereas the partial residual plot shows a significant trend for the ith variable. Furthermore, the degree of multicollinearity can be measured by the VIF statistic in a MLR model, and the VIF statistic for each predictor variable is displayed on the title statement of the VIF plot. The VIF plots for all six predictors showed a very low level of multicollinearity, whereas the crowding of the partial leverage values was not evident in the VIF plots.

Checking for model specification errors: Failure to include significant quadratic or interaction terms results in model specification errors in MLR. The APR plot discussed in the regression diagnostic plots section is effective in detecting the quadratic trend but not the interaction effects. The REGDIAG2 macro also utilizes the RSREG procedure and tests the overall significance of linear, quadratic, and two-factor cross products using the sequential *SS* (Type I). If the regression data contain replicated observations, *a lack-of-fit* test could be performed to check for model specification errors by testing the significance of the deviation from the regression using the *experimental error* as the error term.

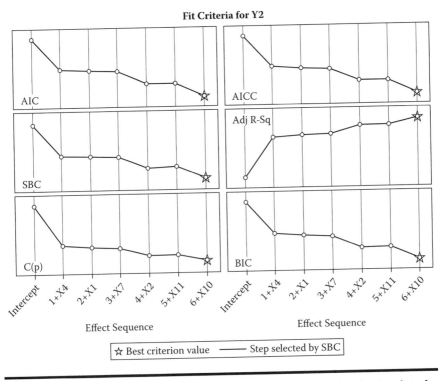

Figure 5.11   Model selection using SAS macro REGDIAG2: Fit criteria plots for the selected best-candidate model derived from using the ODS graphics feature in GLMSELECT procedure.

The statistical significance of model specification errors is presented in Table 5.4. The linear model accounted for 80% of the variation in response and was highly significant (p-value <0.0001). The quadratic was not significant. But at least one of the two-factor interaction was significant (p-value <0.0096) If either quadratic or cross product effects are significant and you want to further select the specific significant higher-order terms, PROC RSREG also includes a table of significance levels for all possible higher-order terms. If the higher-order terms are significant, and you would like to determine the most significant variable, or estimate the amount of reduction in the model SS when you drop one variable completely, RSREG procedure includes a table with useful information (Table 5.5). For each variable, the SS accounts for the linear, quadratic, and all-possible combination of two-factor interactions. Even though all six variables are highly significant, dropping the *X1* and *X11* variables results in a significant reduction in the model SS. Thus, the most significant factor in determining the median price is the number of air bags (*X1*).

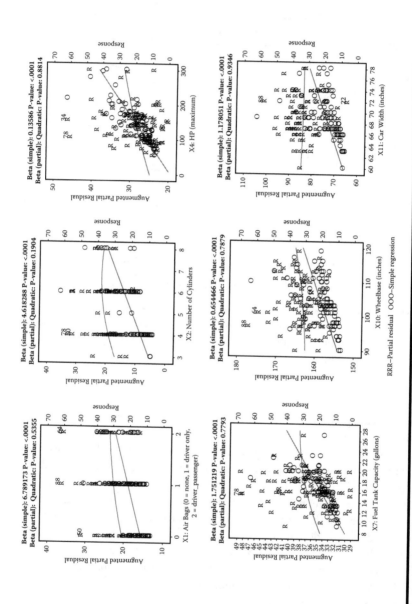

Figure 5.12 Regression diagnostic plot using SAS macro REGDIAG2: Overlay plot of simple linear regression and augmented partial residual plots for all selected predictor variables.

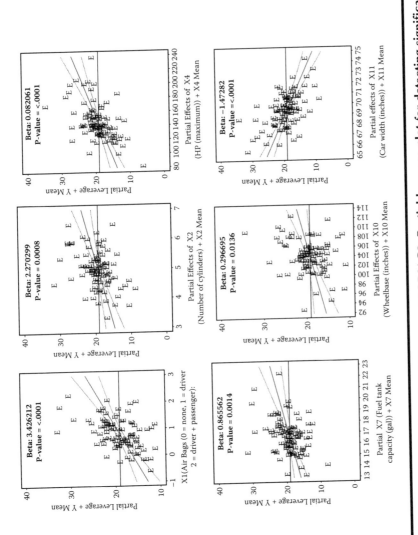

Figure 5.13   Regression diagnostic plot using SAS macro REGDIAG2: Partial leverage plot for detecting significant linear effect for all selected predictor variables.

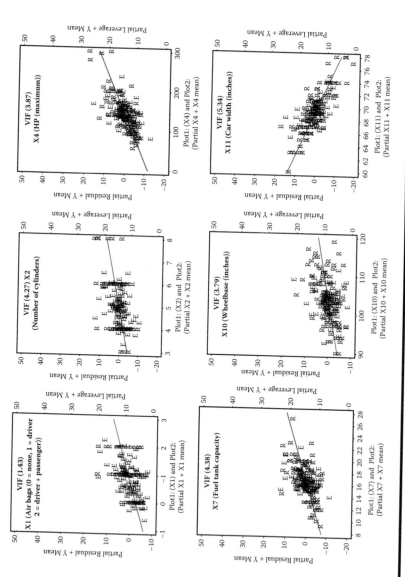

Figure 5.14 Regression diagnostic plot using SAS macro REGDIAG2: VIF plot for detecting multicollinearity—Overlay plots of augmented partial residual and partial leverage plots for all selected predictor variables.

Table 5.4   Macro REGDIAG2—Model Specification Error Summary Table Produced by the RSREG Procedure

Regression	DF	Type I Sum of Squares	R-Square	F Value	Pr > F
Linear	6	5303.527563	0.8054	73.93	<.0001
Quadratic	6	106.545905	0.0162	1.49	0.1977
Cross product	15	421.357618	0.0640	2.35	0.0096
Total model	27	5831.431087	0.8856	18.07	<.0001

The significance of all possible two-factor interactions is verified graphically in a 3-D plot by examining a regression plot between full model residuals + interaction term in question (z-axis) with the two predictor variables in X and Y axis. If you see a significant trend in the partial interaction plot, then the interaction term is considered significant. The significance level of the interaction term is also displayed in the title. One of the interaction terms ($X2*X4$) was significant (Figure 5.15) and the interaction effect was clearly shown in the 3-D plot.

5.12.1.3 Step 3: Fitting the Regression Model and Checking for the Violations of Regression Assumptions

Open the REGDIAG2.SAS macro-call file in the SAS EDITOR window and click RUN to open the REGDIAG2 macro-call window (Figure. 5.4). Input the

Table 5.5   Macro REGDIAG2—Overall Significance (Linear, Quadratic, and Cross Products) of All Predictor Variables

Factor	DF	Sum of Squares	Mean Square	F Value	Pr > F	Label
X1	7	403.903767	57.700538	4.83	0.0002	Air bags (0 = none, 1 = driver only, 2 = driver and passenger)
X2	7	315.088591	45.012656	3.77	0.0018	Number of cylinders
X4	7	321.348416	45.906917	3.84	0.0015	HP (maximum)
X7	7	321.776511	45.968073	3.84	0.0015	Fuel tank capacity (gallons)
X10	7	138.550574	19.792939	1.66	0.1365	Wheelbase (inches)
X11	7	397.000327	56.714332	4.74	0.0003	Car width (inches)

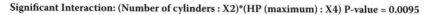

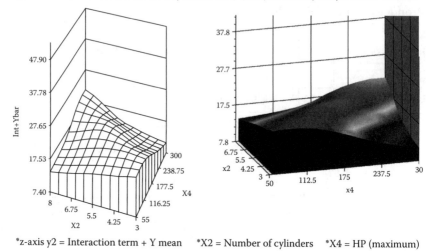

*z-axis y2 = Interaction term + Y mean   *X2 = Number of cylinders   *X4 = HP (maximum)

Figure 5.15 Regression diagnostic plot using SAS macro REGDIAG2: Three-dimensional plot for detecting significant interaction.

appropriate macro-input values by following the suggestions given in the help file (Appendix 2). Include the linear predictors and the selected cross product (interaction term) in macro field #7. Leave the group variable option blank since all the predictors used are continuous. Leave the macro field #14 BLANK to skip regression diagnostics and to run MLR.

Regression model fitting: Simple correlation estimates among the predictor variables and the response is given in Table 5.6. None of the correlation was greater than 0.9. Usually, high degree of correlation (>0.95) may cause some degree of multicollinearity in the model estimates. The observed correlation between the all predictors and the response was very small.

The overall regression model fit was highly significant based on the *F-test* in the ANOVA model (Table 5.7). This result indicates that at least one of the regression coefficient slopes was not equal to zero. The R^2 and the *adjusted* R^2 estimates were almost high, (0.81 and 0.79) indicating that no redundant predicted variable existed in the fitted regression model. The *RMSE* estimate is the error standard deviation and is a very useful indicator for model selection using the same data. The overall response mean (median price), its percentage of coefficient variation, (*CV*) sum of all residuals, and the SS residual are also presented in Table 5.8.

Problems with rounding off and numerical accuracy in the matrix inversion can also be detected if the sum of all residuals is not equal to zero, and/or SSE and SS residuals are not equal. Some degree of differences between the SS residual and the PRESS statistic indicate the presence of significant influential observations in the

Table 5.6   Macro REGDIAG2—Simple Linear Correlations Coefficients among All Predictors and the Response Variable

Variable	Label	X1	X2	X4	X7	X10	X11	Y2
X1	Air bags (0 = none, 1 = driver only, 2 = driver and passenger)	1.0000	0.3744	0.4674	0.3028	0.3720	0.4165	0.5568
X2	Number of cylinders	0.3744	1.0000	0.7940	0.7013	0.6571	0.7976	0.7059
X4	HP (maximum)	0.4674	0.7940	1.0000	0.7102	0.5255	0.6729	0.8155
X7	Fuel tank capacity (gallons)	0.3028	0.7013	0.7102	1.0000	0.7801	0.8066	0.6636
X10	Wheelbase (inches)	0.3720	0.6571	0.5255	0.7801	1.0000	0.8166	0.5496
X11	Car Width (inches)	0.4165	0.7976	0.6729	0.8066	0.8166	1.0000	0.5280
Y2	Midrange price (in $1000)	0.5568	0.7059	0.8155	0.6636	0.5496	0.5280	1.0000

Table 5.7   Macro REGDIAG—Testing for Overall Regression Model Fit by ANOVA

Source	DF	Sum of Squares	Mean Square	F Value	Pr > F
Model	7	5359.93146	765.70449	51.89	<.0001
Error	83	1224.69843	14.75540		
Corrected total	90	6584.62989			

Table 5.8   Macro REGDIAG2—Testing for Overall Regression Model Fit Indicators

Root MSE	3.84128	**R-square**	0.8140
Dependent mean	18.90110	**Adj R-sq**	0.7983
Coeff Var	20.32303		
Sum of residuals	0		
Sum of squared residuals	1224.69843		
Predicted residual SS (PRESS)	1514.93741		

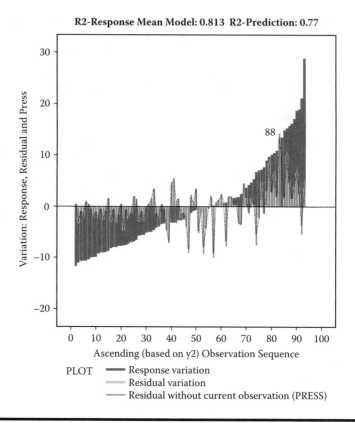

R2-Response Mean Model: 0.813 R2-Prediction: 0.77

PLOT ▬▬ Response variation
     ▪▪▪▪ Residual variation
     ▬▬ Residual without current observation (PRESS)

Figure 5.16   Assessing the MLR model fit using SAS macro REGDIAG2: Explained variation plot showing predicated R².

data. The predictive potential of the fitted model can be determined by estimating the $R^2(prediction)$ by substituting *PRESS* for SSE in the formula for the R^2 estimation. The predictive power of the estimated regression model is considered high if the $R^2(prediction)$ estimate is large and closer to the model R^2.

Figure 5.16 graphically shows the total and the unexplained variation in the response variable in the regression model. The ordered and the centered response variable versus the ordered sequence displays the total variability in the response. This ordered response shows a linear trend with sharp edges at the both ends because the response variable has a positively skewed distribution. The unexplained variability in the response variable is given by the residual distribution. The residual variation shows a random distribution with some sudden peaks, illustrating that the regression model assumptions are violated. The differences between the total and residual variability show the amount of variation in the response accounted for by the regression model and are estimated by the R^2 statistic. If the data contains replicated observations, the deviation from the model includes both *pure error* and *deviation from the regression*. The R^2 estimates can be computed from a regression model using the means of the replicated observations as the response. Consequently, the R^2 is computed based on the means ($R^2_{(mean)}$) is also displayed in the title statement. If there is no replicated data, the $R^2_{(mean)}$ and the R^2 estimate reported by the PROC REG will be identical.

The estimates of $R^2_{(mean)}$ and the $R^2_{(prediction)}$ described previously are also displayed in the title statement. These estimates and the graphical display of explained and unexplained variation help to judge the quality of the model fit.

The overall model fit is illustrated in Figure 5.17 by displaying the relationship between the observed response and the predicted values. The N, R^2, $R^2_{(adjusted)}$, and

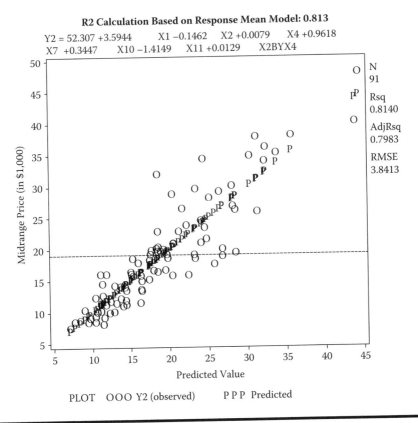

R2 Calculation Based on Response Mean Model: 0.813

Y2 = 52.307 +3.5944   X1 −0.1462   X2 +0.0079   X4 +0.9618
X7 +0.3447   X10 −1.4149   X11 +0.0129   X2BYX4

N	91
Rsq	0.8140
AdjRsq	0.7983
RMSE	3.8413

Midrange Price (in $1,000)

Predicted Value

PLOT   O O O   Y2 (observed)   P P P   Predicted

Figure 5.17   Assessing the MLR model fit using SAS macro REGDIAG2: Overall model fit plot.

RMSE statistics that were useful in comparing regression models and the regression model are also included on the plot.

The regression parameter estimates, *se*, and their *p*-values for testing the parameters = 0 are given in Table 5.9. The *p*-values are derived based on the partial SS (Type II SS). Parameter estimates for all six predictor variables were highly significant. (Ignore the *p*-values associated with the main effects (*X2* and *X4*) when their interaction is significant.) The estimated regression model for predicting the mean median price ($) is 52.30688 + 3.59437 *X1* − 0.14624 *X2* + 0.00794*X4* + 0.96185*X7* + 0.34473*X10* − 1.41489*X11* + 0.01285 *X2*X4*. The magnitude and the sign of the regression parameters provided very useful information relevant to the problem. For example, after controlling, for all other predictor variables, a one unit (1 in.) increase in car width on average decreases the median price by $1414. The information provided in Table 5.10 could be used to find the 95% confidence interval estimates for the parameter. With the 95% level of confidence, we can conclude that on average every 1 inch increase in car width has the potential of decreasing the median price by $924 to $1905 (Table 5.10).

Table 5.9 also provides additional statistics like *Type I* sequential SS, standardized regression coefficient estimates (STB), and the *Type II* partial correlations (PCII). *Type I SS* statistics are influenced by the order of variables entered in the model. This is useful in testing parameter significance in higher-order polynomial models or testing for the significance of parameters in models where the order of entry is important in deciding the significance. The STB and PCII statistics are useful in determining the relative importance of the predictor variables used in the model in terms of accounting for the variation in the response variable. In this example, *X1* is the main variable responsible for the variation in the response. VIF estimates larger than 10 usually provide an indication of multicollinearity. Variables *X2* and *X4* have larger VIF estimates because of the inclusion of their interaction term. However, the impact of multicollinearity on parameter estimates and their significance levels was minimum because both parameters are highly significant, have the right signs (positive), and their estimates are sensible. The impact of multicollinearity on parameter estimates is usually severe in small samples.

Table 5.11 shows a partial list of 10% of the bottom observations and Table 5.12 shows a partial list of 10% of the top observations of the sorted predicted values, which includes the ID, predictor variables used in the model, observed response, predicted response, *se* for predicted values and residuals, and the standardized residual (*student*). If you have any missing values for the response or predictor variables, the whole observation will be excluded from the model fit. However, for any observation, if the response value is missing but all predictor variables are available, the model estimates the predicted value and reports it in this table. Additionally, the residual will be missing for those observations.

Table 5.9   Macro REGDIAG2—Regression Model Parameter Estimates and Their Significance Levels

Variable	Label	DF	Parameter Estimate	Standard Error	t Value	Pr > |t|	Type I SS	Standardized Estimate	Squared Partial Type II Corr
Intercept	Intercept	1	52.30688	12.64441	4.14	<.0001	32510	0	.
i1	Air bags (0 = none, 1 = driver only, 2 = driver and passenger)	1	3.59437	0.69070	5.20	<.0001	2041.67985	0.29725	0.24601
X2	Number of cylinders	1	−0.14624	1.39198	−0.11	0.9166	1894.51444	−0.02235	0.00013296
X4	HP (maximum)	1	0.00794	0.04100	0.19	0.8470	761.71839	0.04733	0.00045108
X7	Fuel tank capacity (gallons)	1	0.96185	0.26207	3.67	0.0004	71.78201	0.37005	0.13963
X10	Wheelbase (inches)	1	0.34473	0.11829	2.91	0.0046	0.47781	0.27463	0.09283
X11	Car width (inches)	1	−1.41489	0.24676	−5.73	<.0001	533.35506	−0.63194	0.28373
X2BYX4	Interaction	1	0.01285	0.00657	1.96	0.0539	56.40389	0.70059	0.04403

Table 5.10 Macro REGDIAG2—Regression Model Parameter Estimates and Their Confidence Interval Estimates

Variable	Label	DF	Variance Inflation	95% Confidence Limits	
Intercept	Intercept	1	0	27.15765	77.45610
X1	Air bags (0 = none, 1 = driver only, 2 = driver and passenger)	1	1.45600	2.22059	4.96815
X2	Number of cylinders	1	20.19934	−2.91482	2.62235
X4	HP (maximum)	1	26.69212	−0.07362	0.08949
X7	Fuel tank capacity (gallons)	1	4.53642	0.44060	1.48310
X10	Wheelbase (inches)	1	3.96271	0.10946	0.58000
X11	Car width (inches)	1	5.42047	−1.90568	−0.92409
X2BYX4	Interaction	1	57.29908	−0.00022230	0.02593

Checking for violations of model assumptions:

i. *Overall check for model assumption violations*: Fit diagnostic plot generated by the ODS graphics feature in PROC REG is presented in Figure 5.18. These plots are useful in visually checking for heteroscedasticity, presence of influential observations, and normality of the residuals. Individual residual plots (residual versus each predictor variable) are presented in Figure 5.19. All these diagnostic plots clearly show that the regression model assumptions, homoscedasticity, absence of influential outliers, and normally distributed residuals are violated here.

ii. *Presence of autocorrelation*: Figure 5.20 shows the trend plot of the residual over the observation sequence. Extreme observations are also identified in the residual plot. If the data is time series data, we can examine the residual plot for a cyclic pattern when there is a sequence of positive residuals following negative residuals. This cyclical pattern might be due to the presence of first-order autocorrelation where the ith residual is correlated with the *lag1* residual. The Durbin–Watson (DW) d statistic measures the degree of first-order autocorrelation. An estimate of the DW statistic and the significance of *1*st order autocorrelation are estimated using the PROC REG and displayed on the title statement. The observed value of the first-order autocorrelation was

Table 5.11   Macro REGDIAG2—Predicted Values, Residuals, Standard Errors, and Outliers for Bottom 10% of the Data

ID	Air Bags (0 = none, 1 = driver only, 2 = driver and passenger)	Number of Cylinders	HP (maximum)	Fuel Tank Capacity (gallons)	Wheelbase (inches)	Car Width (inches)
54	0	4	63	10	90	63
66	0	3	55	10.6	93	63
84	0	3	70	10.6	93	63
51	0	4	92	13.2	98	67
11	0	4	82	13.2	97	66
74	0	4	74	13.2	99	66
45	0	3	73	9.2	90	60
12	0	4	92	13.2	98	66
41	0	4	92	13.2	98	66

Midrange Price (in $1000)	Predicted Value of Y2	Residual	Lower Bound of 95% C.I. for Mean	Upper Bound of 95% C.I. for Mean	Lower Bound of 95% C.I. (Individual Predicted)	Upper Bound of 95% C.I. (Individual Predicted)
7.4	6.9676	0.43244	4.60205	9.3331	−1.03040	14.9655
8.4	7.5434	0.85659	4.86340	10.2234	−0.55315	15.6400
8.6	8.2408	0.35917	5.95910	10.5226	0.26724	16.2144
10.3	8.8649	1.43510	7.39591	10.3339	1.08481	16.6450
8.3	9.3416	−1.04157	7.85878	10.8244	1.55887	17.1243
9	9.5563	−0.55626	7.95545	11.1571	1.75021	17.3623
8.4	10.2442	−1.84418	7.78085	12.7075	2.21674	18.2716
12.2	10.2798	1.92021	8.91842	11.6412	2.51930	18.0403
9.2	10.2798	−1.07979	8.91842	11.6412	2.51930	18.0403

Table 5.12 Macro REGDIAG2—Predicted Values, Residuals, Standard Errors, and Outliers for Top 10% of the Predicted Data

ID	Air Bags (0 = none, 1 = driver only, 2 = driver and passenger)	Number of Cylinders	HP (maximum)	Fuel Tank Capacity (gallons)	Wheelbase (inches)	Car Width (inches)
78	1	8	278	22.5	113	72
77	2	8	295	20	111	74
34	1	8	300	20	96	74
18	2	6	225	20.6	106	71
50	2	8	210	20	117	77
37	2	6	200	18	115	71
87	1	6	300	19.8	97	72
16	2	6	172	21.1	106	70
33	1	8	200	18	114	73

Midrange price (in $1000)	Predicted value of Y2	Residual	Lower bound of 95% C.I. for mean	Upper bound of 95% C.I. for mean	Lower bound of 95% C.I. (individual prediction)	Upper bound of 95% C.I. (individual prediction)
47.9	44.2475	3.65246	40.1087	48.3864	35.5584	52.9367
40.1	43.8010	−3.70097	39.5195	48.0824	35.0430	52.5590
38	35.5894	2.41061	30.8596	40.3192	26.6037	44.5751
35.2	33.6544	1.54563	31.4230	35.8857	25.6951	41.6137
36.1	32.2101	3.88994	29.4858	34.9343	24.0988	40.3214
33.9	32.1298	1.77018	29.5191	34.7405	24.0559	40.2037
25.8	31.1521	−5.35211	27.0542	35.2500	22.4823	39.8219
37.7	31.0423	6.65770	28.1273	33.9573	22.8650	39.2196
34.7	30.2097	4.49027	27.3377	33.0817	22.0476	38.3718

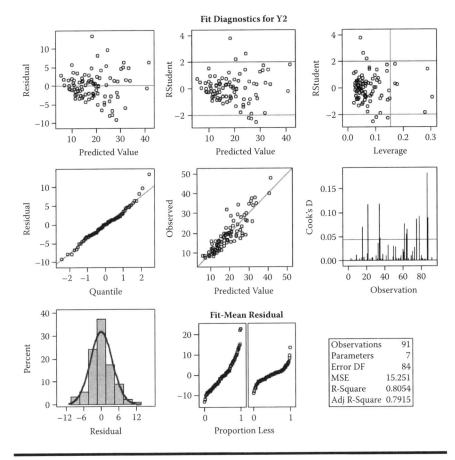

Figure 5.18   Assessing the MLR model fit using SAS macro REGDIAG2: Fit diagnostic plots derived from PROC REG ODS GRAPHICS.

very small and not significant, based on an approximate test (Figure 5.20). Because the dataset used in the case study was simulated, checking for autocorrelation using the DW test is inappropriate.

iii. *Significant outlier/influential observations*: The REGDIAG2 macro automatically excludes extreme influential outliers if the user input is "YES" to macro input #9. Thus, observation number 25 was excluded from the analysis based on the dffits > 1.5 or R student > 4 criteria (Table 5.13). The impact of the Observation number 25 can be seen clearly in outlier detection plot (Figure 5.21), and its impact on normality is also evident in Figure 5.21. However, after excluding extreme influential outliers, when you run the REGDIAG2 macro, additional observation can become outliers if the absolute

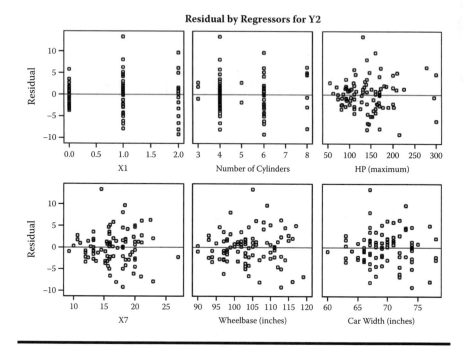

Figure 5.19   Assessing the MLR model fit using SAS macro REGDIAG2: Residual plot by predictor variables derived from PROC REG ODS GRAPHICS.

STUDENT value exceeds 2.5. Also, observations are identified as influential if the DFFITS statistic value exceeds 1.5. Table 5.13 lists two observations (ID #64 and 88) as outlier since the STUDENT value exceeds 2.5. This outlier also showed up in the outlier detection plot (Figure 5.22). Because this outlier has a small DIFFTS value and the STUDENT value falls on the borderline, we can conclude that the impact of this outlier on the regression model was minimum.

iii. *Checking for heteroscedasticity in residuals.* The results of the *Breusch–Pagan* test shows the *heteroscedasticity* exists (Figures 5.21–5.22) before and after excluding the influential outlier 25. Because the test results are significant at the borderline *p*-value (*p*-value > 0.01), this violation of assumption is ignored here. However, regression modeling can be carried out using transformed response variable based on Box–Cox transformation.

iv. *Checking for normality of the residuals*: The residuals appeared to have right-skewed distribution, based on the *p*-values for testing the hypothesis that the skewness and the kurtosis values equal zero and by the *DAGOSTINO–PEARSON OMNIBUS NORMALITY* test before excluding the extreme

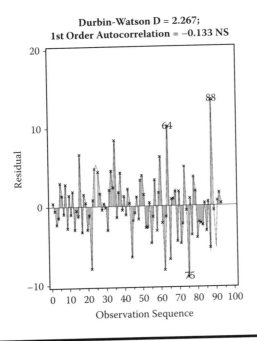

Figure 5.20 Regression diagnostic plot using SAS macro REGDIAG2: Checking for first-order autocorrelation trend in the residual.

outlier 25. This is further confirmed by the symmetrical distribution of the residual histogram and the normally distributed pattern in the normal probability plot (Figures 5.21–5.22).

5.12.1.4 Remedial Measure: Robust Regression to Adjust the Regression Parameter Estimates to Extreme Outliers

The REGDIAG2 macro automatically performs robust regression based on M estimation using SAS PROC ROBUSTREG procedure when the user input "YES" to the REGDIAG2 macro input #9. Robust regression procedure identifies the two extreme data points #25 and #88 as extreme outliers (Table 5.14). The robust regression parameter estimates, their *se* and 95% confidence interval estimates based on M-estimation are presented in Table 5.15. The *se* of the parameter estimates became somewhat smaller when compared with the *se* estimated by the PROC REG and reported in Table 5.9. These robust parameter estimates are recommended over ordinary least-square estimates when the data contains extreme influential data.

Table 5.13 Macro REGDIAG2—List of Influential Outliers (Student ≥ 2.5 or DFFITS > 1.5 or hat ratio > 2)

(a) List of influential outliers excluded from the analysis by the REGDIAG2 macro based on the dffits >1.5 or R student >4 criteria.

ID	X1	X2	X4	X7	X10	X11	DFFITS	R Student
25	2	6	217	18.5	110	69	2.25611	5.36877

(b) List of outliers identified by the REGDIAG2 macro after excluding extreme influential outliers.

ID	Air Bags (0 = none, 1 = driver only, 2 = driver and passenger)	Number of Cylinders	HP (maximum)	Fuel Tank Capacity (gallons)	Wheelbase (inches)	Car Width (inches)	Midrange Price (in $1000)
21	0	6	178	18.5	97	66	23.3
79	2	6	160	15.5	101	75	17.7
32	1	4	208	21.1	109	69	30
34	1	8	300	20	96	74	38
52	0	6	165	27	111	78	16.6
78	1	8	278	22.5	113	72	47.9
77	2	8	295	20	111	74	40.1
87	1	6	300	19.8	97	72	25.8
64	2	6	160	18.4	109	73	34.3
88	1	4	130	14.5	105	67	31.9

Residual	Studentized Residual	Standard Influence on Predicted Value	Hat-Ratio = (hat/hat mean)	Outlier
−1.1200	−0.32608	−0.16238	2.27990	.
1.8628	0.54092	0.26614	2.23239	.
2.0955	0.60471	0.28815	2.11819	.
2.4106	0.79909	0.62853	4.35945	.
−2.7377	−0.85811	−0.57452	3.52852	.
3.6525	1.13121	0.73029	3.33814	.
−3.7010	−1.16329	−0.78880	3.57215	.
−5.3521	−1.65088	−1.06037	3.27247	.
10.0858	2.70677	0.70586	0.67170	*
13.5220	3.60848	0.88042	0.54994	***

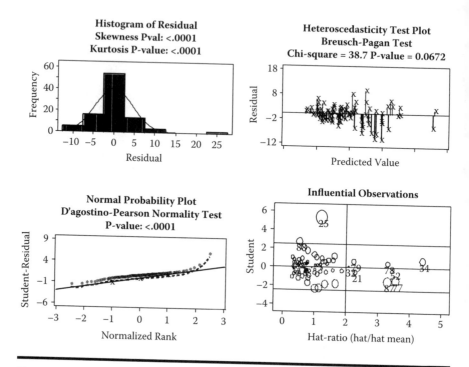

Figure 5.21 Regression diagnostic plot using SAS macro REGDIAG2: Checking for MLR model assumptions—(a) normal probability plot; (b) histogram of residual; (c) residual plot checking for heteroscedasticity; (d) outlier detection plot for response variable (before deleting the outliers).

5.13 Case Study 2: If-Then Analysis and Lift Charts

Perform IF-THEN analysis and construct a lift chart to estimate the differences in the predicted response when one of the continuous predictor variables is fixed at a given value. A following simulate data is used to demonstrate the if-then and lift analysis.

Table 5.14   Macro REGDIAG2—Extreme Observation Identified as Outliers by the SAS ROBUSTREG Procedure

ID	Standardized Robust Residual	Outlier
25	7.0521	*
88	3.6118	*

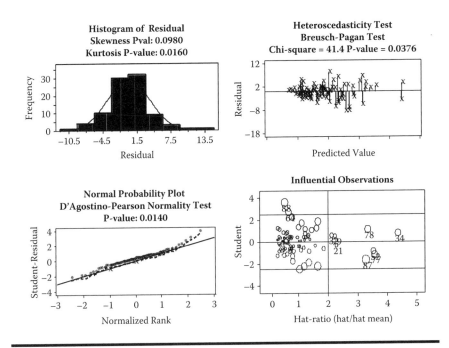

Figure 5.22   Regression diagnostic plot using SAS macro REGDIAG2: Checking for MLR model assumptions—(a) normal probability plot; (b) histogram of residual; (c) residual plot checking for heteroscedasticity; (d) outlier detection plot for response variable (after deleting the outliers).

Table 5.15   Macro REGDIAG2—Robust Regression Parameter (M-estimation) Estimates

Parameter	DF	Estimate	Standard Error	95% Confidence Limits		Chi-Square	Pr > ChiSq
Intercept	1	46.8729	11.4159	24.4981	69.2476	16.86	<.0001
X1	1	3.4379	0.6362	2.1910	4.6848	29.20	<.0001
X2	1	−0.6742	1.2529	−3.1299	1.7814	0.29	0.5905
X4	1	−0.0088	0.0366	−0.0805	0.0630	0.06	0.8109
X7	1	0.9518	0.2445	0.4726	1.4310	15.16	<.0001
X10	1	0.3590	0.1099	0.1435	0.5744	10.66	0.0011
X11	1	−1.3174	0.2247	−1.7577	−0.8770	34.38	<.0001
X2BYX4	1	0.0159	0.0059	0.0044	0.0274	7.40	0.0065
Scale	0	3.8542					

5.13.1 Data Descriptions

Data name	SAS dataset "Sales"
Response (Y) and predictor variables (X)	Y: Response (Product sales in $1000/year))
	X1: 1(Advertising cost/year);
	X2: Advertising cost ratio between competing product/own product.
	X3: Personal disposable income /year;
	ID Year (1 to 116)
Number of observations	116
Source:	Simulated data with a controlled trend and minimum amount of error

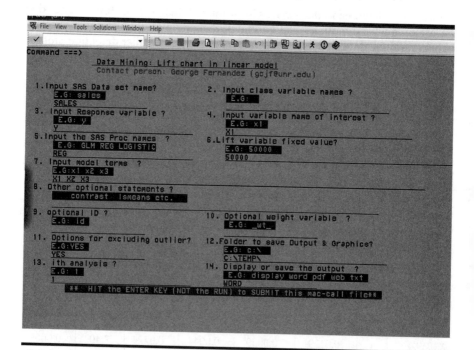

Figure 5.23   Screen copy of LIFT2 macro-call window showing the macro-call parameters required for performing lift chart in linear regression.

Performing if–then analysis and producing the LIFT chart. To perform an if-then analysis and then to create a lift chart showing what happens to the total sales if management spent $50,000 annually on advertising, open the LIFT2.SAS macro-call file in the SAS EDITOR window. Click RUN to open the LIFT2 macro-call window (Figure. 5.23). Input SAS dataset (sales) and response (*Y*) variable names, model statement (*X*1, *X*2, *X*3), the variable name of interest (*X*1), fixed value ($50,000), and other the appropriate macro-input values by following the suggestions given in the help file (Appendix 2). Submit the LIFT2 macro, and you will get the lift chart (Figure 5.24) showing the differences in the predicted values between the original full model and

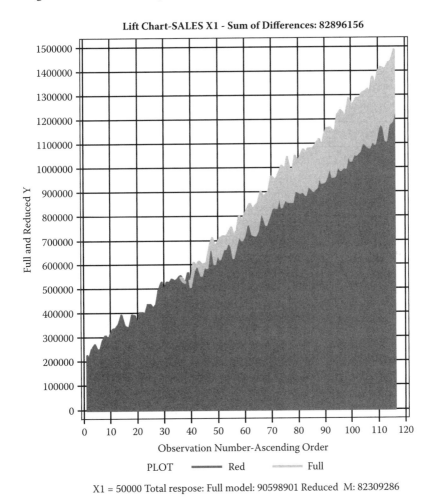

Figure 5.24 Lift chart generated by using SAS macro LIFT2: The differences in predicted values between the original and the reduced model with fixed level of advertising expenses.

Table 5.16   Macro REGDIAG2—Output from IF–THEN Analysis

ID	Predicted Value, Original	x1	Predicted Value (reduced model)	New-value	Original Reduced Model
35	529793.22	48210	536607.59	50000	−6814.37
36	547198.40	48797	551778.12	50000	−4579.72
42	540740.48	52818	530012.61	50000	10727.88
38	526083.16	49933	526338.22	50000	−255.06
39	548175.69	46138	562877.98	50000	−14702.29
41	520534.15	53520	507133.82	50000	13400.33
Partial List					
111	1433029.57	117895	1174559.27	50000	258470.30
113	1444985.02	117501	1188014.64	50000	256970.38
116	1484121.27	118161	1224638.33	50000	259482.94

the new reduced model, where you kept the advertising cost at $50,000/year continuously. On the lift chart, display, the sum of total differences ($8,289,616) between the predicted values of the original model and the predicted value for the reduced model is also displayed. In addition to the lift chart, the if-then analysis outputs a table showing the predicted values of the original, the reduced models, and their differences. A partial list of the if-then analysis output is presented in Table 5.16.

Predicting the response scores for a new dataset. After finalizing the regression model, and saving the regression parameter estimates in a SAS dataset REGEST, new expected responses could be predicted to any new SAS data containing all the required predictor variables.

Open the RSCORE2.SAS macro-call file in the SAS EDITOR window and click RUN to open the RSCORE2 macro-call window (Figure 5.25). Input the name of the new SAS dataset (predsale), regression parameter estimate data (REGEST), response variable name (optional), model statement (*X1 X2 X3*) and other appropriate macro-input values by following the suggestions given in the help file (Appendix 2). Submit the RSCORE2 macro, and you will get an output table showing the predicted scores for your new dataset. If you also have observed response values in the new data, then the RSCORE2 macro performs model validation and produces addition plots showing the success of the validation.

The predicted sales amount for a new input data containing varying levels of advertising (X_1), advertising ratio (X_2), and fixed level of personal disposable income (X_3) are shown in Table 5.17.

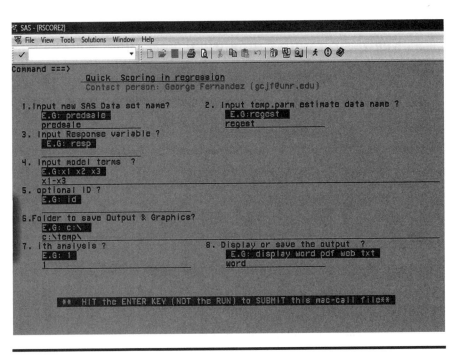

Figure 5.25 Screen copy of RSCORE2 macro-call window showing the macro-call parameters required for estimating predicted scores from an independent dataset.

Table 5.17 Macro RSCORE2—Scoring New Dataset

Advertising ($1000s)	Advertising Ratio (comp/this)	Personal Disposable Income	P_score
118161	0.3	11821	1484121.27
119000	0.3	11820	1487230.39
117500	0.3	11820	1481520.02
119000	1	11820	1456038.61
117500	1	11820	1450328.24
119000	1.3	11820	1442670.70
117500	1.3	11820	1436960.33

5.14 Case Study 3: Modeling Multiple Linear Regression with Categorical Variables

5.14.1 Study Objectives

1. **Model selection**: Using the LASSO method in PROC GLMSELECT to confirm the few predictor variables used in model building is important.

2. **Data exploration using diagnostic plots**: For each predictor variable, check for linear and nonlinear relationships and significant outliers (augmented partial residual plot), significant regression relationship (partial leverage plot), and multicollinearity (VIF plot). Also check for interactions between categorical X predictor in scatter plots, and check the data for very extreme influential observations. If desirable, fit the regression model after excluding these extreme cases.

3. **Regression model fitting and prediction**: Fit the regression model, perform hypothesis testing on overall regression, and on each of the parameter estimates, estimate confidence intervals for parameter estimates, predict scores, and estimate their confidence intervals.

4. **Checking for any violations of regression assumptions**: Perform statistical tests and graphical analysis to detect influential outliers, heteroscedasticity in residuals, and departure from normality.

5. **Save _SCORE_ and REGEST datasets for future use**: These two datasets are created and saved as temporary SAS datasets in the work folder and are also exported to Excel worksheets and saved in the user-specified output folder. The _SCORE_ data contain observed responses and predicted scores, including those observations with missing response value, residuals, and confidence interval estimates for the predicted values. This dataset could be used as the base for developing the scorecards for each observation. Also, after SAS version 9.0, the parameter estimates dataset called REGEST created by the SAS ODS output option could be used (after slight modification) in the RSCORE2 macro for scoring different datasets containing the same variables.

6. **If–then analysis and lift charts**: Perform an if-then analysis and construct a lift chart to estimate the differences in the predicted response when one of the continuous or binary predictor variable is fixed at a given value.

5.14.2 Data Descriptions

Background Information: A small convenience store owner suspects that one of the store managers (Mr. X) has stolen a small amount of money ($50 to 150/day) from the cash register by registering false voids over a 3-year period. The owner caught Mr. X twice in action making fraudulent voids. To prove his claim in court, the owner hired a data-mining specialist to investigate the 3-year daily sales/transaction records. The stores average amount of sales varied from $1000–2000 per day. A total voids amount greater than $400 at any given day was treated as a genuine error. The descriptions of the data and the data mining methods used are presented below:

Data name: (3-year daily sales/ transaction data)	SAS dataset "fraud." Two temporary datasets, fraud_train and fraud_valid, are randomly selected using the RANSPLT2 macro
Response and predictor variables	Voids (total $ amount of voids/day) — Response variable
	Transac (total number of daily transactions)
	Days (*i*th day of business during the 3-year period)
	Manager (Indicator variable with two levels: 0 all other managers; 1; Mr. X)
Number of observations	Total dataset: 896
	Training: 598
	Validation: 298
Source:	Actual daily sales data from a small convenience store

Open the REGDIAG2.SAS macro-call file in the SAS EDITOR window and click RUN to open the REGDIAG2 macro-call window (Figure 5.26). Use the suggestions provided given in the help file (Appendix 2) to input the appropriate macro input values and fit the regression models with indicator variables. Only selected output and graphics produced by the REGDIAG2 macro are shown below.

Step 1: Model selection: LASSO, the new model selection method implemented in the new SAS procedure GLMSELECT is also utilized in REGDIAG2 macro for screening all listed predictor variables and examine and visualize the contribution of each predictor in the model selection. Two informative diagnostic plots (Figure 5.27) generated by the ODS graphics feature in the GLMSELECT can be used to visualize the importance of the predictor variables. The fit criteria plot (Figure 5.27) displays the trend plots of six model selection criteria versus the number of model parameters, and in this example all six criteria identify the three-parameter model (full model) as the best model. The coefficient progression plot displayed in Figure 5.27 shows the stability of the standardized regression coefficients as a result of adding new variables in each model selection step. All included predictor variables significantly contributing to the model fit. The problem of

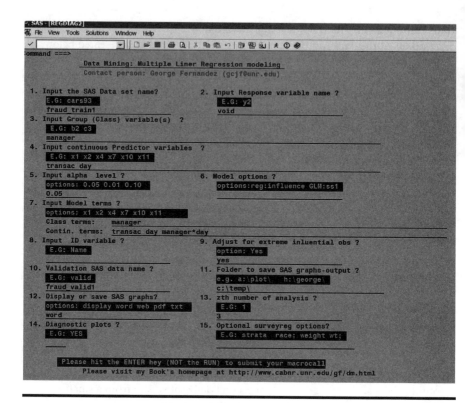

Figure 5.26 Screen copy of REGDIAG2 macro-call window showing the macro-call parameters required, for performing MLR with categorical variables.

multicollinearity among the predictor variables was also not evident since all the standardized regression coefficients have values less than ±1.

Exploratory analysis/diagnostic plots: Open the REGDIAG2.SAS macro-call file in the SAS EDITOR window and click RUN to open the REGDIAG2 macro-call window (Figure 5.26). Use the suggestions provided given in the help file (Appendix 2) to input the appropriate macro input values and fit the regression models with indicator variables. Input dataset name (fraud_train1), response variable (void), group variables (mgr), continuous predictor variables (transac and days), model terms (transac, days mgr), and other appropriate macro-input values. To perform data exploration and to create regression diagnostic plots, input YES in macro field #14. Submit the REGDIAG2 macro, and the following regression diagnostic plots are created:

- Partial regression plots.
- Simple scatter plots by the manager for each predictor variables.
- Box plot of voids by manger.
- Interaction plots between continuous and categorical predictors are produced.

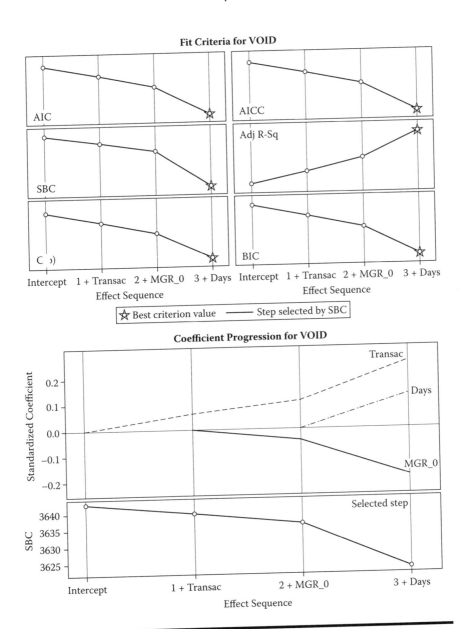

Figure 5.27 Model selection using SAS macro REGDIAG2: Fit criteria plots for the selected model derived from using the ODS graphics feature in GLMSELECT procedure.

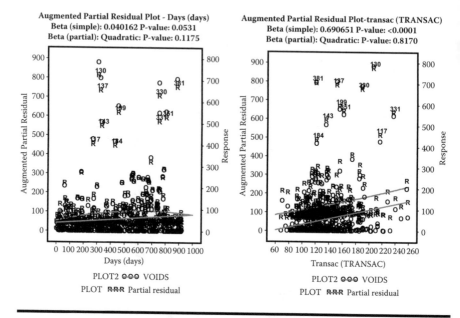

Figure 5.28   Regression diagnostic plot using SAS macro REGDIAG2: Overlay plot of simple linear regression and augmented partial residual plots for two continuous predictor variables in MLR with categorical predictor variables.

Simple linear regression and augmented partial residual (APR) plots for two predictor variables are presented in Figure 5.28. The linear/quadratic regression parameter estimates for the simple and multiple linear regressions and their significance levels are also displayed in the titles of the APR plots. The simple linear regression line describes the relationship between the response and a given predictor variable in a simple linear regression. The APR line shows the quadratic regression effect of the ith predictor on the response variable after accounting for the linear effects of other predictors on the response. The APR plot is very effective in detecting significant outliers and nonlinear relationships. Significant outliers and/or influential observations are identified and marked on the APR plot if the absolute STUDENT value exceeds 2.5 or the DFFITS statistic exceeds 1.5. These influential statistics are derived from the GLM model involving all predictor variables.

The APR plots for the two-predictor variables showed significant linear relationships between the two predictors and void amount. Quadratic effects of two predictor variables on voids were not significant at the 5% level. Many significant outliers were also detected in these APR plots.

Partial leverage plots (PL) for two predictor variables are presented in Figure 5.29. The PL display shows three curves: (a) the horizontal reference line that goes through the response variable mean; (b) the partial regression line, which quantifies the slope of the partial regression coefficient of the ith variable in the

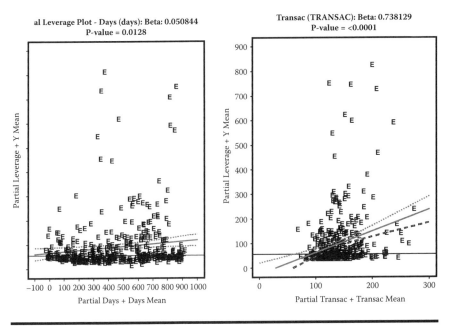

Figure 5.29   Regression diagnostic plot using SAS macro REGDIAG2: Partial leverage plots for testing significant linear effects of two continuous predictor variables in MLR with categorical predictor variables.

MLR; and (c) the 95% confidence band for partial regression line. The partial regression parameter estimates for the ith variable in the multiple linear regression and their significance levels are also displayed in the titles. The slope of the partial regression coefficient is considered statistically significant at the 5% level if the response mean line (the horizontal reference line) intersects the 95% confidence band. If the response mean line lies within the 95% confidence band without intersecting it, then the partial regression coefficient is considered not significant. The PL plots for the two-predictor variables showed significant linear positive relationships between the two predictors and the void.

The VIF plots for two predictor variables are presented in Figure 5.30. The VIF plot displays two overlaid curves: (a) The first curve shows the relationship between partial residual + response mean and the ith predictor variable, and (b) the second curve displays the relationship between the partial leverage + response mean and the partial ith predictor value + mean of ith predictor value. The slope of both regression lines should be equal to the partial regression coefficient estimate for the ith predictor. When there is no high degree multicollinearity, both the partial residual (Symbol R) and the partial leverage (Symbol E) values should be evenly distributed around the regression line. However, in the presence of severe multicollinearity, the partial leverage values, E, shrink and are distributed around the mean of the ith

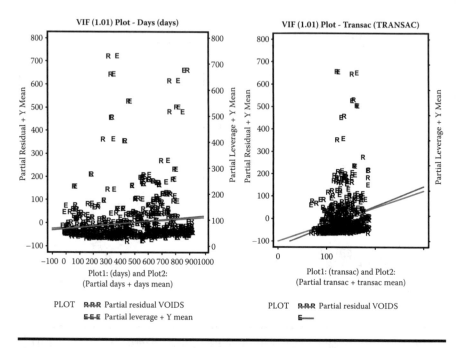

Figure 5.30 Regression diagnostic plot using SAS macro REGDIAG2: VIF plots for testing significant multicollinearity between two continuous predictor variables in MLR in the presence of categorical predictor variables.

predictor variable. Also, the partial regression for the ith variable shows a nonsignificant relationship in the partial leverage plots, whereas the partial residual plot shows a significant trend for the ith variable. Furthermore, the degree of multicollinearity can be measured by the VIF statistic in a MLR model, and the VIF statistic for each predictor variable is displayed on the title statement of the VIF plot. The VIF plots for two predictors showed no multicollinearity where the crowding of the partial leverage values was not evident in the VIF plots.

The interaction between transaction and manager was not statistically significant at the 5% level (figure is not shown). However the interaction between day and manager was statistically significant at the 5% level. A positive linear trend is observed between voids and the ith days of transaction for Mr. X (manager = 1), and no trend is observed for other manager group (manager = 0) (Figure 5.31). This differential trend confirms the presence of significant interaction for the manager x days interaction term. Figure 5.32 shows the box plot display of overall variation for voids by the two-manager group. The median and the third quartile void amount for Mr. X are higher than the other manager group. The void dollar amount that is approximately greater than $75 is identified as outlier for the other manager group, whereas a void dollar amount greater than $225 is identified as the outlier for Mr. X.

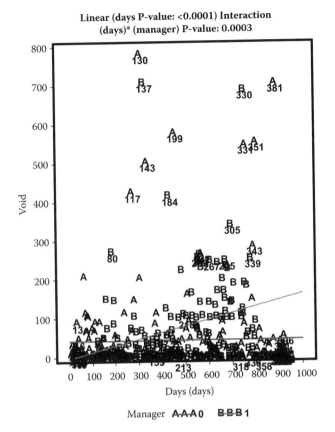

Figure 5.31 Regression diagnostic plot using SAS macro REGDIAG2: Box plot display of significant categorical predictor variables in MLR in the presence of significant covariates.

The interaction effect between the *transaction x days* on voids after accounting for all other predictors was not significant at the 5% level (Figure 5.33).

These regression diagnostic plots clearly revealed that differential trend was observed for the two manager groups, which supports the convenience store manager's hypothesis. This finding could be further confirmed by fitting a multiple regression model, and treating the manager as the categorical variable after excluding the extreme outliers. The results of the regression analysis are presented next.

Fitting regression model and validation: Open the REGDIAG2.SAS macro-call file in the SAS EDITOR window and click RUN to open the REGDIAG2 macro-call window (Figure 5.26). Use the suggestions given in the help file (Appendix 2) to input the appropriate macro input values and fit the regression models with indicator variables. Input dataset name (fraud_train1), response variable (void), group

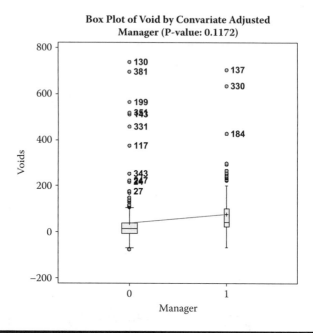

Figure 5.32 Regression diagnostic plot using SAS macro REGDIAG2: Interaction between a categorical and a continuous variable.

variables (manager), continuous predictor variables (transac and days), model terms (transac, days manager and day*manager), and other appropriate macro-input values. To skip outputting regression diagnostic plots, leave macro field #14 BLANK. To exclude extreme outliers/influential observations, input YES in macro field #15. Submit the REGDIAG2 macro to exclude outliers, fit regression with indicator variables, estimate predicted scores, check the regression model assumptions, and validate the model using the independent valid dataset.

Extreme outliers (*student* > *4*) and/or influential observations (*diffits* >1.5) are identified (Table 5.18) and excluded from the regression model fitting. The number and descriptions of categorical variable levels fitted in the model are presented in Table 5.19. Always examine the class level information presented in this table to verify that all categorical variables used in the model are coded correctly. The overall model fit is highly significant (*p*-value <0.0001) (Table 5.20). Verify that the total and the model *df* are correct for the training dataset. The R^2 estimate indicates that about 23% of the variation in the voids could be attributed to the specified model (Table 5.21). The RMSE value reported in Table 5.21 is the smallest estimate among the many regression models tried, indicating that this model is the best.

Figure 5.34 illustrates the total and the unexplained variation in voids after accounting for the regression model. The ordered and the centered response variable

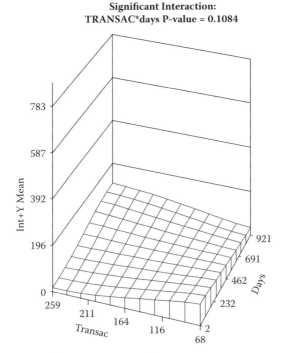

Significant Interaction:
TRANSAC*days P-value = 0.1084

* z-axis void = Interaction term + Y mean
* X-axis days = days
* Y-axis transac = TRANSAC

Figure 5.33   Regression diagnostic plot using SAS macro REGDIAG2: Three-dimensional plot for detecting significant interaction.

Table 5.18   Macro REGDIAG2—List of Influential Outliers Excluded (dffits > 1.5 or Rstudent > 4)

ID	MANAGER	TRANSAC	DAYS	Dffits	R-Student
146	1	125	150	0.61611	5.00296
290	0	173	295	0.35443	4.77228
543	0	186	563	0.36400	4.07533
633	1	104	656	0.38209	4.24605
661	1	150	685	0.47330	4.59364
668	1	174	692	0.50952	4.20830
758	0	156	788	0.41122	4.60275

Table 5.19 Macro REGDIAG2—Class-Level Information

Class	Levels	Values
MANAGER	2	0 1

Table 5.20 Macro REGDIAG2—Overall Model Fit

Source	DF	Sum of Squares	Mean Square	F Value	Pr > F
Model	4	380478.011	95119.503	44.01	<0.0001
Error	586	1266590.125	2161.417		
Corrected total	590	1647068.137			

versus the ordered observation sequence display the total variability in the voids. This total variation plot shows a right skewed trend with very sharp edges at the right end. The unexplained variability in voids is illustrated by the residual distribution. The residual variation shows an unequal variance pattern because the residuals in the higher end are relatively larger than the residual in the lower end, thus illustrating that the equal variance and normally distributed error assumptions are violated. The differences between the total and residual variability shows only a small portion of the variation in the voids was accounted for by the regression model. The predictive ability of the regression model is given by the estimate for $R^2(prediction)$ and is displayed in the title statement. A very small difference (2%) between the model $R^2(mean)$ *and* $R^2(prediction)$ indicates that the impact of the outliers on model prediction is minimum (Figure 5.34).

All terms included in the final regression models are highly significant, based on the partial SS (Type III). The observed positive trend in voids is significant for transac. The significant interaction for *days*manager* confirms the differential slope estimates for days by manager. The regression model parameters, their *se*, and the significance levels produced by the *solution* option in PROC GLM are presented in Table 5.22. These regression estimates could be used to construct the regression equation for both managers as outlined in Table 5.23. The estimated regression model confirms that after adjusting for variations in the number of transactions,

Table 5.21 Macro REGDIAG2—Model-Fit Statistics

R-Square	CV	Root MSE	VOIDS Mean
0.231003	124.6285	46.49104	37.30369

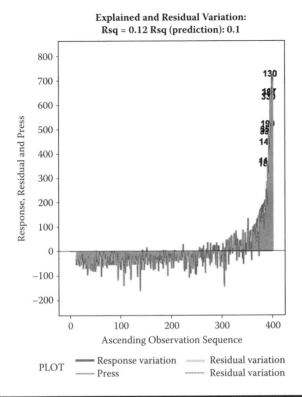

Figure 5.34   Assessing the MLR model fit using SAS macro REGDIAG2: Explained variation plot showing predicated R^2.

the rate of increase in daily void amount for Mr. X is 0.133 whereas for other managers the rate was almost zero (−0.01). These results clearly show that there is a very small chance (<0.0001) that the observed differences in the daily voids between Mr. X and other managers are just due to chance. These regression models could be used to predict the expected void amount for both manager groups.

Table 5.22   Macro REGDIAG2—Significance of Model Effects

Source	DF	Type III SS	Mean Square	F Value	Pr > F
TRANSAC	1	42456.7671	42456.7671	19.64	<0.0001
DAYS	1	113345.3946	113345.3946	52.44	<0.0001
MANAGER	1	19635.1425	19635.1425	9.08	0.0027
DAYS*MANAGER	1	145128.0467	145128.0467	67.14	<0.0001

Table 5.23   Macro REGDIAG2—Parameter Estimates and Their Significance

Parameter	Estimate		Standard Error	t Value	Pr > t
Intercept	−39.74797271	B	10.97681793	−3.62	0.0003
TRANSAC	0.27548794		0.06215819	4.43	<0.0001
DAYS	0.13356213	B	0.01513612	8.82	<0.0001
MANAGER 0	27.13630425	B	9.00332467	3.01	0.0027
MANAGER 1	0.00000000	B	.	.	.
DAYS*MANAGER 0	−0.14130982	B	0.01724511	−8.19	<0.0001
DAYS*MANAGER 1	0.00000000	B	.	.	.

Derived regression model:

Manager(1): −39.74797271 + 0.27548794 Transac + 0.13356213 days

Manager(0):(−39.74797271 + 27.13630425) + 0.27548794 Transac + (0.13356213 − 0.14130982) days

The least square mean estimates for two levels of manager groups and their 95% confidence intervals are presented in Table 5.24. For the manager X, the average void amount varies between $63 to $95/day and for all other managers, the average void amount varied between $24 to $53. The estimated predicted scores and their confidence and prediction intervals are given in Tables 5.25 and 5.26, respectively. These predicted scores and the confidence interval estimates could be used to build scorecards for observations in the dataset.

    i. *Checking for regression model violations: (1) autocorrelation:* Figure 5.35 shows the trend plot of the residual over the observation sequence. No cyclical pattern is evident in the residual plot. Estimates of DW statistic and the first-order autocorrelation displayed on the title statement indicate that the influence of autocorrelation is small and not significant.

    ii. *Significant outlier/influential observations:* Table 5.27 lists several observations as outliers since the STUDENT value exceeds 2.5. These outliers also showed

Table 5.24   Macro REGDIAG2—Adjusted Leas Square Means

MGR	VOID LSMEAN	Standard Error	H0:LS Mean1 = LS Mean2 Pr > |t|	95% Confidence Limits	
0	37.981	6.950	0.0002	24.316	51.647
1	78.593	7.973		62.916	94.271

Table 5.25 Macro REGDIAG2—Partial List of Predicted Scores

ID_	TRANSAC	DAYS	MANAGER	MONTH	YEAR	VOIDS	Predicted	Residual
2	134	3	0	January	1998	9.00	29.703	−20.703
105	126	108	0	April	1998	91.00	23.498	67.502
235	125	239	0	September	1998	29.10	−6.742	35.842
62	191	64	0	March	1998	6.00	54.754	−48.754
771	115	801	0	April	2000	20.00	0.744	19.256
236	118	240	0	September	1998	0.00	−8.851	8.851
				(Partial list)				
414	137	426	0	March	1999	12.00	18.367	−6.367
346	93	355	1	January	1999	4.00	39.486	−35.486
234	126	238	1	September	1998	0.00	14.723	−14.723
701	129	727	0	February	2000	0.50	30.682	−30.182

up in the outlier detection plot (Figure 5.36). Because all these outlier has small DIFFTS value, we can conclude that the impact of this outlier on the regression model estimate was minimum.

iii. *Checking for heteroscedasticity in residuals*: The results of the Breusch–Pagan test and the fan shaped pattern of the residuals in the residual plot (Figure 5.36) both confirm that the residuals have unequal variance. Remedial measures, such as Box–Cox transformation or heterogeneity regression estimates are recommended to combat heteroscedasticity problems associated with the hypothesis testing and interval estimates.

Table 5.26 Macro REGDIAG2—Partial List of Confidence (CI) and Prediction (PI) Interval

ID	Mean lower 95% CI	Mean upper 95% CI	Lower 95% PI	Upper 95% PI
2	6.442	52.963	−62.787	122.193
105	3.253	43.743	−68.280	115.276
235	−27.936	14.453	−98.734	85.251
62	32.387	77.122	−37.515	147.024
771	−23.534	25.022	−92.007	93.495

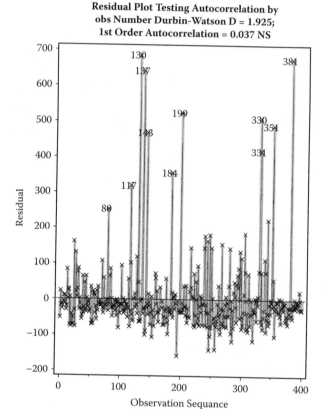

Figure 5.35 Regression diagnostic plot using SAS macro REGDIAG2: Checking for first-order autocorrelation trend in the residual.

iv. *Checking for normality of the residuals*: The residuals appeared to have strong right-skewed distribution based on the *p*-values for testing the hypothesis that the skewness and the kurtosis values equal zero and by the *DAGOSTINO-PEARSON OMNIBUS NORMALITY* TEST (Figure 5.36). This is further confirmed by the right-skewed distribution of the residual histogram and curved upward trend in the normal probability plot (Figure 5.36). The presence of extreme outliers and heteroscedasticity could be the main cause for the right-skewed error distribution, and the remedial measures suggested to combat outliers and heteroscedasticity in this chapter could reduce the impact of right-skewed residuals.

Model validation. To validate the obtained regression model, the regression parameter estimates obtained from the training dataset ($n = 378$) are used to estimate the

Table 5.27   Macro REGDIAG2—List of Outliers (student ≥ 2.5 or DFFITS > 1.5 or hat-ratio > 2)

ID_	TRANSAC	DAYS	VOIDS	RESIDUAL	STUDENT	DIFFITS	OUTLIER
230	118	234	115.00	112.904	2.54910	0.13450	*
645	185	669	248.50	113.665	2.57263	0.31528	*
649	114	673	234.50	120.374	2.71158	0.24465	*
505	112	525	166.13	122.467	2.77639	0.12028	*
888	125	922	153.00	128.344	2.89300	0.24059	*
368	100	378	137.00	128.182	2.90668	0.17994	*
271	146	276	148.00	132.092	2.96556	0.13801	*
574	149	594	149.00	131.630	2.98418	0.13726	*
753	169	783	149.50	131.940	2.99926	0.19670	*
397	120	409	188.00	137.430	3.10464	0.16129	**
614	120	636	242.00	137.164	3.12403	0.23700	**
546	119	566	230.00	140.080	3.14918	0.19772	**
532	136	552	242.25	147.835	3.32934	0.21116	**
751	237	781	190.99	153.354	3.50910	0.39615	***
296	191	301	204.00	163.933	3.73538	0.25869	***
464	129	481	223.50	167.550	3.76830	0.18117	***
533	115	553	257.00	168.717	3.80004	0.23183	***
539	137	559	264.00	169.043	3.80117	0.24216	***
664	117	688	300.00	184.460	4.16745	0.36280	***
50	67	52	202.00	197.924	4.48596	0.53484	***

predicted voids for the validation dataset ($N = 294$). Both training and validation regression slopes and the R^2 values appeared to be somewhat similar in some cases (Figure 5.37). The residuals from the training and validation datasets also show a similar distribution pattern in the residual plot (Figure 5.37). The presence of a large void amount in the validation data could be reason for the differences we observed in the model validation. Thus, we could conclude that the regression model obtained from the training data provides valid estimates to predict the void amount in this investigation.

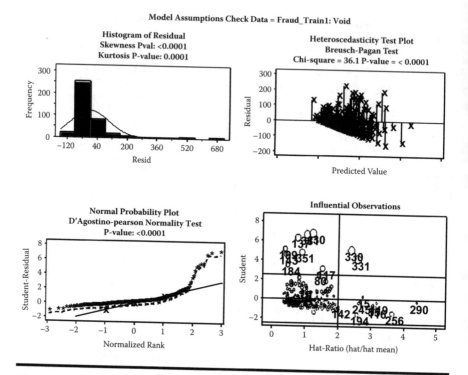

Figure 5.36 Regression diagnostic plot using SAS macro REGDIAG2: Checking for MLR model assumptions: (a) normal probability plot (b) histogram of residual, (c) residual plot checking for heteroscedasticity, (d) outlier detection plot for response variable.

Remedial measure: *Robust regression to adjust the regression parameter estimates to extreme outliers.* The REGDIAG2 macro automatically performs robust regression based on M estimation using the SAS PROC ROBUSTREG procedure when the user input "YES" to the REGDIAG2 macro input #9. Robust regression procedure identifies several extreme data points extreme outliers (Table 5.28). The robust regression parameter estimates, their *se*, and 95% confidence interval estimates based on M-estimation are presented in Table 5.29. The *se* of the parameter estimates became somewhat smaller when compared with the *se* estimated by the PROC REG and reported in Table 5.23. These robust parameter estimates are recommended over ordinary least square estimates when the data contains extreme influential data.

Performing if-then analysis and producing the LIFT chart. The next objective after finalizing the regression model is to perform an if-then analysis and then create a lift chart showing what happens to the total void amount if Mr. X is replaced by any other manager during the 3-year period. Open the LIFT2.SAS macro-call file in

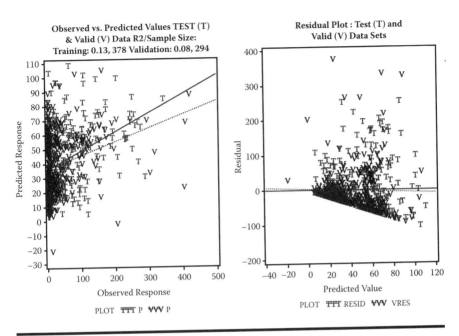

Figure 5.37 Regression model validation plot using SAS macro REGDIAG2: Validating the regression model developed using the training data with independent validation data.

Table 5.28 Macro REGDIAG2—Extreme Observation Identified as Outliers by the SAS ROBUSTREG Procedure

ID	Standardized Robust Residual	Outlier
295	7.0676	*
293	7.1517	*
265	7.3338	*
236	7.3345	*
278	7.4279	*
337	7.6673	*
237	7.9583	*
241	7.9729	*
245	8.2428	*

(continued)

Table 5.28   Macro REGDIAG2—Extreme Observation Identified as Outliers by the SAS ROBUSTREG Procedure (Continued)

ID	Standardized Robust Residual	Outlier
341	9.0737	*
303	10.3973	*
183	13.1271	*
117	13.5113	*
143	16.7036	*
329	17.517	*
349	18.2262	*
198	19.1019	*
328	22.1476	*
137	23.1349	*
379	23.6634	*
130	25.7132	*
155	25.811	*
167	27.0114	*
356	27.4395	*
192	27.8959	*
336	33.5661	*
195	35.423	*
316	40.3722	*
216	46.2156	*

the SAS EDITOR window and click RUN to open the LIFT2 macro-call window (Figure 5.38). Input SAS dataset name (fraud_train1) and response variable name (void), model statement (transac day manager manager*day), the variable name of interest (manager), fixed level ('0'), and any other appropriate macro-input values by following the suggestions given in the help file (Appendix 2). Submit the lift macro, and you will get the lift chart (Figure 5.39) showing the differences in the predicted values between the original full model and the new reduced model where you fixed

Table 5.29 Macro REGDIAG2—Robust Regression Estimates

Parameter		DF	Estimate	Standard Error	95% Confidence Limits		Chi-Square	Pr > ChiSq
Intercept		1	−27.5935	8.7405	−44.7246	−10.4624	9.97	0.0016
Manager	0	1	17.0338	6.9687	3.3754	30.6922	5.97	0.0145
Manager	1	0	0.0000	.	.	.	.	.
DAYS		1	0.0767	0.0119	0.0533	0.1000	41.38	<.0001
TRANSAC		1	0.2046	0.0475	0.1114	0.2978	18.52	<.0001
DAYS*Manager	0	1	−0.0833	0.0136	−0.1099	−0.0567	37.64	<.0001
DAYS*manager	1	0	0.0000	.	.	.	.	.

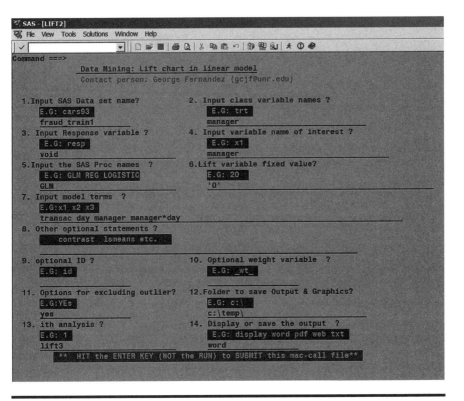

Figure 5.38 Screen copy of LIFT2 macro-call window showing the macro-call parameters required for performing lift chart in linear regression with categorical variables.

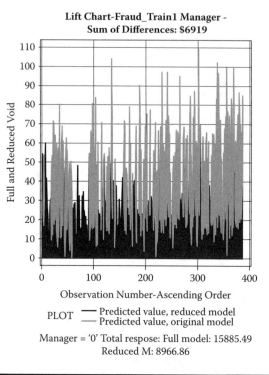

Figure 5.39 Lift chart generated by using SAS macro LIFT2: The differences in predicted *voids* between the original and the reduced model with the same level of *manager*.

the manager variable at 0 meaning other managers. On the lift chart display, the sum of differences ($6919) between the predicted values of the original model and the predicted value for the reduced model is also displayed. In addition to the lift chart, the if-then analysis output table shows the predicted values of the original and the reduced models, and the differences in the predicted values. A partial list of the if-then analysis output is presented in Table 5.30. We can conclude that this convenience store lost about $16,543 due to fraudulent voids during the 3-year period.

5.15 Case Study 4: Modeling Binary Logistic Regression

5.15.1 Study Objectives

1. *Best candidates model selection using AICC and SBC*: Perform sequential stepwise model selection and determine the subset where AICC and SBC are at minimum. Then run all possible combination of models within the optimum subset range and select the best candidate models.

Table 5.30 Macro LIFT2—Partial List of Differences between Original and Reduced (IF–THEN) Model Predicted Scores

MANAGER	id	Predicted (Original)	Predicted. (Reduced)	If-Then	Difference in Predicted Values
		Sum of the Difference between the Original and the Reduced Model Score = $6919			
0	393	51.749	51.749	0	0.0000
1	398	71.673	39.387	0	32.2860
1	401	50.975	18.280	0	32.6947
1	402	33.705	0.874	0	32.8309
0	412	53.306	53.306	0	0.0000
1	416	51.398	16.660	0	34.7382
1	418	105.838	70.827	0	35.0107
0	419	72.088	72.088	0	0.0000
1	427	45.611	9.374	0	36.2368
(Partial list of observation)					
1	490	70.785	25.965	0	44.8196
1	735	98.233	20.035	0	78.1972
1	715	98.913	23.440	0	75.4725

2. *Data exploration using diagnostic plots*: Check for the significance of each continuous predictor variable and multicollinearity among the predictors by studying simple logit and delta logit plots.

3. *Logistic regression model fitting and prediction*: Fit the logistic regression model, perform hypothesis testing on overall regression and on each parameter estimate, estimate confidence intervals for parameter estimates and odds ratios, predict probability scores and estimate their confidence intervals, produce classification tables, and false positive and false negative estimates.

4. *Checking for any violations of logistic regression assumptions*: Perform statistical tests to detect over dispersion, and graphical analysis to detect influential outliers.

5. *Computing goodness-of-fit tests and measures for evaluating the fit*: Obtain Hosmer and Lemeshaw goodness-of-fit statistics and model adequacy estimates, generalized and adjusted generalized R^2, Brier scores, and ROC curves.

6. *Save _score_ and estim datasets for future use*: These two datasets are created and saved as temporary SAS datasets in the work folder and also exported to excel work sheets and saved in the user-specified output folder. The _score_ data contains the observed variables, predicted probability scores, and confidence interval estimates. This dataset could be used as the base for developing the scorecards for each observation. Also, the parameter estimate dataset called ESTIM could be used in the LSCORE2 macro for scoring different datasets containing the same predictor variables.

7. *If–then analysis and lift charts*: Perform an if-then analysis, and construct a lift chart to estimate the differences in predicted probabilities when one of the continuous or binary predictor variables is fixed at a given value.

5.15.2 Data Descriptions

Background Information: To predict the probability of going bankrupt for financial institutions, a business analyst collected the four financial indicators from 200 institutions 2 years prior to bankruptcy. He also recorded the same financial indicator from 200 financially sound firms. To develop a model predicting the probability of going bankrupt, the analysis used a logistic regression. The descriptions of the data and the data mining methods used are presented below:

Data name: bank	SAS dataset "bank." Two temporary datasets, bank_tr1 and bank_v1, created randomly using the RANSPLT2 macro.
Response variable	Y: (0: financially sound; 1 bankrupt) — Response variable
Predictor variables	X1 (CF_TD cash flow/total debt)
	X2 (NI_TA net income/total assets)
	X3 (CA_CL current assets/current liabilities)
	X4 (CA_NS current assets/net sales)
Number of observations	Total dataset: 400
	Training: 360
	Validation: 40
Source:	Simulated data

5.15.2.1 Step 1: Best Candidate Model Selection

Open the LOGIST2.SAS macro-call file in the SAS EDITOR window and click RUN to open the LOGIST2 macro-call window (Figure 5.40). Input dataset name (bank_tr1), binary response variable (*y*), continuous predictor variables (*X*1–*X*4), model terms (*X*1–*X*4), and other appropriate macro-input values. To perform model selection, leave the macro field #6 blank. Follow the instruction given in Appendix 2.

The first step in best candidate model selection is to perform sequential stepwise selection. In the sequential stepwise selection, the stepwise variable selection is carried forward until all the predictor variables are entered by changing the default SLENTER and SLSTAY *p*-values from 0.15 to 1. Using the ODS GRAPHICS feature in the PROC LOGISTIC, an overlaid ROC plot where each ROC curve represents the best model in the sequential stepwise selection is generated (Figure 5.41). Because, the ROC statistics is always large for the full model, this criterion is not useful in selecting the optimum subset. However, in each sequential stepwise selection step, both AIC and SBC statistics are recorded, and the best subsets with the minimum AIC and minimum SBC are identified (Figure 5.42). Thus, based on minimum AIC, a three-variable subset model and, based on minimum SBC, a two-variable subset are identified as optimum subsets.

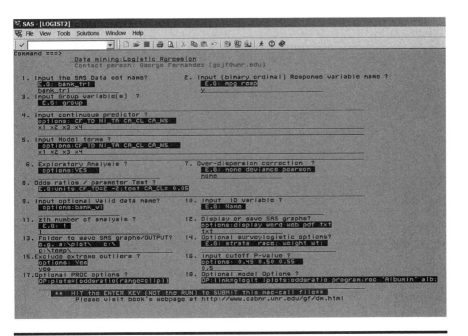

Figure 5.40 Screen copy of LOGIST2 macro-call window showing the macro-call parameters required for performing BLR.

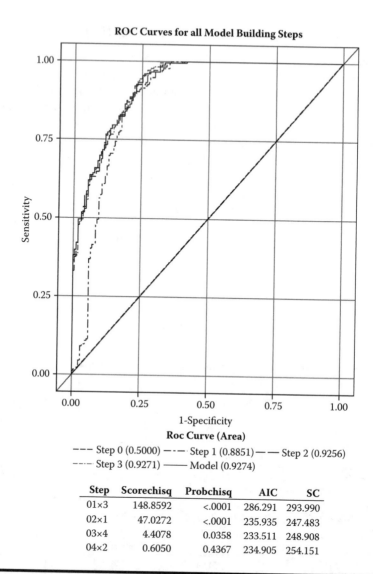

Figure 5.41 includes the following table:

Step	Scorechisq	Probchisq	AIC	SC
01×3	148.8592	<.0001	286.291	293.990
02×1	47.0272	<.0001	235.935	247.483
03×4	4.4078	0.0358	233.511	248.908
04×2	0.6050	0.4367	234.905	254.151

Figure 5.41 Comparative ROC curves generated by SAS macro LOGIST2: Comparing the area under the ROC in sequential model building when no categorical variable is included. The plot is generated by the ODS Graphics feature.

In the second step, all possible combinations of multiple logistic regressions models are performed between (SBC: 2-1) one-variable subset and (AIC: 3+1) four-variable subsets and the best candidate models based on delta AICC and delta SBC are identified (Figure 5.43). A three-variable model ($X1$, $X3$, and $X4$) and a two-variable model ($X1$ and $X3$) were identified as the best models based on delta AICC

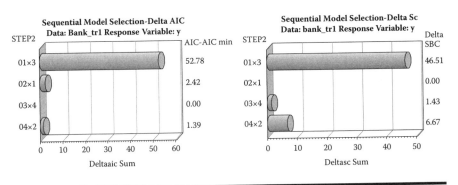

Figure 5.42   Delta AIC and Delta SC plots generated by SAS macro LOGIST2: Comparing Delta AIC and SC in sequential model-building steps.

and delta SBC criteria, respectively. However, because the full model with all four predictors is also identified as one of the best candidate model, the next exploratory analysis step was performed using all four predictors.

5.15.2.2 Step 2: Exploratory Analysis/Diagnostic Plots

Open the LOGIST2.SAS macro-call file in the SAS EDITOR window, and click RUN to open the LOGIST2 macro-call window (Figure 5.44). To perform data exploration and to create regression diagnostic plots, input YES in macro field #6. (The standard logistic regression analysis is suppressed when performing exploratory graphical analysis). Use the help file (Appendix 2) to input the appropriate macro input values to produce exploratory graphs in the BLR. Once the LOGIST2 macro has been submitted, overlay plots of simple and delta logit for each predictor

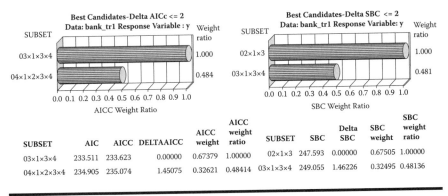

Figure 5.43   Delta AICC and Delta SBC plots generated by SAS macro LOGIST2: Comparing Delta AICC and SBC values for the best candidates based on all possible model selection within optimum subsets.

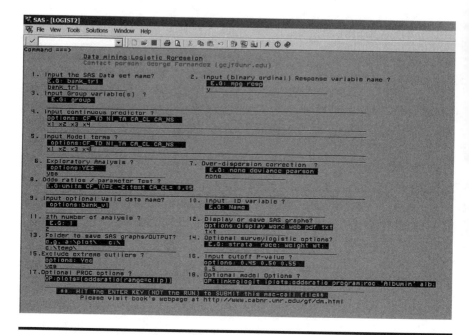

Figure 5.44 Screen copy of LOGIST2 macro-call window showing the macro-call parameters required for performing exploratory diagnostic plots in BLR.

are produced. Only selected output and graphics produced by the LOGIST2 macro are shown below.

Exploratory analysis/Diagnostic plots: Overlay plots of simple and partial delta logit plots for all four predictors are presented in Figures 5.45–5.48. The $X1$ (CF_TD) variable showed a significant negative trend in both simple and delta logit plots (Figure 5.45). A moderate degree of multicollinearity was evident, because the delta logit data points were clustered near the mean of CF_TD. The quadratic trend of $X1$ was not statistically significant.

The $X2$ (NI_TA) variable showed a significant quadratic delta logit plot (Figure 5.46). A moderate degree of multicollinearity was evident because the delta logit data points were clustered near the mean of NI_TA. However, the effect of adding a quadratic terms needs to be explored in the next step.

The $X3$ (CA_CL) variable showed a significant negative and quadratic trend in both simple and delta logit plots (Figure 5.47). No evidence of multicollinearity was observed because the delta logit data points spread along the partial logit regression line of CA_CL.

The $X4$ (CA_NS) variable showed a significant positive trend in both simple and delta logit plots. The positive trend improved from the simple logit to the partial delta curve. No evidence of multicollinearity was observed because the delta logit data points spread along the partial logit regression line of CF_TD (Figure 5.48).

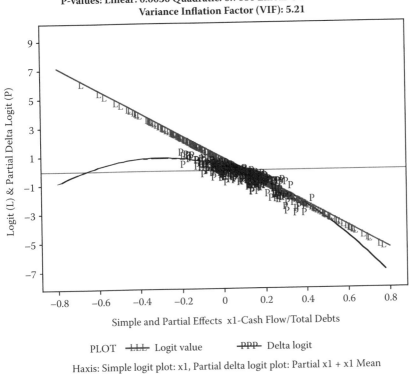

Logit & Partial Delta Logit Plots data: bank_tr1 Resp = y vs Cash Flow/Total Debts
P-values: Linear: 0.0030 Quadratic: 0.7080 Linear Odds Ratio = 0.0061
Variance Inflation Factor (VIF): 5.21

PLOT ⎯LLL⎯ Logit value   ⎯PPP⎯ Delta logit

Haxis: Simple logit plot: x1, Partial delta logit plot: Partial x1 + x1 Mean

Figure 5.45   Logistic regression diagnostic plot using SAS macro LOGIST2. Overlay plot of simple logit and partial delta logit plots for the *CF/TD ratio* is shown.

Among all the possible two variable interactions among the four variables tested, only one two-factor interaction between $X3*X4$ appeared to be significant at the 5% level (Figure 5.49). The significance of this interaction needs to be verified in the presence of all linear and quadratic terms in the next step.

5.15.2.3 Step 3: Fitting Binary Logistic Regression

Open the LOGIST2.SAS macro-call file in the SAS EDITOR window and click RUN to open the LOGIST2 macro-call window (Figure 5.50). Input training dataset name (bank_tr1), validation dataset name (bank_v1), binary response variable (Y), four continuous predictor variables ($X1$-$X4$) and model terms including quadratic and interaction terms ($X1$-$X4$ $X2*X2$ $X3*X3$ $X3*X4$). To skip data exploration and create regression diagnostic plots leave macro field #6 blank. To exclude extreme outliers, input YES to macro field #14. Input other appropriate macro

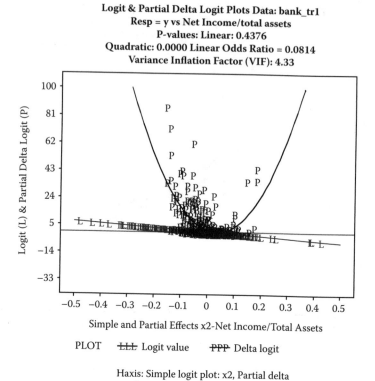

Figure 5.46 Logistic regression diagnostic plot using SAS macro LOGIST2. Overlay plot of simple logit and partial delta logit plots for the *NI/TA* ratio is shown.

parameters and submit the LOGIST2 macro to fit the binary logistic regression. In the initial run, X3*X4 interaction came out to be not significant. Therefore, the step was repeated only using X1-X4 X2*X2 X3*X3 terms. Only selected output and graphics produced by the LOGIST2 macro are shown below.

The characteristics of the training dataset, number of observations in the dataset, type of model fit, and the frequency of event and nonevent are given in Table 5.31. The LOGIST2 macro models the probability of the event. If you need to model the probability of the nonevent, recode the nonevent to 1 in your data before running the macro.

Table 5.32 shows the model convergence status and statistics for testing the overall model significance. The *AIC* and *SC* statistics for the intercept and the covariates are useful for selecting the best model from different logistic models. Lower values are desirable. The −2 Log *L* criteria is the −2 Log likelihood statistic for models with only an intercept and the full model.

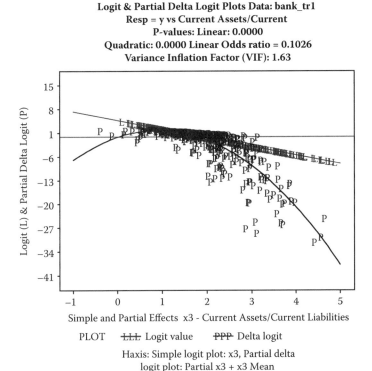

Logit & Partial Delta Logit Plots Data: bank_tr1
Resp = y vs Current Assets/Current
P-values: Linear: 0.0000
Quadratic: 0.0000 Linear Odds ratio = 0.1026
Variance Inflation Factor (VIF): 1.63

Simple and Partial Effects x3 - Current Assets/Current Liabilities

PLOT   L̶L̶L̶ Logit value   P̶P̶P̶ Delta logit

Haxis: Simple logit plot: x3, Partial delta
logit plot: Partial x3 + x3 Mean

Figure 5.47   **Logistic regression diagnostic plot using SAS macro LOGIST2. Overlay plot of simple logit and partial delta logit plots for the *CA/CL* ratio is shown.**

The results of testing the global null hypothesis that all regression coefficients are equal to zero are given in Table 5.33. All three tests indicate the overall logistic model is highly significant, and at least one of the parameter estimates is significantly different from zero. The parameter estimates, their *se* using maximum likelihood methods, and significance levels of each variable using the Wald *chi-square* are given in Table 5.34. The estimated logistic regression model for estimating the *log (odds of bankruptcy)* $= -4.9659 - 49.5493X2 + 408.6X2^2 + 13.5103\ X3 - 6.5072\ X3^2 - 16.8370\ X4 + 12.4635\ X3{*}X4$ (The final model was reestimated after dropping the nonsignificant *X1* variable from the model).

All model terms parameter estimates are statistically significant (Table 5.34). These parameter estimates estimate the change in the log odds of bankruptcy when one unit changes a given predictor variable while holding all other variables constant. For any observation in the data, the probability of the event could be predicted by the BLR parameter estimates. However, in the presence of significant quadratic and interaction terms, interpreting the parameter coefficient or the linear

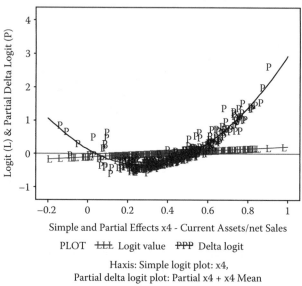

Logit & Partial Delta Logit Plots Data: bank_tr1
Resp = y vs Current assets/net S
P-values: Linear: 0.0288 Quadratic: 0.0754
Linear Odds Ratio = 7.1381
Variance Inflation Factor (VIF): 1.15

Simple and Partial Effects x4 - Current Assets/net Sales

PLOT ⎯L⎯L⎯L⎯ Logit value PPP Delta logit

Haxis: Simple logit plot: x4,
Partial delta logit plot: Partial x4 + x4 Mean

Figure 5.48 **Logistic regression diagnostic plot using SAS macro LOGIST2. Overlay plot of simple logit and partial delta logit plots for the *CA/NS* ratio is shown.**

odds ratio is not feasible. The reference probability of 0.41 indicates the probability of a company going bankruptcy is 0.41 when all the predictor variables are at the mean level (Table 5.35). However, the standardized marginal probabilities for the significant variables reported in Table 5.35 are not interpretable when the quadratic and interactions terms are included in the model.

By default, the LOGIST2 macro outputs the odds ratio estimate for one-unit increase in all linear predictor variables while holding all other variables constant. These odds ratios and their confidence intervals are computed by exponentiating the parameter estimates and their confidence intervals. Because both quadratic and interaction terms are significant for all three $X2$–$X4$ predictor variables, the odds ratio table is not produced. Therefore, to get the odds-ratio estimates, the model was refitted again with only linear predictors, and the odds ratio estimates (Table 5.36), odds ratio plot (Figure 5.51) and Table 5.36 shows the odds ratio estimates and the profile likelihood confidence interval estimates are generated. The interpretation of the odds ratio is valid for a linear continuous or binary predictor when one unit change in the predictor variable is relevant. If interaction terms are included in the model, the interpretation of the odds ratio for the given variable is not valid.

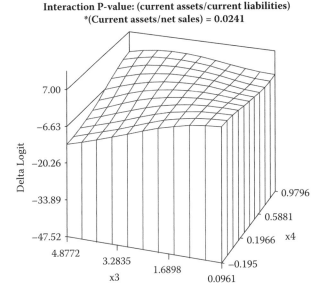

Interaction P-value: (current assets/current liabilities)
*(Current assets/net sales) = 0.0241

*X-axis x4 = Current assets/net sales
*Y-axis x3 = Current assets/current liabilities

Figure 5.49   Binary logistic regression diagnostic plot using SAS macro LOGIST2: Three-dimensional plot for detecting significant interaction between two continuous variables.

An odds-ratio value equal to 1 implies no change in the odds of the event when you increase the predictor variable by one unit. The chance of the bankruptcy increases multiplicatively when the odds ratio of a variable is greater than *one* and vice versa.

Sometimes a change in the default 1 unit increase in a predictor variable may not be meaningful. In that case, the customized odds ratio can be estimated for a 0.1 unit change in $X1$ (Table 5.37). Since an increase in one unit in the CF_TD ratio is unrealistic, we could estimate the customized odds-ratio estimates and their confidence intervals in the LOGIST2 macro by specifying the values in the macro input #8. The customized odds-ratio estimates and their profile confidence interval estimates for a ± 0.1 unit change in $X1$ is given in Table 5.37. While holing other predictor variables at a constant level, an increase of 0.1 unit in the $X1$ reduces the odds of bankruptcy by 0.42 to 0.83 times, and a decrease of 0.1 unit in the $X1$ increases the odds of bankruptcy by 1.20 to 2.36 times. There is a 20%–136% increase, or 17%–58% decrease, in the chance of bankruptcy when the $X1$ ratio decreases or increases 0.1 units (Table 5.37).

Formal tests results for checking for model adequacy are presented in Table 5.38. The *p*-value for the Hosmer and Lemeshow goodness-of-fit test using the *Pearson*

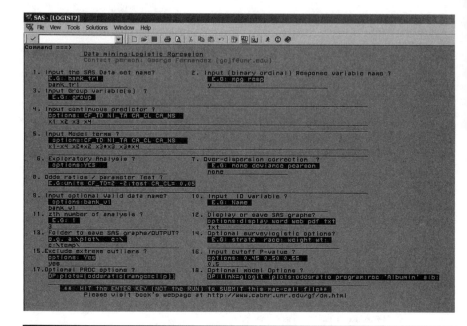

Figure 5.50   Screen copy of LOGIST2 macro-call window showing the macro-call parameters required for performing final BLR model-building step.

chi-square test is not significant at a 5% level. The estimated R^2 and the max-rescaled R^2 statistics computed from the -2 *log likelihood estimates* are moderately high. Both deviance and Pearson goodness-of-fit statistics are not significant, indicating that the variance of the binary response variable does not exceed the expected nominal variance. Thus, overdispersion is not a problem in this logistic regression model.

The biased adjusted classification table created by the *CTABLE* option in the SAS LOGIST2 procedure is presented in Table 5.39. In computing the biased adjusted classification table, SAS uses an approximate *jackknifing* method using the observed sample proportion as the prior probability estimates. The classification table uses the estimated logistic model based on cross-validation techniques to classify each observation either as events or nonevents at different probability cutpoints. Each observation is classified as event if its estimated probability is greater than or equal to a given probability cutpoint. Otherwise, the observation is classified as nonevent. The classification table provides several measures of predictive accuracy at various probability cutpoints.

For each probability cutpoint, the correct and incorrect columns provide the frequency of events and nonevents correctly and incorrectly classified. For example, at 0.5 probability cutpoint, the LOGISTIC procedure correctly classifies 160 events and 157 nonevents. It misclassifies 23 events and 17 nonevents (Table 5.39). The next five columns provide predictive accuracy of the model at various cutpoint probabilities.

Table 5.31   Macro LOGIST2—Model Information

Model Information		
Dataset	WORK.DATA	
Response variable	Y: Bankrupt 1 = yes 0 = no	
Number of response levels	2	
Model	binary logit	
Optimization technique	Fisher's scoring	
Number of observations read		400
Number of observations used		357
	Ordered Value Y	*Total Frequency*
	1: 1	177
	0: 2	180

Table 5.32   Macro LOGIST2—Model-Fit Statistics

Model Convergence Status		
Convergence Criterion (GCONV=1E-8) Satisfied		
Model-Fit Statistics		
Criterion	*Intercept Only*	*Intercept and Covariates*
AIC	496.882	147.994
SC	500.760	179.016
−2 Log *L*	494.882	131.994

Table 5.33   Macro LOGIST2—Testing Global Null Hypothesis: BETA = 0

Testing Global Null Hypothesis: BETA = 0			
Test	*Chi-Square*	*DF*	*Pr > ChiSq*
Likelihood ratio	362.8876	7	<0.0001
Score	167.3661	7	<0.0001
Wald	31.8048	7	<0.0001

Table 5.34 Macro LOGISTIC—Analysis of Maximum Likelihood Estimates and Their Significance Levels: Final Model

Parameter	DF	Full Model					Label
		Estimate	Standard Error	Wald Chi-Square	Pr > ChiSq	Standardized Estimate	
Intercept	1	-4.9884	2.3653	4.4478	0.0349		Intercept: Y = 1
X1	1	0.2646	2.1599	0.0150	0.9025	0.0377	Cash flow/total debts
X2	1	-50.2264	12.8048	15.3857	<.0001	-3.5246	Net income/total assets
X3	1	13.5096	3.2689	17.0799	<.0001	7.5416	Current assets/current liabilities
X4	1	-16.7935	5.8456	8.2533	0.0041	-1.8129	Current assets/net sales
X2*X2	1	409.8	91.0880	20.2402	<.0001		Net income/total assets * net income/total assets
X3*X3	1	-6.5060	1.3404	23.5593	<0.0001		Current assets/current liabilities * current assets/current liabilities
X3*X4	1	12.4583	3.6923	11.3850	0.0007		Current assets/current liabilities * Current assets/net sales

ii) Reduced Model after Dropping Nonsignificant X1 Variable

Parameter	DF	Estimate	Standard Error	Wald Chi-Square	Pr > ChiSq	Standardized Estimate	Label
Intercept	1	−4.9659	2.3541	4.4499	0.0349		Intercept: Y = 1
X2	1	−49.5493	11.5188	18.5038	<.0001	−3.4771	Net income/total assets
X3	1	13.5103	3.2680	17.0912	<.0001	7.5420	Current assets/current liabilities
X4	1	−16.8370	5.8313	8.3369	0.0039	−1.8176	Current assets/net sales
X2*X2	1	408.6	90.4947	20.3834	<.0001		Net income/total assets * net income/total assets
X3*X3	1	−6.5072	1.3412	23.5394	<.0001		Current assets/current liabilities * current assets/current liabilities
X3*X4	1	12.4635	3.6892	11.4133	0.0007		Current assets/current liabilities * Current assets/net sales

Table 5.35 Macro LOGIST2—Reference Probability = 0.418. Standardized Marginal Probabilities of Significant Variables When Prob chisq ≤ 0.20

Variable	Standardized Beta	Pr > Chi-Square	Standardized Marginal Probability
X2	−6.6817	<0.0001	−2.8
X3	−7.2098	<0.0001	−3.0
X4	1.5444	0.0002	0.65
X2*X2	6.5902	<0.0001	2.75
X3*X3	−6.6332	<0.0001	−2.8
X3*X4	2.4541	0.0007	1.03

- *Correct*: The overall percentage of correct classification (total frequency of correct classification/sample size)
- *Sensitivity*: Measure of accuracy of classifying events or true positives (number of correctly classified events/total number of events)

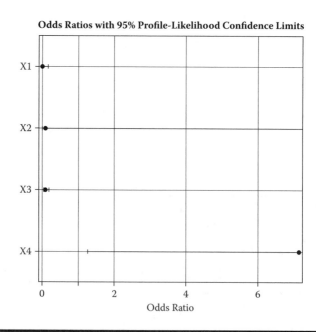

Figure 5.51 Assessing the BLR fit: Predicted probability plot for a given continuous variable and the odds ratio plot generated by SAS ODS graphics using the SAS macro LOGIST2.

Table 5.36   Macro LOGIST2—Standard Odds-Ratio Estimates and Their Confidence Interval Estimates for Linear Predictors

Effect	Unit	Estimate	95% Confidence Limits	
X1	1.0000	0.006	<0.001	0.161
X2	1.0000	0.081	<0.001	44.856
X3	1.0000	0.103	0.052	0.184
X4	1.0000	7.138	1.256	43.348

- *Specificity*: Measure of accuracy of classifying nonevents or true negatives (number of correctly classified nonevents/total number of nonevents)
- *False positive*: Measure of error in classification of nonevents (number of falsely classified nonevents as events/total number of events)
- *False negative*: Measure of error in classification of events (number of falsely classified events as nonevents/total number of nonevents)

Table 5.37   Macro LOGIST2–Customized Odds Ratio Estimates and Their Confidence Interval Estimates for Linear Predictors

Effect	Unit	Estimate	95% Confidence Limits	
X1	0.1000	0.601	0.423	0.833
X1	−0.1000	1.665	1.201	2.366

Table 5.38   Macro LOGIST2—Model Adequacy Tests

Hosmer and Lemeshow Goodness-of-Fit Test				
Chi-square	DF	Pr > Chi-sq		
0.8579	7	0.9968		
R-square	0.6381	Max-rescaled R-square	0.8509	
Deviance and Pearson Goodness-of-Fit Statistics				
Criterion	Value	DF	Value/DF	Pr > Chi-sq
Deviance	131.9943	349	0.3782	1.0000
Pearson	118.2825	349	0.3389	1.0000

Table 5.39 Macro LOGIST2—Classification Table

				Classification Table					
	Correct		Incorrect			Percentages			
Prob Level	Event	Non event	Event	Non event	Correct	Sensitivity	Specificity	False POS	False NEG
0.050	177	126	54	0	84.9	100.0	70.0	23.4	0.0
0.100	176	131	49	1	86.0	99.4	72.8	21.8	0.8
0.150	176	137	43	1	87.7	99.4	76.1	19.6	0.7
0.200	175	138	42	2	87.7	98.9	76.7	19.4	1.4
0.250	172	142	38	5	88.0	97.2	78.9	18.1	3.4
0.300	172	147	33	5	89.4	97.2	81.7	16.1	3.3
0.350	168	150	30	9	89.1	94.9	83.3	15.2	5.7
0.400	164	152	28	13	88.5	92.7	84.4	14.6	7.9
0.450	163	155	25	14	89.1	92.1	86.1	13.3	8.3
0.500	160	157	23	17	88.8	90.4	87.2	12.6	9.8
0.550	156	161	19	21	88.8	88.1	89.4	10.9	11.5
0.600	145	164	16	32	86.6	81.9	91.1	9.9	16.3
0.650	143	169	11	34	87.4	80.8	93.9	7.1	16.7
0.700	142	170	10	35	87.4	80.2	94.4	6.6	17.1
0.750	139	172	8	38	87.1	78.5	95.6	5.4	18.1
0.800	132	174	6	45	85.7	74.6	96.7	4.3	20.5
0.850	127	176	4	50	84.9	71.8	97.8	3.1	22.1
0.900	120	179	1	57	83.8	67.8	99.4	0.8	24.2
0.950	118	180	0	59	83.5	66.7	100.0	0.0	24.7
1.000	0	180	0	177	50.4	0.0	100.0	.	49.6

The ROC (receiver operating characteristic) curve presented in Figure 5.52 provides the measure of predicted accuracy of the fitted logistic regression model for the training and validation data. Since the ROC curve rises quickly, the fitted model has a relatively high predictive accuracy. The higher predictive ability of this logistic regression model is confirmed by the large area under the ROC

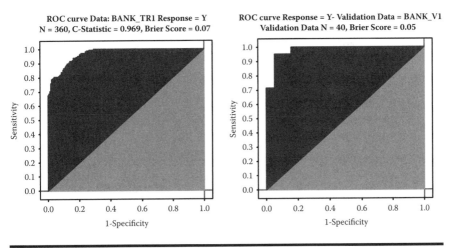

ROC curve Data: BANK_TR1 Response = Y
N = 360, C-Statistic = 0.969, Brier Score = 0.07

ROC curve Response = Y- Validation Data = BANK_V1
Validation Data N = 40, Brier Score = 0.05

Figure 5.52 Assessing the BLR fit: Comparing the area under the ROC curves generated for the training and validation data using the SAS macro LOGIST2.

curve (0.97) that is displayed in the figure title. The *C-statistic* estimated by the LOGISTIC procedure is equal to the area under the ROC curve. The interpretation of the area under the ROC curve is as follows: The probability of picking an event from a pair of events and nonevents is 0.50. However, if the predicted probability scores based on the logistic regression model are given to these two events, and the event is asked to be picked now, the probability of making the right choice will be equal the area under the ROC curve. Very large values for area under the ROC curve for both the training and validation data validate the fitted logistic regression model.

Figure 5.53 shows the trend between the percentages of correct classification and the false positive and false negatives versus different cutpoint probability values. If the objective is to find the cutpoint probability that gives overall predictive accuracy, select the cutpoint probability where the false positive and false negative curve intersects, and the overall classification percentage is high. Approximately at the cutpoint probability equal to 0.575, both false positive and false negatives estimates are low.

The estimated predicted probability scores and their confidence intervals are given in Tables 5.40 and 5.41. These predicted scores and the confidence interval estimates could be used to build scorecards for each financial institution, which helps to identify financially troubled firms.

Significant outlier/influential observations: Table 5.42 lists several observations as influential outliers based on the larger DIFDEV statistic (DIFDEV > 4). The DIFDEV statistic measures the change in the model deviance statistic when these observations are excluded individually. These outliers also showed up in the outlier detection plot (Figure 5.54) that illustrates the leverage of each observation (hat-value). The diameter

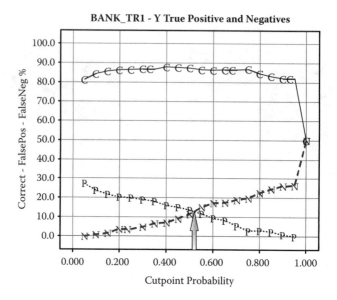

Figure 5.53   Assessing the BLR fit: Overlay plot showing false-positive and false-negative percentages versus different cutpoint probabilities using the SAS macro LOGIST2.

of the bubbles in the outlier detection plot is proportional to the *cbar* statistic, which is an influential statistic that quantifies the change in the parameter confidence interval estimates when each observation is excluded. Because some of the outliers have relatively larger *cbar* statistics, to investigate the impact of these outliers, this analysis should be repeated by selecting macro option #14 to exclude extreme influential points.

Model validation: To validate the obtained logistic regression model estimates, the regression parameter estimates obtained from the training dataset ($n = 360$) are used to predict the bankruptcy for the validation dataset ($N = 40$), and the results are presented in Table 5.43. The Brier score for this validation dataset is relatively small: 0.048 (Table 5.43) similar to the Brier score (0.07) obtained from the training data (Figure 5.52). The misclassification percentage is 5% (2 out of 40 is misclassified). Even though the validation results are satisfactory, the impact of the extreme outliers on model validation could be verified by refitting the model by excluding the outliers from the analysis.

Lift or gain charts: The cumulative lift charts are very useful informative display that portrait the success of applying predictive probability scores in

Table 5.40 Macro LOGIST2—Partial List of the Bottom 10% of the Predicted Probability and Their Confidence Intervals

ID	Bankrupt 1 = yes 0 = no	PREDICTS	Net Income/ Total Assets	Current Assets/Current Liabilities	Current Assets/net Sales	Estimated Probability	Lower 95% Confidence Limit	Upper 95% Confidence Limit
299	0	0	0.0332	4.8772	0.3614	1.7219E-22	4.8498E-31	6.1137E-14
316	0	0	0.0684	4.7802	0.4554	2.9545E-21	2.4226E-29	3.6032E-13
300	0	0	0.0917	4.7775	0.3474	3.6211E-21	3.1785E-29	4.1252E-13
323	0	0	0.065	4.5875	0.3779	4.6569E-19	3.1627E-26	6.8571E-12
257	0	0	0.0731	4.5594	0.3631	1.0173E-18	9.3574E-26	1.1059E-11
213	0	0	0.096	4.5804	0.4105	1.0828E-18	8.5298E-26	1.3745E-11
339	0	0	0.1901	4.7133	0.4914	2.235E-17	9.1699E-25	5.4473E-10
222	0	0	0.0201	4.4335	0.642	8.9036E-17	3.5088E-23	2.2593E-10

Table 5.41 Macro LOGIST2—Partial List of the Top 10% of the Predicted Probability and Their Confidence Intervals

ID	Bankrupt 1 = yes 0 = no	PREDICTS	Net Income/Total Assets	Current Assets/Current Liabilities	Current Assets/Net Sales	Estimated Probability	Lower 95% Confidence Limit	Upper 95% Confidence Limit
106	1	1	-0.27	0.7595	0.1622	1	1.00000	1
118	1	1	-0.282	0.4723	0.1358	1	1.00000	1
126	1	1	0.392	2.109	0.4626	1	1.00000	1
131	1	1	-0.327	1.4098	0.3312	1	1.00000	1
134	1	1	-0.369	1.4488	0.3924	1	1.00000	1
162	1	1	-0.276	1.3407	0.2739	1	1.00000	1
171	1	1	-0.313	0.7975	0.6694	1	1.00000	1
174	1	1	-0.254	1.7005	0.1698	1	1.00000	1
27	1	1	0.3877	2.2275	0.7125	1	1.00000	1
36	1	1	-0.28	0.7267	0.3056	1	1.00000	1
39	1	1	-0.399	1.5107	0.4094	1	1.00000	1
5	1	1	-0.434	0.5271	0.4018	1	1.00000	1
62	1	1	-0.319	1.0858	0.4174	1	1.00000	1
67	1	1	0.4279	2.0578	0.3875	1	1.00000	1
72	1	1	-0.294	1.4769	0.6884	1	1.00000	1

79	1	1	-0.315	0.9303	0.1253	1	1.00000	1
92	1	1	-0.288	0.7051	0.3679	1	1.00000	1
95	1	1	-0.265	0.9153	-0.016	1	1.00000	1
98	1	1	-0.481	0.2874	-0.066	1	1.00000	1
108	1	1	-0.247	1.3503	0.3698	1	1.00000	1
94	1	1	-0.245	0.9064	0.567	1	1.00000	1
130	1	1	-0.238	1.2637	0.4831	1	1.00000	1
105	1	1	-0.233	1.2902	0.7173	1	1.00000	1
170	1	1	-0.227	1.4691	0.7809	1	1.00000	1

Table 5.42 Macro LOGISTIC—Partial List of Influential Observations

ID	Net Income/Total Assets	Current Assets/Current Liabilities	Current Assets/Net Sales	Bankrupt 1 = yes 0 = no	Estimated Probability	Deviance Residual	Hat-Ratio = (hat/hatmean)	Confidence Interval Displacement CBar	One Step Difference in Deviance
107	0.1147	1.0744	0.2137	1	0.53766	1.11403	3.11487	0.04736	1.28844
115	0.0706	1.7882	0.1238	1	0.21097	1.76410	2.44574	0.15985	3.27191
117	0.0413	2.0944	0.3344	1	0.11557	2.07743	1.19630	0.15657	4.47230
12	-0.038	2.3201	0.6194	1	0.69268	0.85695	5.81887	0.04794	0.78231
13	0.0135	0.8969	0.1306	1	0.21992	1.74039	2.70629	0.16853	3.19749
132	0.0959	1.3823	0.2253	1	0.49887	1.17934	2.32932	0.04081	1.43164
133	0.1333	1.4966	0.3586	1	0.88555	0.49305	2.11203	0.00474	0.24784
136	0.0827	1.9096	0.7931	1	0.55627	1.08305	3.20731	0.04531	1.21832
139	0.0081	1.0179	0.3308	1	0.48897	1.19620	2.01067	0.03645	1.46733
146	0.1714	0.8084	-0.195	1	0.86068	0.54779	9.13848	0.02928	0.32935
148	-0.01	1.7177	0.0553	1	0.61527	0.98559	4.58668	0.05207	1.02347

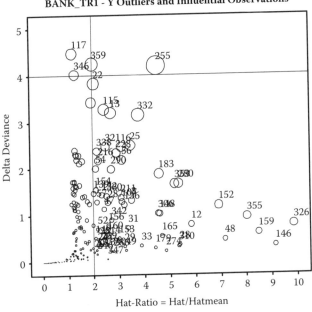

Figure 5.54 BLR Diagnostic plot: Checking for influential outliers in BLR using the SAS macro LOGIST2.

identifying the events. The horizontal line in the lift charts shows the probability of selecting an event randomly from the population. For example, if you randomly select a company, the probability that the company is going to be bankrupt is around 0.52 (Figure 5.55). However, if we use the predicted probability scores from the logistic regression and pick the top 30% the companies,

Table 5.43 Macro LOGIST2—Classification Table of Observed versus Predicted Probability at the Cutoff *p*-value: 0.5—Validation Data

Table of OBSERVED by PREDICTS			
N = 40 Brier Score 0.048			
OBSERVED	PREDICTS		
Frequency	0	1	Total
0	17	2	19
1	0	21	21
Total	17	23	40

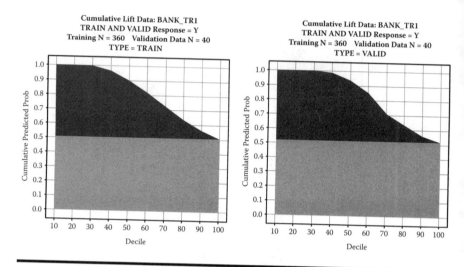

Figure 5.55 Assessing the BLR fit: Comparing the LIFT chart performances generated for the training and validation data using the SAS macro LOGIST2.

then there is almost 98% chance that these companies can go bankrupt. Thus, the accuracy of prediction based on logistic regression is increased by 46%. Furthermore, if we select the top 50% of the companies, the probability of going bankrupt is still above 90%. This finding clearly shows the success of applying logistic regression in identifying the event. Because the validation data success rate in the lift chart is also very similar to training data, the model validation can also be confirmed.

Performing if-then analysis: After finalizing the regression model, the next objective is to perform an *if-then* analysis and create a lift chart showing what happens to the predictive probability if the $X3$ is held constant at 1.0. (Figure 5.56). Open the LIFT2.SAS macro-call file in the SAS EDITOR window and click RUN to open the LIFT2 macro-call window. Input the SAS dataset (bank_trl) and response (y) variable name, model statement ($X2$ $X3$ $X4$ $X2*X2$ $X3*X3$ $X3*X4$), the variable name of interest ($X3$), fixed level (1), and other appropriate macro-input values by following the suggestions given in the help file (Appendix 2). Submit the lift macro, and you will get the lift chart showing the differences in the predicted probability scores between the original full model and the new reduced model where you fixed the $X3$ ratio at 1. On the lift chart display, the mean difference (-0.067) between the predicted probability values of the original model and the mean predicted probability value for the reduced model is also displayed. In addition to the lift chart, the *if–then* analysis output table shows, the predicted probability values of the original and the reduced models, and the differences in the predicted probability values are also produced. A partial list of the *if-then* analysis output is presented in Table 5.44. Thus, we could conclude that the average probability of bankruptcy drops by 6.7%

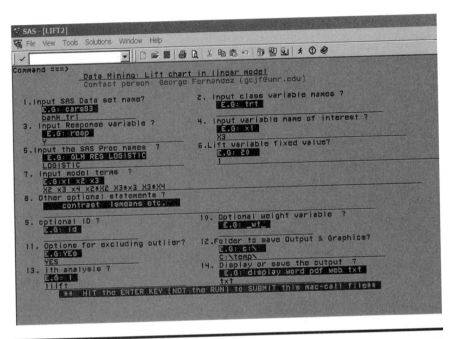

Figure 5.56 Screen copy of LIFT2 macro-call window showing the macro-call parameters required for performing lift chart in BLR.

Table 5.44   Macro LIFT2—A Partial List of Difference in Mean Predicted Probability between Original and the Reduced Model = −0.067

Current Assets/Current Liabilities	ID	OBS	Predicted Value, Original	Predicted Value, Reduced Model	NEWVAR	Difference in Probability Values
1.7747	39	122	0.96862	0.02839	1	0.94024
1.4185	94	123	0.97835	0.92279	1	0.05556
1.8677	318	124	0.96264	0.02137	1	0.94127
1.4966	132	125	0.95516	0.83882	1	0.11634
2.0759	181	126	0.95121	0.46101	1	0.49019
2.1367	301	127	0.96540	0.14500	1	0.82039
1.6507	336	128	0.96473	0.45748	1	0.50725
1.4004	73	129	0.97125	0.64301	1	0.32823
1.4944	352	130	0.94978	0.98840	1	−0.03862

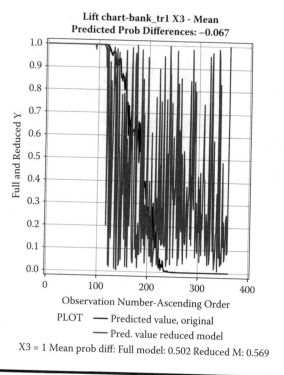

X3 = 1 Mean prob diff: Full model: 0.502 Reduced M: 0.569

Figure 5.57 Lift chart generated by using SAS macro LIFT2: The differences in predicted probability for *bankruptcy* between the original and the reduced model if the *CA/CL* ratio was held constant at 2.5.

if the CA_CL (**current assets/current liabilities**) ratio is held constant at 1 without any change in other financial indicators (Figure 5.57).

5.16 Case Study: 5 Modeling Binary Multiple Logistic Regression

5.16.1 Study Objectives

1. *Best-candidate model selection using AICC and SBC*: Perform sequential stepwise model selection and determine the subset where AICC and SBC are at a minimum. Then run all possible combination of models within the optimum subset range and select the best-candidate models.

2. *Data exploration using diagnostic plots*: Check for the significance of the selected continuous predictor variable and multicollinearity among the predictors by studying simple logit and delta logit trends.

3. *Logistic regression model fitting and prediction*: Fit the logistic regression model, perform hypothesis testing on overall regression and on each parameter estimate, estimate confidence intervals for parameter estimates and odds ratios, predict probability scores and estimate their confidence intervals, and produce classification tables, and false-positive and false-negative estimates.

4. *Checking for any violations of logistic regression assumptions*: Perform statistical tests to detect overdispersion and graphical analysis to detect influential outliers.

5. *Save _score_ and estim datasets for future use*: These two datasets are created and saved as temporary SAS datasets in the work folder and also exported to excel work sheets and saved in the user-specified output folder. The _score_ data contains the observed variables, predicted probability scores, and confidence-interval estimates. This dataset could be used as the base for developing the scorecards for each observation. Also, the parameter estimate dataset called ESTIM could be used in the LSCORE2 macro for scoring different datasets containing the same predictor variables.

5.16.2 Data Descriptions

Background Information: To predict the probability of getting coronary artery disease, a health analyst collected the 8 continuous and 1 binary health indicators from 462 patients. To develop a model predicting the probability of getting coronary artery disease, the analysis used a logistic regression. The descriptions of the data, and the data-mining methods used, are presented below:

Data name: Coronary	SAS dataset "coronary." Two temporary datasets, train and valid, created randomly using the RANSPLT2 macro.	
Response variable		
Predictor variables	Y: (0: NO; 1 yes) — Coronary artery disease	
	X1	Systolic blood pressure
	X2	Cumulative tobacco used in kg
	X3	Low-density lipoprotein cholesterol
	X4	Adiposity
	X5	Type A behavior
	X6	BMI
	X7	Current alcohol consumption
	X8	Age at onset

	C1	Family history of heart disease (Absent/present)
Number of observations	Total dataset: 462	
	Training: 300	
	Validation: 162	
Source:	South African Hearth Disease Data[39]	

Step 1: Best-candidate model selection: Open the LOGIST2.SAS macro-call file in the SAS EDITOR window and click RUN to open the LOGIST2 macro-call window (Figure 5.58). Input dataset name (coronary_train1), binary response variable (*Y*), continuous predictor variables (*X*1–*X*8), group variable (*C*1) model terms (*C*1 *X*1–*X*8), and other appropriate macro-input values. To perform model selection, leave the macro field #6 blank. Follow the instruction given in Appendix 2.

The first step in best-candidate model selection is to perform sequential stepwise selection. In sequential stepwise selection, stepwise variable selection

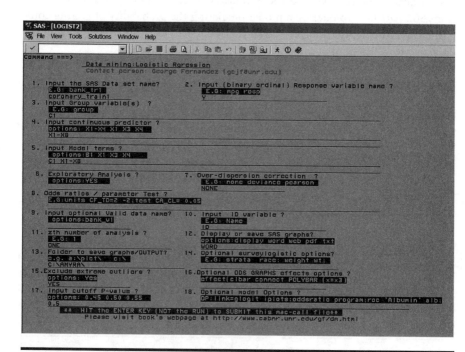

Figure 5.58 Screen copy of LOGIST2 macro-call window showing the macro-call parameters required for performing BLR.

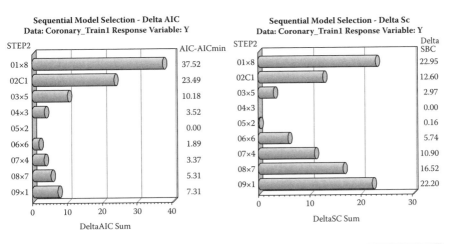

Figure 5.59 Delta AICC and Delta SC plots generated by SAS macro LOGIST2: Comparing Delta AICC and SC in sequential model-building steps.

is carried forward until all the predictor variables are entered by changing the default SLENTER and SLSTAY *p*-values from 0.15 to 1. The ODS GRAPHICS feature available in the PROC LOGISTIC, an overlaid ROC plot where each ROC curve represents the best model in the sequential stepwise selection, is generated but is not available when categorical predictors are included. In each sequential stepwise selection step, both AIC and SBC statistics are recorded, and the best subsets with the minimum AIC and minimum SBC are identified (Figure 5.59). Thus, based on minimum AIC, a five-variable subset model, and, based on minimum SBC, a four-variable subset are identified as optimum subsets.

Before performing all possible combination of multiple logistic regression models, the levels of categorical variables are automatically converted to dummy variables (D1, D2, etc.) by the GLMMOD procedure and treated as continuous variables in the subsequent model-selection step. In the second step, all possible combination of multiple logistic regression models are performed between the three-variable subsets (SBC: 4–1) and the six-variable subsets (AIC: 5+1), and the best-candidate models based on delta AICC and delta SBC are identified (Figure 5.60). A five-variable model (*D2* (categorical), *X*2, *X*3, and *X*5, and *X*8) and a four-variable model (*D2* (categorical), *X*3, and *X*5, and *X*8) were identified as the best models based on delta AICC and delta SBC criteria, respectively. However, because the *X*6 predictor was also identified as one of the variables by AICC as the best-candidate models, the next exploratory analysis step was performed using all six predictors.

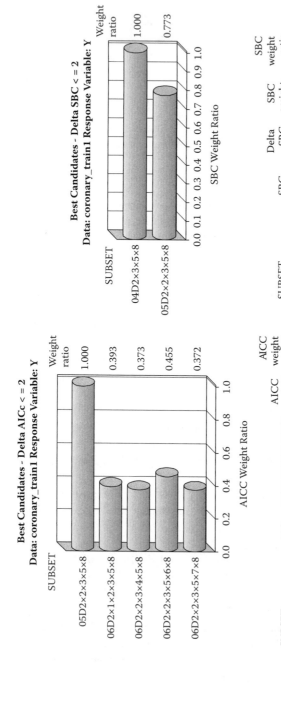

Best Candidates - Delta SBC < = 2
Data: coronary_train1 Response Variable: Y

SUBSET	SBC	Delta SBC	SBC weight	SBC weight ratio
04D2×3×5×8	311.888	0.00000	0.56416	1.00000
05D2×2×3×5×8	312.404	0.51613	0.43584	0.77254

Best Candidates - Delta AICc < = 2
Data: coronary_train1 Response Variable: Y

SUBSET	AIC	AICC	DELTAAICC	AICC weight	AICC weight ratio
05D2×2×3×5×8	290.181	290.468	0.00000	0.38567	1.00000
06D2×2×3×5×6×8	291.660	292.043	1.57527	0.17545	0.45492
06D2×2×3×5×7×8	292.060	292.443	1.97547	0.14363	0.37242
06D2×1×2×3×5×8	291.955	292.338	1.87028	0.15139	0.39253
06D2×2×3×4×5×8	292.057	292.440	1.97237	0.14386	0.37300

Figure 5.60   Delta AICC and Delta SC plots generated by SAS macro LOGIST2: Comparing Delta AICC and SC values for the best candidates based on all possible model selection within optimum subsets.

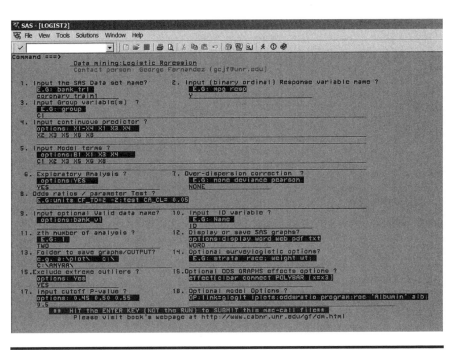

Figure 5.61 Screen copy of LOGIST2 macro-call window showing the macro-call parameters required for performing exploratory diagnostic plots in BLR.

Step 2: Open the LOGIST2.SAS macro-call file in the SAS EDITOR window and click RUN to open the LOGIST2 macro-call window (Figure 5.61). Input dataset name (coronary_train1), binary response variable (*Y*), group predictor (*C*1), continuous predictor variables (*X*2 *X*3 *X*5 *X*6 *X*8), model terms (*C*1 *X*2 *X*3 *X*5 *X*6 *X*8) and other appropriate macro-input values. To perform data exploration and to create regression diagnostic plots, input YES in macro field #6. Once the LOGIST2 macro has been submitted, overlay plots of simple and delta logit for each predictor variable is produced. Only selected output and graphics produced by the LOGIST2 macro are shown in the figure.

Exploratory analysis/diagnostic plots: Overlay plots of simple and partial delta logit plots for two selected continuous variables are presented in Figures 5.62 and 5.63. The *X*5 (type A behavior) showed a significant positive trend in both simple and partial delta logit plots. A significant quadratic effect was also evident in the partial delta logit plot. No evidence for multicollinearity was evident in the delta logit plot (Figure 5.62). Similarly, the *X*6 (BMI) showed a significant positive trend in the simple logit plot. However, highly significant quadratic effect was evident in the partial delta logit plot. No evidence for multicollinearity was evident in

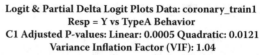

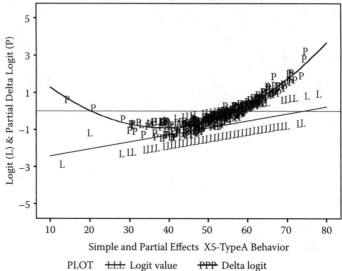

Haxis: Simple logit plot: X5, Partial delta logit plot: Partial X5 + X5 mean

Figure 5.62 **Logistic regression diagnostic plot using SAS macro LOGIST2. Overlay plot of simple logit and partial delta logit plots for *type A behavior* is shown.**

the delta logit plot (Figure 5.63). No significant quadratic effects or problems with multicollinearity were observed in other three predictors (*X*2, *X*3, and *X*8) (plots not shown).

The 3-D interaction plot clearly revealed the interaction trend between *X*2 (cumulative tobacco use) and *X*5 (Figure 5.64). The box plot display of the categorical variable *C*1 (presence of family history of heart attack) confirms that it was a significant risk factor in contributing to coronary heart disease (Figure 5.65). The increasing level of low-density lipoprotein cholesterol becomes a significant risk factor only with the patients with family history of heart disease, and this was clearly shown by the significant interaction plot between *C*1 and *X*3 in Figure 5.66.

Step 3: Fitting binary logistic regression

Open the LOGIST2.SAS macro-call file in the SAS EDITOR window and click RUN to open the LOGIST2 macro-call window (Figure 5.67). Input the training dataset name (coronary_train1), validation dataset name (coronary_valid1), binary response variable (*Y*), group predictor

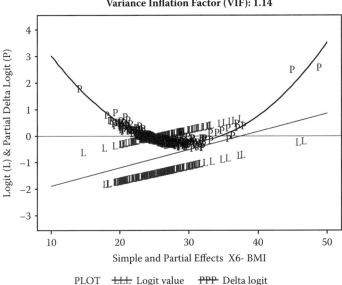

Logit & Partial Delta Logit Plots Data: Coronary_Train1 Resp = Y vs BMI
C1 Adjusted P-values: Linear: 0.7426 Quadratic: 0.0333
Variance Inflation Factor (VIF): 1.14

PLOT ~~LLL~~ Logit value ~~PPP~~ Delta logit

Haxis: Simple logit plot: X6, Partial delta logit plot: Partial X6 + X6 mean

Figure 5.63 Logistic regression diagnostic plot using SAS macro LOGIST2. Overlay plot of simple logit and partial delta logit plots for *bmi* is shown.

(*C*1), continuous predictor variables (*X*2 *X*3 *X*5 *X*6 *X*8), and model terms (*C*1 *X*2 *X*3 *X*5 *X*6 *X*8 *X*5**X*5 *X*6**X*6 *X*2**X*5 *C*1**X*3). To skip data exploration and create regression diagnostic plots, leave macro field #6 blank. To exclude extreme outliers, input YES to macro field #14. Input other appropriate macro parameters and submit the LOGIST2 macro to fit the binary logistic regression. Only selected output and graphics produced by the LOGIST2 macro are shown below.

The characteristics of the training dataset, number of observations in the dataset, type of model fit, and the frequency of event and nonevent are given in Table 5.45. The LOGIST2 macro models the probability of the event (presence of coronary artery disease). If you need to model the probability of the nonevent, record the nonevent to 1 in your data before running the macro. Table 5.46 shows the model convergence status and statistics for testing the overall model significance. The *AIC* and *SC* statistics for the intercept and the covariates are useful for selecting the best model from different logistic models using the same data and the same response variables but with different model terms. Lower values are desirable. The −2 Log *L* criterion

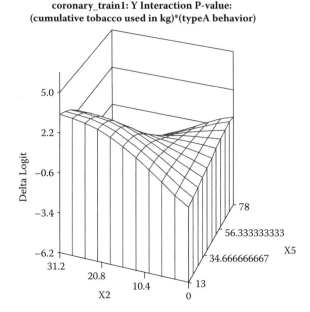

coronary_train1: Y Interaction P-value:
(cumulative tobacco used in kg)*(typeA behavior)

*X-axis X5 = TypeA behavior
*Y-axis X2 = Cumulative tobacco used in kg

Figure 5.64 **Binary logistic regression diagnostic plot using SAS macro LOGIST2:** Three-dimensional plot for detecting significant interaction between two continuous variables.

is the −2 Log likelihood statistics for models with only an intercept and for the full model.

The results of testing the global null hypothesis that all regression coefficients are equal to zero are given in Table 5.47. All three tests indicate that the overall logistic model is highly significant, and that at least one of the parameter estimates is significantly different from zero.

The parameter estimates, their *se* using maximum likelihood methods, and significance levels of each variable using the Wald *chi-square* are given in Table 5.48. The estimated logistic regression model for estimating the $\log$ (*odds of coronary disease*) $= 3.1874 + 1.6786(C1:$ absent$) + 0.4849\ X2 + 0.7139X3 − 0.1998X5 − 0.6386\ X6 + 0.0869\ X8 + 0.00300\ \boldsymbol{X5*X5} + 0.0107\ \boldsymbol{X6*X6} − 0.00757\ \boldsymbol{X2*X5} − 0.6288\ \boldsymbol{X3*C1}.$

All models terms included are statistically significant at 5% level. Except for one linear predictor variable ($X8$), all other variables are significantly involved in either a quadratic effect or in interaction. Therefore, interpreting the odds ratio or standardized regression coefficient becomes

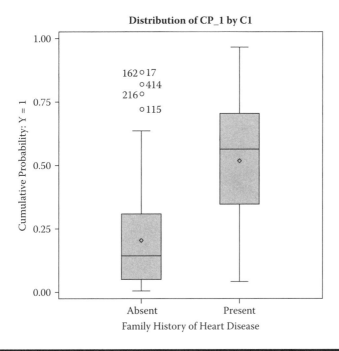

Figure 5.65   Binary logistic regression diagnostic plot using SAS macro LOGIST2: Box-plot display of predicted probabilities by covariate adjusted categorical variable.

problematic. The reference and marginal probability for standardized predictor variables (zero mean and 1 standard deviation) are presented in Table 5.49. The reference probability of 0.246 indicates that the probability of getting coronary artery disease is 0.25 when all the predictor variables are at the mean level. However, the standardized marginal probabilities for the significant variables reported in Table 5.49 are not interpretable when the quadratic and interactions terms are included in the model.

By default, the LOGIST2 macro outputs the odds ratio estimate for one-unit increase in all predictor variables while holding all other variables constant. These odds ratios and their confidence intervals are computed by exponentiating the parameter estimates and their confidence intervals. Table 5.50 shows the odds-ratio estimates and the profile likelihood confidence interval estimates for one of the predictor variables. The interpretation of the odds-ratio is valid for a linear continuous or binary predictor when one unit change in the predictor variable is relevant. If interaction terms are included in the model, the interpretation of the odds ratio for given variable is not valid.

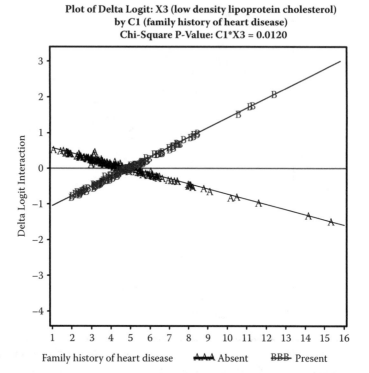

**Plot of Delta Logit: X3 (low density lipoprotein cholesterol)
by C1 (family history of heart disease)
Chi-Square P-Value: C1*X3 = 0.0120**

Family history of heart disease   ▲▲▲ Absent   B̶B̶B̶ Present

Figure 5.66   Binary logistic regression diagnostic plot using SAS macro LOGIST2: Two-dimensional plot for detecting significant interaction between a continuous and a categorical variable.

An odds ratio value equal to 1 implies no change in the odds of the event when you increase the predictor variable by one unit. The chance of getting coronary artery disease increases multiplicatively when the odds ratio of a variable is greater than one, and vice versa. For example, one unit increase in the X8 (age at onset) while holding all other predictors constant multiplies the odds of getting heart attack by 1.057 to 1.13 times (Table 5.50) and this is illustrated in the odds-ratio plot (Figure 5.68).

The interaction between the categorical predictor (C1: family history of heart disease) and the cumulative use of tobacco usage on the predicted probability and its 95% confidence interval scores of getting heart disease at the mean level of other three attributes is displayed in Figure 5.69. The new ODS graphics feature was used to generate this informative graph. The probability of being diagnosed with coronary artery disease is about 10%–12% higher if a subject has a prior family history of heart attack, and the health risk increases much higher if the patient also consumes more tobacco products.

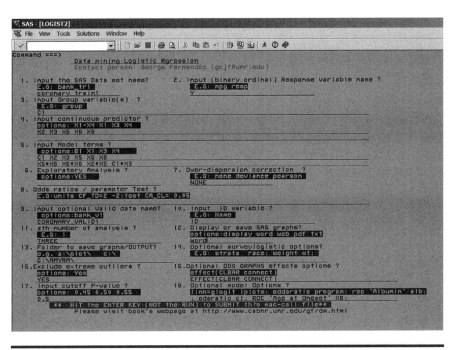

Figure 5.67   Screen copy of LOGIST2 macro-call window showing the macro-call parameters required for performing the final BLR model-building step.

Formal test results for checking for model adequacy are presented in Table 5.51. The *p*-value for the Hosmer and Lemeshow goodness-of-fit test using the *Pearson chi-square* test is not significant at a 5% level. The estimated R^2 and the max-rescaled R^2 statistics computed from the *−2 log likelihood estimates* (0.50) are at an acceptable level. A smaller Brier score closer to 0 indicates that the predictive power of the logistic regression model is high. Both deviance and Pearson goodness-of-fit statistics are not significant, indicating that the variance of the binary response variable does not exceed the expected nominal variance. Thus, overdispersion is not a problem in this logistic regression model.

The biased adjusted classification table created by the *CTABLE* option in the SAS LOGIST2 procedure is presented in Table 5.52. In computing the bias-adjusted classification table, SAS uses an approximate *jackknifing* method using the observed sample proportion as the prior probability estimates. The classification table uses the estimated logistic model based on cross-validation techniques to classify each observation either as events or nonevents at different probability cutpoints. Each observation is classified as event if its estimated probability is greater than or equal to

Table 5.45   Macro LOGIST2—Model Information

Model Information	
Dataset	Work.data
Response Variable	Y (0, 1) coronary heart disease
Number of Response Levels	2
Model	binary logit
Optimization technique	Fisher's scoring
Number of observations read	462
Number of observations used	291
Y (ordered value)	Total frequency
1 (1)	96
0 (2)	195

Table 5.46   Macro LOGIST2—Model-Fit Statistics

Model-Fit Statistics		
Convergence Criterion (GCONV=1E–8) Satisfied		
Criterion	Intercept Only	Intercept and Covariates
AIC	371.049	259.070
SC	374.723	299.476
−2 Log L	369.049	237.070

Table 5.47   Macro LOGIST2—Testing Global Null Hypothesis: BETA = 0

Test	Chi-Square	DF	Pr > Chi-sq
Likelihood ratio	131.9797	10	<0.0001
Score	106.3651	10	<0.0001
Wald	63.8599	10	<0.0001

Table 5.48 Macro LOGIST2—Analysis of Maximum Likelihood Estimates and Their Significance Levels—Final Model

Parameter		DF	Estimate	Standard Error	Wald Chi-Square	Pr > Chi-sq	Standardized Estimate	Label
Intercept		1	3.1874	4.9467	0.4152	0.5194		Intercept: Y=1
C1	Absent	1	1.6786	0.9934	2.8552	0.0911		Family history of heart disease absent
X2		1	0.4849	0.2104	5.3130	0.0212	1.1983	Cumulative tobacco used in kg
X3		1	0.7139	0.1754	16.5701	<.0001	0.8109	Low-density lipoprotein cholesterol
X5		1	−0.1998	0.1122	3.1684	0.0751	−1.0727	Type A behavior
X6		1	−0.6386	0.2919	4.7872	0.0287	−1.4761	BMI
X8		1	0.0869	0.0169	26.3974	<.0001	0.7119	Age at onset
X5*X5		1	0.00300	0.00111	7.3177	0.0068		Type A behavior * type A behavior
X6*X6		1	0.0107	0.00506	4.4661	0.0346		BMI * BMI
X2*X5		1	−0.00757	0.00376	4.0551	0.0440		Cumulative tobacco used in kg * type A behavior
X3*C1	Absent	1	−0.6288	0.1957	10.3273	0.0013		Family history of heart disease Absent * low-density lipoprotein cholesterol

Table 5.49 Macro LOGIST2—Reference Probability = 0. 269 Standardized Marginal Probabilities (Prob > chi-sq <=0.20)

Variable	Standardized Beta	Pr > chi-SQUARE	Standardized Marginal Probability
C1	−1.2429	0.0002	−0.33
X2	0.3723	0.0254	0.10
X3	1.4814	<.0001	0.40
X5	0.8845	<.0001	0.24
X6	−0.3452	0.0838	−0.09
X8	1.2859	<.0001	0.35
X5*X5	0.2797	0.0068	0.08
X6*X6	0.1866	0.0346	0.05
X2*X5	−0.3244	0.0440	−.09
X3*C1	−1.3048	0.0013	−.35

a given probability cutpoint. Otherwise, the observation is classified as a nonevent. The classification table provides several measures of predictive accuracy at various probability cutpoints.

For each probability cutpoint, the correct and incorrect columns provide the frequency of events and nonevents correctly and incorrectly classified. For example, at 0.5 probability cutpoint, the LOGISTIC procedure correctly classifies 60 events and 170 nonevents. It misclassifies 25 events and 36 nonevents. The next five columns provide predictive accuracy of the model at various cutpoint probabilities:

■ *Correct*: The overall percentage of correct classification (total frequency of correct classification/sample size)

■ *Sensitivity:* Measure of accuracy of classifying events or true positives (number of correctly classified events/total number of events)

Table 5.50 Macro LOGIST2—Profile Likelihood Confidence Interval for Adjusted Odds Ratios

Effect	Unit	Estimate	95% Confidence	Limits
X8	1.0000	1.091	1.057	1.130

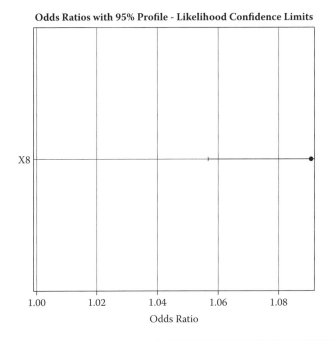

Odds Ratios with 95% Profile - Likelihood Confidence Limits

Figure 5.68 **Assessing the BLR fit: The odds ratio with 95% CI plot generated by SAS ODS graphics using the SAS macro LOGIST2.**

- *Specificity*: Measure of accuracy of classifying nonevents or true negatives (number of correctly classified nonevents/total number of nonevents)
- *False positive*: Measure of error in classification of nonevents (number of falsely classified nonevents as events/total number of events)
- *False negative*: Measure of error in classification of events (number of falsely classified events as nonevents/total number of nonevents)

The ROC (receiver operating characteristic) curve presented in Figure 5.70 provides the measure of predicted accuracy of the fitted logistic regression model. Since the ROC curve rises quickly, the fitted model has a relatively high predictive accuracy. The effect of dropping the $X8$ variable from the model is visualized by the change in the area under the curve in Figure 5.70. The higher predictive ability of this logistic regression model is confirmed by the large area under the ROC curve (0.87). The *C-statistic* estimated by the LOGISTIC procedure is equal to the area under the ROC curve. Figure 5.71 shows the trend between the percentages of correct classification, false positive, and false negative versus different cutpoint probability values. If the objective is to find the cutpoint probability that gives overall predictive

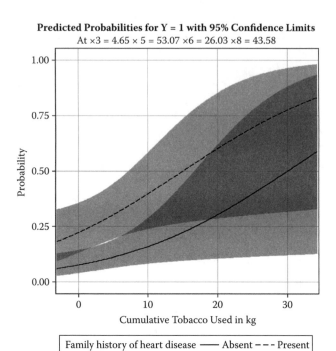

Figure 5.69 Assessing the BLR fit: Predicted probability plot for a given continuous variable generated by SAS ODS graphics using the SAS macro LOGIST2.

accuracy, select the cutpoint probability where the false-positive and false-negative curves intersect, and the overall classification percentage is high. Approximately at the cutpoint probability equal to 0.575, both false-positive and false-negative estimates are low.

Table 5.51 Macro LOGIST2—Model Adequacy Tests

Hosmer and Lemeshow Goodness-of-Fit Test				
Chi-Square	*DF*	*Pr > Chi-sq*		
6.9089	8	0.5465		
R-Square	0.3646	Max-rescaled R-Square	0.5074	
Deviance and Pearson Goodness-of-Fit Statistics				
Criterion	*Value*	*DF*	*Value/DF*	*Pr > Chi-sq*
Deviance	237.0698	280	0.8467	0.9705
Pearson	234.4148	280	0.8372	0.9780

Table 5.52 Macro LOGIST2—Classification Table

	Correct		Incorrect		Percentages				
Problem Level	Event	Non-event	Event	Non-event	Correct	Sensitivity	Specificity	False POS	False NEG
0.050	96	57	138	0	52.6	100.0	29.2	59.0	0.0
0.100	95	83	112	1	61.2	99.0	42.6	54.1	1.2
0.150	87	102	93	9	64.9	90.6	52.3	51.7	8.1
0.200	82	119	76	14	69.1	85.4	61.0	48.1	10.5
0.250	79	133	62	17	72.9	82.3	68.2	44.0	11.3
0.300	74	148	47	22	76.3	77.1	75.9	38.8	12.9
0.350	66	154	41	30	75.6	68.8	79.0	38.3	16.3
0.400	63	161	34	33	77.0	65.6	82.6	35.1	17.0
0.450	62	167	28	34	78.7	64.6	85.6	31.1	16.9
0.500	60	170	25	36	79.0	62.5	87.2	29.4	17.5
0.550	52	174	21	44	77.7	54.2	89.2	28.8	20.2
0.600	48	177	18	48	77.3	50.0	90.8	27.3	21.3
0.650	42	183	12	54	77.3	43.8	93.8	22.2	22.8
0.700	38	187	8	58	77.3	39.6	95.9	17.4	23.7
0.750	32	189	6	64	75.9	33.3	96.9	15.8	25.3
0.800	28	189	6	68	74.6	29.2	96.9	17.6	26.5
0.850	22	192	3	74	73.5	22.9	98.5	12.0	27.8
0.900	17	195	0	79	72.9	17.7	100.0	0.0	28.8
0.950	10	195	0	86	70.4	10.4	100.0	0.0	30.6
1.000	0	195	0	96	67.0	0.0	100.0	.	33.0

The estimated predicted probability scores and their confidence intervals are given in Tables 5.53 and 5.54. These predicted scores and the confidence interval estimates could be used to build risk scores for each patient, which helps to identify potentially unhealthy patients.

Significant outlier/influential observations: Table 5.55 lists several observations as influential outliers based on the larger DIFDEV statistic (DIFDEV > 4). The

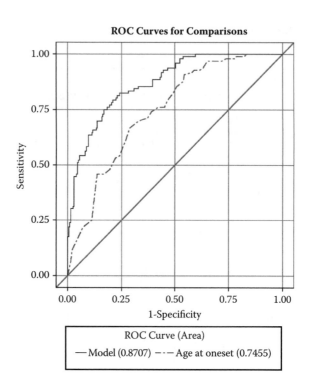

Figure 5.70 Assessing the BLR fit: Comparing ROC curves for a full model and with single continuous predictor variable generated by SAS ODS graphics using the SAS macro LOGIST2.

DIFDEV statistic measures the change in the model deviance statistic when these observations are excluded individually. These outliers also showed up in the outlier detection plot (Figure 5.72) that illustrates the leverage of each observation (hat-value). The diameter of the bubbles in the outlier detection plot is proportional to the *cbar* statistic, which is an influential statistic that quantifies the change in the parameter confidence interval estimates when each observation is excluded. Because some of the outliers have relatively larger *cbar* statistics, to investigate the impact of these outliers, this analysis should be repeated by selecting macro option #14 to exclude extreme influential points.

   Model validation: To validate the obtained logistic regression model estimates, the regression parameter estimates obtained from the training dataset ($n = 360$) are used to predict the presence of coronary artery disease for the validation dataset ($N = 162$), and the results are presented in Table 5.56. The Brier score for this validation dataset is acceptable, and 0.19 similar to the Brier score obtained from the training data (0.13) in Figure 5.73. The misclassification percentage is 27% (44 out

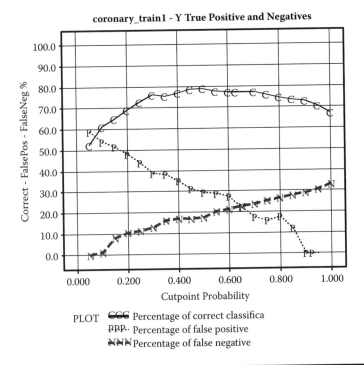

coronary_train1 - Y True Positive and Negatives

PLOT   CCC Percentage of correct classifica
PPP·· Percentage of false positive
NNN Percentage of false negative

Figure 5.71   Assessing the BLR fit: Overlay plot showing false-positive and false-negative percentages versus different cutpoint probabilities using the SAS macro LOGIST2.

of 162 is misclassified). Even though the validation results are somewhat satisfactory based on the Brier score and the ROC curves (Figure 5.73), the impact of the extreme outliers on model validation could be verified by refitting the model by excluding the outliers from the analysis.

Lift or gain charts: The cumulative lift charts are very useful informative display that portrait the success of applying predictive probability scores in identifying the events. The horizontal line in the lift charts shows the probability of selecting an event randomly from the population. For example, if you randomly select a subject, the probability that the subject is going to be diagnosed with coronary artery disease is around 0.35 (Figure 5.74). However, if we use the predicted probability health-risk scores from the logistic regression and picked the top 25% of the subjects, then there is almost 80% chance that these subjects are going to be diagnosed with coronary artery disease. Thus, the accuracy of prediction based on logistic regression is increased by 45%. The performance of the lift chart based on validation data is also very satisfactory (Figure 5.74). This finding clearly shows the success of applying logistic regression in identifying the event.

Table 5.53 Macro LOGIST2—Partial List of the Bottom 10% of the Predicted Probability and Their Confidence Intervals

ID	Coronary Heart Disease	PREDICTS	Cumulative Tobacco Used in kg	Low-Density Lipoprotein Cholesterol	Type A Behavior	BMI	Age at Onset	Estimated Probability	Lower 95% Confidence Limit	Upper 95% Confidence Limit
411	0	0	0	3.1	41	24.8	16	0.002744	.000536598	0.013908
288	0	0	0	3.69	43	27.66	19	0.003323	.000708316	0.015443
289	0	0	0.02	2.8	45	24.82	17	0.003679	.000805076	0.016639
292	0	0	0	1.71	42	22.03	16	0.003750	.000742550	0.018706
434	0	0	0	4	40	21.94	16	0.004218	.000864030	0.020322
165	0	0	0	3.08	43	22.13	16	0.004375	.000945934	0.019987
43	0	0	0	1.07	47	22.15	15	0.004457	.000925119	0.021188
14	0	0	0	1.87	49	23.63	15	0.004561	.001019574	0.020152
267	0	0	0.12	1.96	37	20.01	18	0.005678	.001051050	0.030059
433	0	0	0	1.43	42	19.38	16	0.006140	.001173612	0.031460
435	0	0	0	2.46	47	22.01	18	0.006650	.001629814	0.026717
195	0	0	0.05	2.79	46	21.62	18	0.006804	.001655947	0.027517
439	0	0	0.06	4.15	49	22.59	16	0.007050	.001809269	0.027062
290	0	0	0.05	4.61	51	23.23	16	0.008116	.002190958	0.029591
110	0	0	0	3.04	49	22.04	18	0.008274	.002188199	0.030767

Table 5.54 Macro LOGIST2—Partial List of the Top 10% of the Predicted Probability and Their Confidence Intervals

ID	Coronary Heart Disease	PREDICTS	Cumulative Tobacco Used in kg	Low-Density Lipoprotein Cholesterol	Type A Behavior	BMI	Age at Onset	Estimated Probability	Lower 95% Confidence Limit	Upper 95% Confidence Limit
126	1	1	8.6	11.17	70	33.14	59	0.99924	0.98993	0.99994
18	1	1	10.5	8.29	78	32.73	53	0.99698	0.96229	0.99977
26	1	1	4	12.42	54	23.23	42	0.99255	0.89414	0.99952
99	1	1	3.2	11.32	55	27.07	51	0.98947	0.90344	0.99894
295	1	1	2.6	7.22	71	27.87	56	0.98900	0.94935	0.99769
40	1	1	11.2	5.81	75	27.68	58	0.97879	0.86196	0.99708
19	1	1	2.6	7.46	61	29.3	62	0.96741	0.89610	0.99031
335	1	1	19.45	4.22	28	23.95	59	0.96271	0.44460	0.99880
115	1	1	31.2	3.17	47	19.4	59	0.95892	0.60462	0.99720
372	1	1	8.2	7.75	46	26.53	64	0.95825	0.86316	0.98817
192	1	1	0.7	4.9	72	35.94	49	0.94828	0.78857	0.98903
230	1	1	0	5.47	71	28.99	50	0.94434	0.81301	0.98512
245	1	1	4.6	7.4	57	28.67	60	0.94022	0.83884	0.97940
334	1	1	6.4	8.49	56	28.94	51	0.93996	0.79558	0.98437
463	1	1	0	4.82	62	14.7	46	0.93600	0.65122	0.99135

Table 5.55 Macro LOGIST2—Partial List of Influential Observations

ID	Cumulative Tobacco Used in kg	Low-Density Lipoprotein Cholesterol	Type A behavior	BMI	Age at Onset	Coronary Heart Disease	Estimated Probability	Deviance Residual	Hat-Ratio = (hat/hatmean)	Confidence Interval Displacement CBar	One Step Difference in Deviance
108	7.6	5.5	42	37.41	54	1	0.73850	0.77864	2.1135	0.03075	0.63702
115	31.2	3.17	47	19.4	59	1	0.95892	0.28966	2.0153	0.00353	0.08744
118	0.7	5.91	13	20.6	42	0	0.19447	−0.65765	11.7274	0.19225	0.62475
12	14.1	4.44	65	23.09	40	1	0.58884	1.02917	2.7898	0.08232	1.14150
129	8.8	7.41	35	29.44	60	1	0.44700	1.26901	2.3360	0.11982	1.73021
132	2	3.08	45	31.44	58	1	0.12162	2.05272	0.6150	0.17188	4.38552
142	2.78	4.89	63	19.3	25	1	0.45976	1.24664	2.1117	0.10193	1.65605
160	6.75	5.45	53	25.62	43	1	0.13188	2.01289	0.3212	0.08091	4.13262
162	27.4	3.12	66	27.45	62	1	0.62311	0.97266	10.8640	0.42149	1.36756
17	7.5	15.33	60	25.31	49	0	0.53600	−1.23925	6.6641	0.38899	1.92472
180	15	4.91	41	27.96	56	0	0.52831	−1.22592	3.7011	0.18219	1.68506
193	0.4	3.41	56	23.59	39	1	0.13355	2.00663	0.4958	0.12392	4.15049
208	4	6.65	54	28.4	60	0	0.86854	−2.01447	0.6046	0.15451	4.21259
21	1.61	1.74	74	20.92	20	1	0.39605	1.36104	2.9328	0.19014	2.04258
23	0.3	6.38	62	24.64	50	0	0.87871	−2.05406	0.6518	0.18300	4.40217
273	2.2	4.16	65	37.24	41	1	0.37410	1.40230	2.1738	0.14979	2.11624

297	0	4.19	56	23.65	42	1	0.11500	2.07983	0.2843	0.08359	4.40927
298	0.1	3.28	73	20.42	17	0	0.35915	−0.94335	2.6733	0.06300	0.95291
328	0	6.63	37	29.41	62	0	0.61833	−1.38795	2.0808	0.13831	2.06472
335	19.45	4.22	28	23.95	59	1	0.96271	0.27570	2.9829	0.00492	0.08094
338	7.28	3.56	20	26.8	58	1	0.72338	0.80476	7.5121	0.15165	0.79928
349	5.4	2.36	51	18.36	61	0	0.59512	−1.34475	2.5744	0.15846	1.96682
376	0	2.4	70	30.74	29	0	0.19656	−0.66159	2.3086	0.02339	0.46109
386	19.6	6.03	49	26.99	44	0	0.35209	−0.93167	3.2368	0.07576	0.94376
396	12.5	2.73	48	35.58	48	0	0.25194	−0.76194	2.0773	0.02870	0.60925
408	25.01	3.7	57	30.54	61	1	0.78174	0.70176	3.0142	0.03590	0.52836
414	8.2	14.16	52	28.5	55	1	0.44878	1.26587	4.9644	0.28373	1.88616
415	0.92	2.66	49	20.58	63	1	0.36745	1.41505	2.0883	0.14754	2.14990
45	0	2.99	54	46.58	17	0	0.11859	−0.50246	8.3780	0.06236	0.31482
455	1.6	7.22	36	31.5	51	1	0.57310	1.05518	2.4507	0.07605	1.18945
462	5.4	11.61	64	27.35	40	0	0.34323	−0.91698	2.8117	0.06215	0.90300
48	1.91	7.56	52	30.01	33	1	0.45100	1.26198	2.6457	0.13527	1.72786
53	0.9	9.12	56	28.64	42	1	0.12302	2.04717	0.6949	0.19232	4.38321
58	5.1	2.96	55	25.52	38	1	0.07665	2.26648	0.2996	0.13797	5.27491
82	8.14	4.93	53	45.72	53	1	0.82439	0.62148	9.7120	0.12357	0.50980
87	10.5	4.49	67	19.37	49	1	0.72691	0.79869	2.2068	0.03419	0.67210

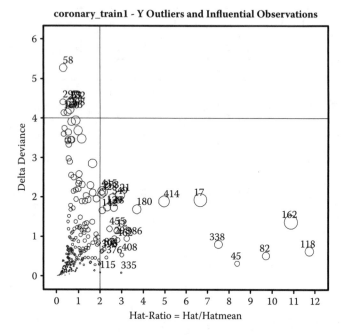

Figure 5.72   BLR Diagnostic plot: Checking for influential outliers in BLR using the SAS macro LOGIST2.

Table 5.56   Macro LOGIST2—Classification Table of Observed versus Predicted Probability at the Cutoff *p*-value: 0.5—Validation Data

Brier score = 0.19310		
Table of OBSERV by PREDICTS		
OBSERV	*PREDICTS*	
Frequency	0	1	Total
0	83	23	106
1	21	35	56
Total	104	58	162

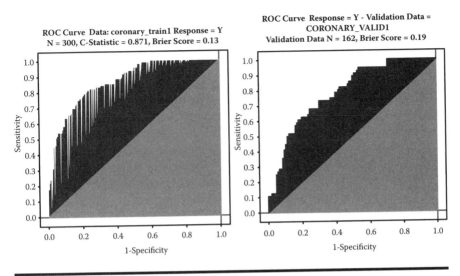

Figure 5.73 Assessing the BLR fit: Comparing the area under the ROC curves generated for the training and validation data using the SAS macro LOGIST2.

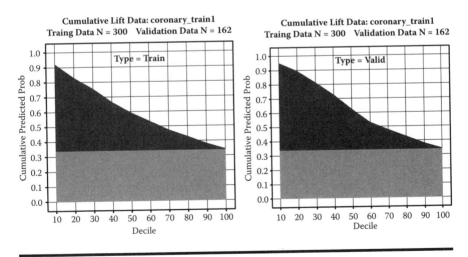

Figure 5.74 Assessing the BLR fit: Comparing the LIFT chart performances generated for the training and validation data using the SAS macro LOGIST2.

5.17 Case Study: 6 Modeling Ordinal Multiple Logistic Regression

5.17.1 Study Objectives

1. *Best-candidate model selection using AICC and SBC*: Perform sequential step-wise model selection and determine the subset where AICC and SBC are at the minimum. Then run all possible combinations of models within the optimum subset range and select the best-candidate models.

2. *Data exploration using diagnostic plots*: Check for the significance of the selected continuous predictor variable and multicollinearity among the predictors by studying simple logit and delta logit trends.

3. *Logistic regression model fitting and prediction*: Fit the logistic regression model, perform hypothesis testing on overall regression and on each parameter estimate, estimate the confidence intervals for the parameter estimates and odds ratios, predict probability scores and estimate their confidence intervals, and produce classification tables, and false positive and false negative estimates.

4. *Save _score_ and estim datasets for future use*: These two datasets are created and saved as temporary SAS datasets in the work folder. The _score_ data contains the observed variables, predicted probability scores, and confidence interval estimates. This dataset could be used as the base for developing the scorecards for each observation. Also, the parameter estimate dataset called ESTIM could be used in the LSCORE2 macro for scoring different datasets containing the same predictor variables.

5.17.2 Data Descriptions

Data name	SAS dataset CARS93	
Multiattributes	Ferank (Fuel efficiency rank): 1 = Not efficient, 2 = intermediate, 3 = fuel efficient	
	$X1$	Air Bags (0 = none, 1 = driver only, 2 = driver and passenger)
	$X2$	Number of cylinders
	$X3$	Engine size (liters)
	$X4$	HP (maximum)
	$X5$	RPM (revs per minute at maximum HP)
	$X6$	Engine revolutions per mile (in highest gear)

	X7 Fuel tank capacity (gallons)
	X8 Passenger capacity (persons)
	X9 Car length (inches)
	X10 Wheelbase (inches)
	X11 Car width (inches)
	X12 U-turn space (feet)
	X13 Rear seat room (inches)
	X14 Luggage capacity (cu ft)
	X15 Weight (pounds)
Number of observations	92
Car93: Data Source	Lock, R. H. (1993)

Step 1: Best-candidate model selection: Open the LOGIST2.SAS macro-call file in the SAS EDITOR window and click RUN to open the LOGIST2 macro-call window (Figure 5.75). Input dataset name (cars93), ordinal response variable (ferank), continuous predictor variables ($X1–X15$), model terms ($X1–X15$), and other appropriate macro-input values. To perform model selection, leave the macro field #6 blank. Follow the instruction given in Appendix 2.

The first step in best-candidate model selection is to perform sequential stepwise selection. In sequential stepwise selection, stepwise variable selection is carried forward until all the predictor variables are entered by changing the default SLENTER and SLSTAY p-values from 0.15 to 1. The ODS GRAPHICS feature in the PROC LOGISTIC, that is, to produce the overlaid ROC plot where each ROC curve represents the best model in the sequential stepwise selection, is not available with ordinal logistic regression. However, in each sequential stepwise selection step, both AIC and SC statistics are recorded, and the best subsets with the minimum AIC and minimum SC are identified (Figure 5.76). Thus, based on minimum AIC, a four-variable subset model, and, based on minimum SBC, a two-variable subset are identified as optimum subsets.

In the second step, all possible combination of multiple logistic regressions models are performed between the one-variable subset (SBC: 2 − 1) and the five-variable subsets (AIC: 4 + 1), and the best-candidate models based on delta AICC and delta SBC are identified (Figure 5.77). A four-variable model ($X15$, $X2$, $X7$, and $X14$) and a two-variable model ($X15$ and $X2$)

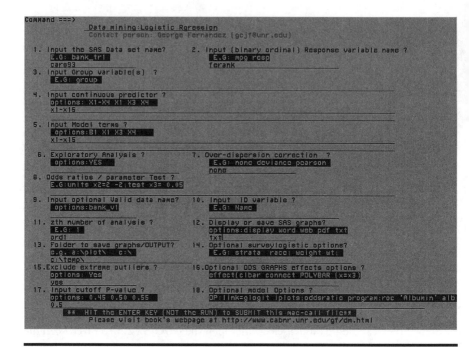

Figure 5.75 Screen copy of LOGIST2 macro-call window showing the macro-call parameters required for performing ordinal LR.

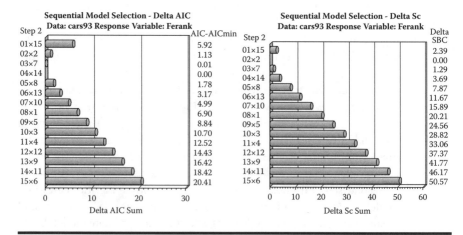

Figure 5.76 Delta AICC and Delta SC plots generated by SAS macro LOGIST2: Comparing Delta AICC and SC in sequential model-building steps.

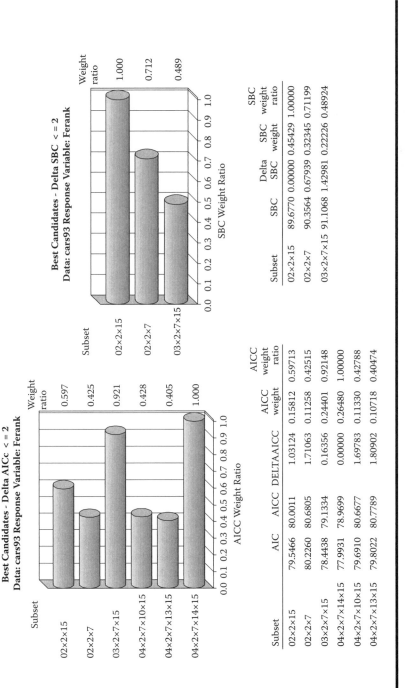

Best Candidates - Delta AICc < = 2
Data: cars93 Response Variable: Ferank

Subset	AIC	AICC	DELTAAICC	AICC weight	AICC weight ratio
02×2×15	79.5466	80.0011	1.03124	0.15812	0.59713
02×2×7	80.2260	80.6805	1.71063	0.11258	0.42515
03×2×7×15	78.4438	79.1334	0.16356	0.24401	0.92148
04×2×7×14×15	77.9931	78.9699	0.00000	0.26480	1.00000
04×2×7×10×15	79.6910	80.6677	1.69783	0.11330	0.42788
04×2×7×13×15	79.8022	80.7789	1.80902	0.10718	0.40474

Best Candidates - Delta SBC < = 2
Data: cars93 Response Variable: Ferank

Subset	SBC	Delta SBC	SBC weight	SBC weight ratio
02×2×15	89.6770	0.00000	0.45429	1.00000
02×2×7	90.3564	0.67939	0.32345	0.71199
03×2×7×15	91.1068	1.42981	0.22226	0.48924

Figure 5.77 Delta AICC and Delta SC plots generated by SAS macro LOGIST2: Comparing Delta AICC and SC values for the best candidates based on all possible model selection within optimum subsets in performing ordinal LR.

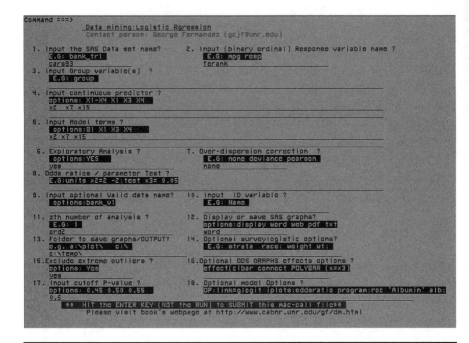

Figure 5.78 Screen copy of LOGIST2 macro-call window showing the macro-call parameters required for performing exploratory diagnostic plots in performing ordinal LR.

were identified as the best models based on the delta AICC and delta SC criteria, respectively. Based on the best-candidate model selection, the three-variable model ($X15$, $X2$, and $X7$) was chosen.

Step 2: Open the LOGIST2.SAS macro-call file in the SAS EDITOR window and click RUN to open the LOGIST2 macro-call window (Figure 5.78). Use the suggestions provided in the help file (Appendix 2) to input the appropriate macro input values and fit the ordinal LR. Only selected output and graphics produced by the LOGIST2 macro are shown in the figure.

Exploratory analysis/Diagnostic plots: Input dataset name (cars93), ordinal response variable (ferank), continuous predictor variables ($X15$, $X2$, $X7$) model terms ($X15$, $X2$, $X7$) and other appropriate macro-input values. To perform data exploration and to create regression diagnostic plots, input YES in macro field #6. Once the LOGIST2 macro has been submitted, overlay plots of simple and delta logit for each predictor variable, and outputs from forward selection model selection are produced.

Overlay plots of simple and partial delta logit plots for a three-variable subset are presented in Figures 5.79–5.81. The $X2$ (number of cylinder) variable

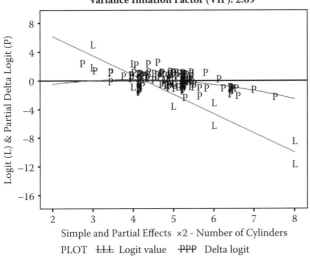

Logit & Partial Delta Logit Plots Resp = Ferank vs Number of Cylinders
P-Values: Linear: 0.0106 Quadratic: 0.7425 Linear Odds Ratio = 0.2369
Variance Inflation Factor (VIF): 2.69

Simple and Partial Effects ×2 - Number of Cylinders
PLOT LLL Logit value PPP Delta logit

Haxis: Simple logit plot: ×2, Partial delta
logit plot: Partial ×2 + ×2 mean

Figure 5.79 Logistic regression diagnostic plot using SAS macro LOGIST2. Overlay plot of simple logit and partial delta logit plots for *number of cylinder* is shown.

showed a significant negative trend in both simple and delta logit plots. The quadratic effect was not statistically significant. No evidence of multicollinearity was observed.

The X7 (fuel tank capacity) showed a negative trend in simple logit plots. But, the partial effects of X7 was not significant at the 0.05 level. A moderate degree of multicollinearity was evident in the full model since the delta logit data points were clustered near the mean of X7. The observed quadratic effect was not significant.

The X15 (weight of the vehicle) showed a significant negative trend in both simple and delta logit plots. A moderate level of multicollinearity was observed in the full model since the delta logit data points clustered near the center of the partial logit regression line of X15.

When tested for all possible combinations two-factor interactions among the three X2, X7, and X15 variables, only the interaction between X7 and X15 was statistically significant (Figure 5.82). Therefore, based on the exploratory graphical analysis, three linear terms (X2, X7, and X15) and one interaction term (X7*X15) were identified as significant models terms; these will be further tested in the next step.

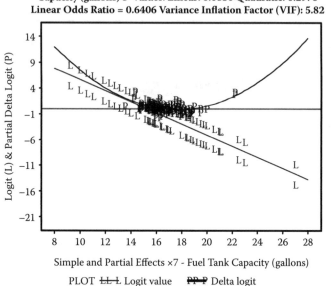

Logit & Partial Delta Logit Plots Resp = Ferank vs Fuel Tank
Capacity (gallons) P-Values: Linear: 0.0830 Quadratic: 0.2976
Linear Odds Ratio = 0.6406 Variance Inflation Factor (VIF): 5.82

Simple and Partial Effects ×7 - Fuel Tank Capacity (gallons)

PLOT ⌐L-L-L Logit value   P-P-P Delta logit

Haxis: Simple logit plot: ×7, Partial delta logit plot: Partial ×7 + ×7 mean

Figure 5.80 Logistic regression diagnostic plot using SAS macro LOGIST2. Overlay plot of simple logit and partial delta logit plots for *fuel tank capacity* is shown.

Step 3: *Fitting binary logistic regression*

Open the LOGIST2.SAS macro-call file in the SAS EDITOR window and click RUN to open the LOGIST2 macro-call window (Figure 5.83). Input training dataset name (cars93), ordinal response variable (Ferank), three significant continuous predictor variables ($X2$, $X7$, $X15$) and model terms ($X2$, $X7$, $X15$, $X7*X15$). To skip data exploration and create regression diagnostic plots, leave macro field #6 blank. To exclude extreme outliers, input YES to macro field #14. Input other appropriate macro parameters and submit the LOGIST2 macro to fit the binary logistic regression. Only selected output and graphics produced by the LOGIST2 macro are shown in the figure.

The characteristics of the training dataset, number of observations in the dataset, type of model fit, and the frequency of ordinal events are given in Table 5.57. The LOGIST2 macro modeled the probabilities that are cumulated over the lower-order values.

Table 5.58 shows the model convergence status and statistics for testing the overall model significance. The *AIC* and *SC* statistics for the intercept

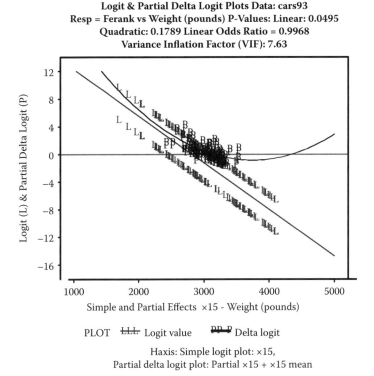

Logit & Partial Delta Logit Plots Data: cars93
Resp = Ferank vs Weight (pounds) P-Values: Linear: 0.0495
Quadratic: 0.1789 Linear Odds Ratio = 0.9968
Variance Inflation Factor (VIF): 7.63

Simple and Partial Effects ×15 - Weight (pounds)

PLOT ᴸᴸᴸ Logit value ᴾᴾ ᴾ Delta logit

Haxis: Simple logit plot: ×15,
Partial delta logit plot: Partial ×15 + ×15 mean

Figure 5.81 Logistic regression diagnostic plot using SAS macro LOGIST2. Overlay plot of simple logit and partial delta logit plots for automobile weight is shown.

and the covariates are useful for selecting the best model from different logistic models. Lower values are desirable. The *–2 Log L* criteria is the *–2 Log likelihood* statistics for models with only an intercept and the full model.

The results of testing the global null hypothesis that all regression coefficients are equal to zero are given in Table 5.59. All three tests indicate that the overall ordinal logistic model is highly significant and that at least one of the parameter estimates is significantly different from zero.

The score chi-square for testing the proportional odds assumption performs a test of the parallel lines assumption with 0.4008, which is not significant with respect to a chi-square distribution with four degrees of freedom. This indicates that the proportional odds assumption is reasonable (Table 5.60).

The parameter estimates, their *se* using maximum likelihood methods, and significance levels of each variable using the Wald *chi-square* are given in Table 5.61.

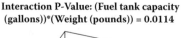

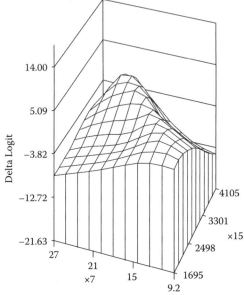

* X-axis ×15 = Weight (pounds)
* Y-axis ×7 = Fuel tank capacity (gallons)

Figure 5.82 Binary logistic regression diagnostic plot using SAS macro LOGIST2: Three-dimensional plot for detecting significant interaction between two continuous variables.

To build proportional odds model, let us consider the probabilities of fuel efficiency ranking:

$\theta_1 = \pi_1$, probability of Ferank = 3 fuel efficient

$\theta_2 = \pi_1 + \pi_2$, probability of Ferank = 3 or Ferank = 2, moderately and above fuel efficient

where
π_1 = Probability of Ferank = 3 fuel efficient
π_2 = Probability of Ferank = 2 moderately fuel efficient
π_3 = Probability of Ferank = 1 fuel inefficient

Then we can construct the cumulative logits:

$$\text{logit}(\theta_1) = \log(\theta_1/(1 - \theta_1)) = \log(\pi_1/(\pi_2 + \pi_3)),$$

$$\text{logit}(\theta_2) = \log(\theta_2/(1 - \theta_2)) = \log((\pi_1 + \pi_2)/(\pi_3)),$$

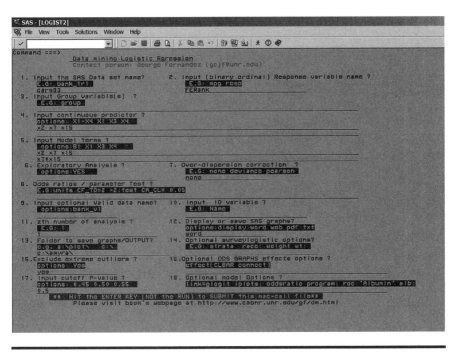

Figure 5.83 Screen copy of LOGIST2 macro-call window showing the macro-call parameters required for performing the final ordinal LR model-building step.

Table 5.57   Macro LOGIST2—Model Information

Dataset	WORK.DATA_I
Response variable	FERANK
Number of response levels	3
Model	Cumulative logit
Optimization technique	Fisher's scoring
Number of observations read	93
Number of observations used	90
FERANK	Total frequency
3	18
2	32
1	40
Probabilities modeled are cumulated over the lower-order values	

Table 5.58   Macro LOGIST2—Model-Fit Statistics

Convergence Criterion (GCONV=1E–8) Satisfied		
Criterion	*Intercept Only*	*Intercept and Covariates*
AIC	192.995	77.630
SC	197.995	92.629
–2 Log L	188.995	65.630

Table 5.59   Macro LOGIST2—Testing Global Null Hypothesis: BETA = 0

Test	Chi-square	DF	Pr > Chi-sq
Likelihood ratio	123.3649	4	<.0001
Score	68.9134	4	<.0001
Wald	38.7851	4	<.0001

The estimated logistic regression model for estimating the proportional odds logistic regression model based on ordinal logistic regression parameters is reported in Table 5.61:

log (π_1 : probability of highly fuel efficient)/probability of [π_2: moderately fuel efficient + π_3 : fuel inefficient]) = 43.2932 –1.4399X2–1.8868 X7–0.0117X15 + 0.000505X7*X15.

log (probability of [π_1:highly fuel efficient + π_2: moderately fuel efficient])/(π_3: fuel inefficient) = 48.8086 –1.4399X2–1.8868 X7–0.0117X15+ 0.000505X7*X15

These parameter estimates calculate the change in the proportional log odds of fuel efficiency when one unit changes in a given predictor variable while all other variables are held constant. All three partial linear logistic regression parameter estimates

Table 5.60   Macro LOGIST2—Score Test for the Proportional Odds Assumption

Chi-square	DF	Pr > Chi-sq
0.4008	4	0.9824

Table 5.61   Macro LOGIST2—Analysis of Maximum Likelihood Estimates and Their Significance Levels—Final Model

Parameter	DF		Estimate	Standard Error	Wald Chi-Square	Pr > Chi-sq	Standardized Estimate	Label
Intercept	3	1	43.2932	14.9741	8.3590	0.0038		Intercept: FERANK=3
Intercept	2	1	48.8086	15.5024	9.9128	0.0016		Intercept: FERANK=2
X2		1	−1.4399	0.5302	7.3769	0.0066	−1.0400	Number of cylinders
X7		1	−1.8868	0.9073	4.3245	0.0376	−3.2130	Fuel tank capacity (gallons)
X15		1	−0.0117	0.00541	4.6728	0.0306	−3.7797	Weight (pounds)
X7*X15		1	0.000505	0.000295	2.9399	0.0864		Fuel tank capacity (gallons)* Weight (pounds)

are negative and statistically significant, confirming that an increase in the number of cylinders or increase in fuel tank capacity or an increase in the weight of the vehicles all decrease the fuel efficiency rankings. For any observation in the data, the probability of the event being fuel efficient could be predicted by the ordinal LR parameter estimates.

By default, the LOGIST2 macro outputs the odds-ratio estimate for one-unit increase in all predictor variables while holding all other variables constant. These odds ratios and their confidence intervals are computed by exponentiating the parameter estimates and their confidence intervals. Table 5.62 shows the odds-ratio estimates and the profile likelihood confidence interval estimates. The interpretation of the odds ratio is valid for a linear continuous or binary predictor when a one-unit change in the predictor variable is relevant. If the interaction terms are

Table 5.62   Macro LOGIST2—Profile Likelihood Confidence Interval for Adjusted Odds Ratios

Effect	Unit	Estimate	95% Confidence Limits	
X2: Number of cylinders	1	0.237	0.073	0.623

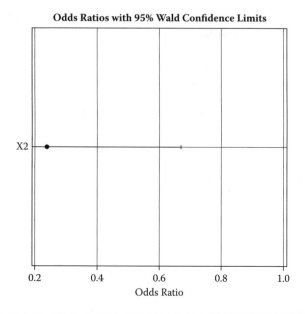

Figure 5.84   Assessing the BLR fit: Odds ratio plot generated by SAS ODS graphics using the SAS macro LOGIST2.

included in the model, the interpretation of the odds ratio for a given variable is not valid.

An odds-ratio value equal to one implies no change in the odds of the event when you increase the predictor variable by one unit. The chance of the fuel efficiency rankings increasing multiplicatively happens when the odds ratio of a variable is greater than one, and vice versa. For example, a one-unit increase in the number of cylinders while holding all other predictors constant lowers the odds of being fuel efficient by 0.073 to 0.623 times (Figure 5.84). Because the variable $X7$ and $X5$ are involved in two-factor interaction, the odds-ratio estimates are not available. Increasing the number of cylinders by one unit has no effect on their fuel efficiency rankings in fuel-efficient and fuel-inefficient cars with a 16 gallon fuel tank and 3050 lb weight. However, increasing the number of cylinders by one unit has a significant negative effect on the moderately fuel efficient average-size cars with 16 gallon fuel tank and 3050 lb weight (Figure 5.85).

The Hosmer and Lemeshow goodness-of-fit test using the Pearson chi-square test results for checking for model adequacy, and the classification table used in computing the ROC curves, are not available when performing ordinal logistic regression. The estimated R^2 and the max-rescaled R^2 statistics computed from the *−2 log likelihood estimates* are high (Table 5.63). Both deviance and Pearson goodness-of-fit

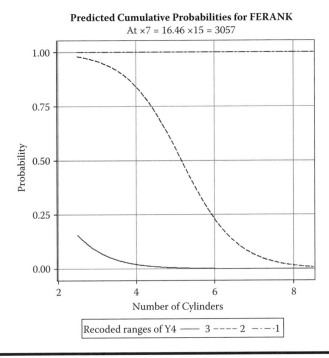

Predicted Cumulative Probabilities for FERANK
At ×7 = 16.46 ×15 = 3057

Figure 5.85 **Assessing the BLR fit: Predicted probability plot for a given continuous variable generated by SAS ODS graphics using the SAS macro LOGIST2.**

statistics are not significant, indicating that the variance of the binary response variable does not exceed the expected nominal variance. Thus, overdispersion is not a problem in this ordinal logistic regression model (Table 5.63).

The estimated predicted probability scores and their confidence intervals are given in Tables 5.64 and 5.65. These predicted scores and the confidence interval estimates could be used to build scorecards for each automobile, which helps to identify the top fuel-efficient and fuel-inefficient automobiles.

Table 5.63 **Macro LOGIST2—Model Adequacy Tests: R^2, Max-Rescaled R^2, and Overdispersion Diagnostic Tests**

R^2	0.7461	Max-rescaled R^2	0.8502	
Deviance and Pearson Goodness-of-Fit Statistics for Testing Overdispersion				
Criterion	*Value*	*DF*	*Value/DF*	*Pr > Chi-sq*
Deviance	65.6301	166	0.3954	1.0000
Pearson	103.7169	166	0.6248	1.0000

Table 5.64   Macro LOGIST2—Partial List of the Bottom 10% (Fuel-Inefficient Automobiles) of the Predicted Probability and Their Confidence Intervals

ID	Fuel Efficiency Rank	Number of Cylinders	Fuel Tank Capacity (gallons)	Weight (pounds)	Estimated Probability	Lower 95% Confidence limit	Upper 95% Confidence limit
25	1	8	20	4055	.000004092	.000000050	.000333426
24	1	8	20	3950	.000004835	.000000066	.000356268
5	1	8	20	3935	.000004951	.000000068	.000360561
8	1	8	22.5	4000	.000006229	.000000038	.001024365
12	1	8	23	3910	.000006703	.000000036	.001240175
4	1	8	18	3620	.000009186	.000000160	.000525839
10	1	8	20	3380	.000011960	.000000158	.000902270
9	1	6	20	4100	.000067848	.000001780	.002580449
21	1	6	19.8	3805	.000107668	.000004966	.002329254

Table 5.65   Macro LOGIST2—Partial List of the Top 10% (Top Fuel-Efficient Automobiles) of the Predicted Probability and Their Confidence Intervals

ID	Fuel Efficiency Rank	Number of Cylinders	Fuel Tank Capacity (gallons)	Weight (pounds)	Estimated Probability	Lower 95% Confidence Limit	Upper 95% Confidence Limit
93	3	3	10.6	1695	1.00000	0.99964	1.00000
90	3	3	9.2	2045	1.00000	0.99943	1.00000
91	3	3	10.6	1965	0.99999	0.99932	1.00000
87	3	4	10	1845	0.99999	0.99832	1.00000
89	3	4	11.9	2055	0.99987	0.99437	1.00000
93	3	3	10.6	1695	0.99973	0.97729	1.00000
76	3	4	11.9	2285	0.99951	0.98997	0.99998
73	2	4	12.4	2240	0.99944	0.98934	0.99997
81	3	4	11.9	2345	0.99931	0.98795	0.99996

Significant outlier/influential observations: Diagnostic statistics useful in identifying influential outliers are not available when performing ordinal logistic regression.

Lift or gain charts: Cumulative lift charts are very useful informative display that portrait the success of applying predictive probability scores in identifying the binary events. However, these charts are is not meaningful when you have an ordinal data.

5.18 Summary

The methods of using supervised predictive models to predict continuous and binary response variables with user-friendly SAS macro applications have been covered in this chapter. The graphical methods of performing diagnostic and exploratory analysis, model assessment and validation, detecting violation of model assumptions, and producing lift charts have all been covered here. The steps involved in employing user-friendly SAS macro applications REGDIAG2 for performing regression modeling and validation, LOGIST2 for performing binary logistic regression and validation, LIFT2 for generating lift charts and IF-THEN analysis, and RSCORE2 (regression) and LSCORE2 (logistic) for predicting scores from new dataset have been presented.

References

1. Neter, J., Kutner, M. H., Nachtsheim, C. J., and Wasserman, W. *Applied Linear Regression Models*. Richard D. Irwin, 1996, chap. 1–5.
2. Montgomery, D. C. and Peck, E. A., *Introduction to Linear Regression Analysis*, 2nd ed. John Wiley & Sons, New York, 1992, chap. 1–4.
3. Freund, R. J. and Littell, R. C., *SAS System for Regression*, 1st ed., SAS Institute, Cary, NC, 1986.
4. SAS Institute, Regression with quantitative and qualitative variables. In SAS ONLINE Documentation at http://support.sas.com/onlinedoc/913/getDoc/en/statug.hlp/reg_sect54.htm (last accessed 06-21-09).
5. Fernandez, G. C. J., Detection of model specification, outlier and multicollinearity in multiple regressions using partial regression/residual plots. P.1246–1251. SAS Users Group International Conference, San Diego, CA, 1997. http://www2.sas.com/proceedings/sugi22/STATS/PAPER269.PDF.
6. Mallows, C. L. 1986. Augmented partial residual plots. *Technometrics* 28: 313–319.
7. Sall, J. 1990. Leverage plots for general linear hypothesis. *The American Statistician* 44: 308–315.
8. Stine, R. A. 1995. Graphical Interpretation of variance inflation factors. *The American Statistician* 49: 53–56.
9. SAS Institute. Model Selection Methods: The REG Procedure. In SAS ONLINE Documentation at http://support.sas.com/onlinedoc/913/getDoc/en/statug.hlp/reg_sect30.htm (last accessed 06-21-09).
10. Neter, J., Kutner, M. H., Nachtsheim, C. J., and Wasserman, W., *Applied Linear Regression Models*. Richard D. Irwin, 1996, chap. 8.

11. Tibshirani, R. 1996. Regression shrinkage and selection via the Lasso. *Journal of the Royal Statistical Society Series B*, 58: 267–288.
12. Cohen, R. A., Introducing the GLMSELECT Procedure for Model Selection, SUGI 31 proceedings, 2006, http://www2.sas.com/proceedings/sugi31/207-31.pdf.
13. SAS Institute, The GLMSELECT procedure, http://support.sas.com/rnd/app/papers/glmselect.pdf.
14. Neter, J., Kutner, M. H., Nachtsheim, C. J., and Wasserman, W. *Applied Linear Regression Models*. Richard D. Irwin, 1996, chap. 3.
15. SAS Institute, SAS/ETS Users guide: The AUTOREG Procedure. In SAS ONLINE Documentation at http://support.sas.com/91doc/getDoc/etsug.hlp/autoreg_sect9.htm (last accessed 06-21-09).
16. SAS Institute. The REG Procedure: Influential diagnostics, In SAS ONLINE Documentation at http://support.sas.com/onlinedoc/913/getDoc/en/statug.hlp/reg_sect39.htm (last accessed 06-21-09).
17. Neter, J., Kutner, M. H., Nachtsheim, C. J., and Wasserman, W., *Applied Linear Regression Models*. Richard D. Irwin, 1996, chap. 10.
18. SAS Institute, The MIXED Procedure: REPEATED statement In SAS ONLINE Documentation at http://support.sas.com/onlinedoc/913/getDoc/en/statug.hlp/mixed_sect18.htm#stat_mixed_mixedrandom (last accessed 06-21-09).
19. DAgostino, R. B., Belanger, A., and DAgostino, R. B. Jr., A suggestion for using powerful and informative tests of normality, *The American Statistician*, 44: 316–321, 1990.
20. SAS Institute, GENMOD procedure: Overview, In SAS ONLINE Documentation at http://support.sas.com/onlinedoc/913/getDoc/en/statug.hlp/genmod_sect1.htm (last accessed 06-21-02).
21. Zelterman D., *Models for Discrete Data*, Oxford Science Publications, Cladrendon Press, Oxford, U.K., 1999, chap. 3.
22. SAS Institute, *Logistic Regression Examples Using the SAS Systems*, version 6, 1st ed., Cary, NC, 1995.
23. SAS Institute, The LOGISTIC Procedure: Overview, In SAS ONLINE Documentation at http://support.sas.com/onlinedoc/913/getDoc/en/statug.hlp/logistic_sect1.htm (last accessed 06-21-09).
24. SAS Institute, The LOGISTIC Procedure: Odds Ratio Estimation In SAS ONLINE Documentation at http://support.sas.com/onlinedoc/913/getDoc/en/statug.hlp/logistic_sect31.htm (last accessed 06-21-02).
25. SAS Institute, The LOGISTIC Procedure: Confidence Intervals for Parameters, In SAS ONLINE Documentation at http://support.sas.com/onlinedoc/913/getDoc/en/statug.hlp/logistic_sect30.htm (last accessed 06-21-09).
26. SAS Institute, The LOGISTIC Procedure: Model fitting. In SAS ONLINE Documentation at http://support.sas.com/onlinedoc/913/getDoc/en/statug.hlp/logistic_sect27.htm (last accessed 06-21-09).
27. SAS Institute, The LOGISTIC Procedure: Score Statistics and Tests. In SAS ONLINE Documentation at http://support.sas.com/onlinedoc/913/getDoc/en/statug.hlp/logistic_sect29.htm (last accessed 06-21-09).
28. SAS Institute, The LOGISTIC Procedure: The Hosmer–Lemeshow Goodness-of-Fit Test, In SAS ONLINE Documentation at http://support.sas.com/onlinedoc/913/getDoc/en/statug.hlp/logistic_sect36.htm (last accessed 06-21-09).

29. SAS Institute, Logistic Regression Examples Using the SAS Systems, version 6, 1st ed., Cary, NC, 1995, chap. 7.
30. SAS Institute, The LOGISTIC Procedure: Generalized Coefficient of Determination, In SAS ONLINE Documentation at http://support.sas.com/onlinedoc/913/getDoc/en/statug.hlp/logistic_sect28.htm (last accessed 06-21-02).
31. SAS Institute, The LOGISTIC Procedure: Classification Table, In SAS ONLINE Documentation at http://support.sas.com/onlinedoc/913/getDoc/en/statug.hlp/logistic_sect34.htm (last accessed 06-21-02).
32. SAS Institute, The LOGISTIC Procedure: Receiver Operating Characteristic Curves, In SAS ONLINE Documentation at http://support.sas.com/onlinedoc/913/getDoc/en/statug.hlp/logistic_sect37.htm (last accessed 06-21-02).
33. SAS Institute, *Logistic Regression Examples Using the SAS Systems*, version 6, 1st ed., Cary, NC, 1995, chap. 6.
34. SAS Institute, The LOGISTIC Procedure: Overdispersion In SAS ONLINE Documentation at http://support.sas.com/onlinedoc/913/getDoc/en/statug.hlp/logistic_sect35.htm (last accessed 06-21-09).
35. SAS Institute, The REG Procedure: Overview. In SAS ONLINE Documentation at http://support.sas.com/onlinedoc/913/getDoc/en/statug.hlp/reg_sect1.htm (last accessed 06-21-09).
36. SAS Institute, The GLM Procedure: Overview. In SAS online Documentation at http://support.sas.com/onlinedoc/913/getDoc/en/statug.hlp/glm_sect1.htm (last accessed 06-21-09).
37. SAS Institute, PROC LOGISTIC Syntax. In SAS ONLINE Documentation at http://support.sas.com/91doc/getDoc/statug.hlp/logistic_sect1.htm (last accessed 06-21-09).
38. Lock, R. H., New car data, *Journal of Statistics Education*, 1: 1, 1993.
39. Rousseauw, J., du Plessis, J., Benade, A., Jordaan, P., Kotze, J., and Ferreira, J., Coronary risk factor screening in three rural communities, *South African Medical Journal*. 64: 430–436, 1983.

Chapter 6

Supervised Learning Methods: Classification

6.1 Introduction

The goal of developing supervised classification models is to fit a model or decision tree that will correctly associate the predictor variables with the nominal group levels. Classification models use nominal response to approximate the probability of class membership as a function of the predictor variables. Decision tree analysis is one of the popular tools used in data mining to predict the membership of a group response by means of a decision tree. In general, classification models allow the analysts to determine the form of the relationship between the nominal response and the predictor variables, and further investigate which predictor variables are significantly associated with categorical response. Two supervised classification model techniques, discriminant analysis (DA) and chi-squared automatic interaction detection methods (CHAID), are discussed in this chapter.

Discriminant analysis, a multivariate statistical technique, assigns subjects (cases, observations) to predetermined groups or categories. The main purpose of discriminant analysis is to predict membership in two or more mutually exclusive predetermined nominal group levels from a set of numerical predictors.

Classification or decision tree analysis is used to predict the membership of cases or observations in the levels of a categorical response from their measurements on one or more predictor variables and then outputs a decision tree. The flexibility of classification trees makes them a very attractive supervised classification option. When stringent distributional assumptions of classical statistical methods are met, the traditional methods may be preferable. However, as an exploratory technique, or as a technique of last resort when traditional methods fail, the potentials of classification trees are unsurpassed.

A brief nonmathematical description and the application of these supervised classification methods are given in this chapter. Readers are encouraged to refer to the following statistics books for a mathematical account of DA: Sharma[1] and Johnson and Wichern;[2] for CHAID, Breiman et al.[3]

6.2 Discriminant Analysis

Discriminant analysis (DA), a multivariate statistical technique, is commonly used to build a predictive or descriptive model of group discrimination based on observed predictor variables and to classify each observation into one of the predetermined group levels. In DA, multiple quantitative attributes are used to discriminate single mutually exclusive classification variables. DA is different from the cluster analysis because prior knowledge of the group membership is required.

DA is a very powerful tool for selecting the predictor variables that allow the discrimination between different group levels, and for classifying cases into different group levels with a better than chance accuracy. The common objectives of DA are:

- To investigate differences between groups of observations
- To discriminate between groups effectively
- To identify important discriminating variables
- To perform hypothesis testing on the differences between the expected groupings
- To classify new observations into preexisting groups

Based on the objectives of the analysis, DA can be classified into three types: stepwise discriminant analysis (SDA), canonical discriminant analysis (CDA), and discriminant function analysis (DFA). SDA is an exploratory component in the discriminant step, where the important discriminating variables are selected in a stepwise fashion. When group membership is known in advance and the purpose of the analysis to describe and highlight the major group differences, CDA is appropriate. In DFA, the focus is on classifying cases into predefined groups or to predict group membership on the basis of predictor variable measures. The statistical theory, methods, and the computational aspects of SDA,[4–6] CDA,[4,5,7] and DFA[4,5,8] are presented in detail in the literature.

6.3 Stepwise Discriminant Analysis

It is an exploratory component in DA commonly used to identify a subset of predictor variables that maximize discrimination among group members. The predictor variables within each group levels are assumed to have a multivariate normal distribution with a common covariance matrix. The SDA analysis is a useful exploratory component to either CDA or DFA. In SDA, continuous predictor variables are chosen to enter or leave the model based on the significance level of an F-test or squared partial correlation from an analysis of covariance, where the variable is already chosen, the group response acts as predictors, and the variable under consideration is considered as the response variable. A moderate significance level, in the range of 0.10 to 0.25, often performs better than the use of a much larger or a much smaller significance level in the model selection.[6]

Three types of SDA methods are available in the SAS systems:[6]

1. *Forward selection* begins as a null model. At each step, the predictor variable that contributes most to the discriminatory power of the model as measured by Wilks' lambda, the likelihood ratio criterion is selected. Once a variable is selected, it cannot be removed from the model. When none of the unselected variables meet the entry criterion, the forward selection process stops.
2. *Backward elimination* begins as a full model. At each step, the predictor variable that contributes least to the discriminatory power of the model as measured by Wilks' lambda is removed. Once a variable is excluded, it cannot be entered into the model. When all remaining variables meet the criterion to stay in the model, the backward elimination process stops.
3. *Stepwise selection* begins, similar to forward selection, as a null model. At each step, the significance of the previously entered predictor variables is compared. The variable that contributes least to the discriminatory power of the model is removed. Otherwise, the variable that is not in the model that contributes most to the discriminatory power is entered. When all variables in the model meet the criterion to stay and none of the other variables meet the criterion to enter, the stepwise selection process stops.

Using stepwise selection methods to evaluate the relative importance of predictor variables or selecting predictor variables for finding the best-fitted DA is not always guaranteed to give the best results. Limitations in model selection methods include the following:

■ Only one variable can be entered into the model at each step. Thus, all possible combinations of predictors are not evaluated.
■ The selection process does not take into account the relationships between variables that have not yet been selected. Thus, some important variables could be excluded in the selection process.
■ Wilks' lambda may not be the best criterion for evaluating the discriminatory power in a specific application.

The problems inherent with stepwise methodologies as outlined earlier can be serious, especially in small samples. Reducing the number of predictor variables to a manageable size is a useful strategy in the exploratory analysis stage. It is important to exclude highly correlated redundant predictor variables if they do not have discriminatory power. Therefore, using SDA as a stand-alone DA method is not recommended. However, SDA can be a treated as a valuable exploratory tool in excluding redundant variables for building a successful discrimination model.

6.4 Canonical Discriminant Analysis

In CDA, linear combinations of the quantitative predictor variables that summarize between-group variation and provide maximal discrimination between the groups are extracted. The methodology used in CDA is very similar to a one-way multivariate ANOVA (MANOVA). In MANOVA, the goal is to test for equality of the mean vector across class levels, while in CDA, the significance of continuous predictor variables in discriminating categorical response is investigated. Thus, the CDA can be conceptualized as the inverse of multivariate MANOVA, and all assumptions for MANOVA apply to CDA.

6.4.1 Canonical Discriminant Analysis Assumptions

■ *Multivariate normal distribution:* It is a requirement for performing hypothesis testing in CDA. It is assumed that the predictor variables within group have a multivariate normal distribution. The validity of multivariate normality assumptions can be performed by first standardizing the predictors within each group level and testing for significant Mardia's multivariate kurtosis.[9] Significant departure from multivariate normality can be examined visually in a Q-Q plot.[9] The results of CDA could be misleading if multivariate assumptions are severely violated. When some of the predictor variables are heavily skewed, performing a Box-Cox-type transformation can reduce the severity of the problem.[9]

■ *Homogeneity of variance–covariance:* It is assumed that the variance–covariance matrices of predictors are homogeneous across groups. Minor departures from homogeneity of variance–covariance are not that important; however, before accepting the results of CDA, it is probably a good idea to verify the equality of within-groups variance–covariance matrices. The problem with the heterogeneous variance–covariance can also be confounded by the lack of multivariate normality. The Box-Cox transformation recommended for reducing the impact of departure from multivariate normality also indirectly corrects the unequal variance–covariance problems. The SAS DISCRIM procedure provides a method for testing the homogeneity of within-error variance-covariance based on the unbiased likelihood ratio statistic adjusted by the Bartlett correction.

■ *Multicollinearity among the predictors:* Another assumption of CDA is that the predictor variables that are used to discriminate group members are not highly correlated. If any one of the predictors is totally dependent on the other variables, then the matrix is said to be ill-conditioned. The results of CDA become unstable and unpredictable when severe multicollinearity is present.

6.4.2 Key Concepts and Terminology in Canonical Discriminant Analysis

■ *Canonical discriminant function (CDF):* It is a latent variable that is created as a linear combination of predictor variables, such that $CDF_1 = a + b_1x_1 + b_2x_2 + \cdots + b_nx_n$, where the b's are raw discriminant coefficients, the x's are predictor variables, and a is a constant. The product of the unstandardized coefficients with the observations yields the discriminant scores. The group centroid is the mean value for the discriminant scores for a given group level. The CDF is considered analogous to multiple regressions, but the b's are discriminant coefficients that maximize the distance between the means of the group members. The first CDF always provides the most overall discrimination between groups; the second CDF provides second most, and so on. The process of extracting CDFs can be repeated until the number of CDFs equals the number of original variables or the number of classes minus one, whichever is smaller. Moreover, the CDFs are independent or orthogonal, similar to principal components; that is, their contributions to the discrimination between groups do not overlap. The CDF score is the value resulting from applying a CDF to the data for a given observation.

■ *Squared canonical correlation (Rc²):* It is the measure of multiple correlations between the discriminating group and the CDF. The first canonical correlation is at least as large as the multiple correlations between the groups and any of the original variables. The second canonical correlation is obtained by finding the linear combination uncorrelated with the first canonical variable that has the highest possible multiple correlations with the groups. If the original variables have high within-group correlations, the first canonical correlation can be large even if all the multiple correlations are small. In other words, the first canonical variable can show substantial differences between the classes, even if none of the original variables do.

■ *MANOVA test of group mean differences:* The *F test* of Wilks lambda shows which variables contributions are significant. Wilks lambda is also used in discriminant analysis to test the significance of the CDFs as a whole. One can test the number of CDFs that add significantly to the discrimination between groups. Only those found to be statistically significant should be used for interpretation; nonsignificant functions (roots) should be ignored. The variables should have an approximately multivariate normal distribution within each class with a common covariance matrix in order for the MANOVA *p*-values to be valid.

■ *Structure coefficients:* They provide the correlations of each predictor variable with each CDF. These simple Pearson correlations are called *structure coefficients* or *discriminant loadings*. The raw discriminant function coefficients denote the unique (partial) contribution of each variable to the discriminant functions, while the structure coefficients denote the simple correlations between the variables and the functions. To assign meaningful names to the discriminant functions, the structure coefficients should be used. However, to learn each predictor variable's unique contribution to the discriminant function, the discriminant function coefficients (weights) should be used.

■ *Biplot display of canonical discriminant analysis:* The biplot[10] display of CDA is a visualization technique for investigating the interrelationships between the group members and the canonical function scores in CDA. The term *biplot* means a plot of two dimensions with the observation and variable spaces plotted simultaneously. In CDF, relationships between CDF scores and CDF structure loadings associated with any two CDFs can be illustrated in a biplot display.[11] The success of CDF in discriminating between the group members can be visually verified in the biplot.

6.5 Discriminant Function Analysis

Another major purpose of DA is to classify observation into predefined group levels. In a dataset containing multiple quantitative predictor variables and a classification variable, discriminant functions derived by the DFA can be used to classify each observation into one of the group levels. The discriminant function, also known as a classification criterion, is determined by a measure of the generalized squared distance. The classification criterion can be determined based on either the individual within-group covariance matrices (yielding a quadratic function) or the pooled covariance matrix (yielding a linear function). This classification criterion also takes into account the prior probabilities of the groups. The calibration information can be stored in a special SAS dataset and applied to other datasets. The derived classification criterion from the training dataset can be applied to a validation dataset simultaneously.

6.5.1 Key Concepts and Terminology in Discriminant Function Analysis

■ *Parametric DFA*: Using a generalized squared distance measure, a parametric method can be used to develop a classification criterion when the distribution within each group level is multivariate normal. The classification criterion can be derived based on either the individual within-group covariance matrices or the pooled covariance matrix that also takes into account the prior probabilities of the classes. Each observation is placed in the class from which it has the smallest generalized squared distance. The posterior probability of an observation belonging to each class can also be computed.

■ *Nonparametric DFA*: When the multivariate normality assumption within each group is not met, nonparametric DFA methods can be used to estimate the group-specific densities. Either a kernel or the *k*-nearest-neighbor method can be used to generate a nonparametric density estimate in each group and to produce a classification criterion. The performance of a discriminant criterion can be evaluated by estimating the probabilities of misclassification of new observations in the validation data.

■ *Classification criterion*: The classification functions can be used to determine which group level each case most likely belongs to. There are as many classification criteria as there are group levels. Each classification criterion allows the computation of classification probability scores for each member in each group level. The Mahalanobis distance estimates are used to determine proximity in computing the classification criterion. The classification criterion can be used to directly compute posterior probability scores for new observations.

■ *Mahalanobis distance*: It is the distance between an observation and the centroid for each group level in *p*-dimensional space defined by *p* variables and their covariance that is used in DFA. Thus, the smaller the Mahalanobis distance, the closer the observation is to the group level centroid and the more likely it is to be assigned to that group level. In SAS, only the pooled covariance matrix can be used to calculate the Mahalanobis distance in the *k*-nearest-neighbor method. With parametric or kernel DFA methods, either the individual within-group covariance matrices or the pooled covariance matrix can be used to calculate the Mahalanobis distance.[8]

■ *Priori probabilities*: Sometimes, we know ahead of time that there are more observations in one group level than in any other level. Thus, we should adjust our prediction according to the prior probability. Prior probabilities are the likelihood of belonging to a particular group level given that no information about the observation is available. Different prior probabilities can be specified, which will then be used to adjust the classification of cases and the computation of posterior probabilities accordingly. The prior probabilities can be set to be proportional to the sizes of the groups in the training data or can be set equal to each group level.[8] The specification of different prior probabilities can greatly affect the accuracy of the prediction.

■ *Posterior probabilities*: The posterior probability is a probability based on knowledge of the values of predictor variables that the respective case belongs to a particular group level. The probability that an observation belongs to a particular group level is approximately proportional to the Mahalanobis distance from that group centroid. The Mahalanobis distance is one of the main components in estimating the group-specific probability densities. Once the posterior probability scores are computed for an observation, it can be used to decide how to classify the observation. In general, a given observation is classified to a predefined group level for which it has the highest posterior probability scores. With the estimated group-specific probability densities and their associated prior probabilities, the posterior probability estimates of group membership for each class can be estimated.

■ *Classification table*: It is used to assess the performance of DFA and is simply a table in which the rows are the observed categories of the group levels and the columns are the predicted categories of the group levels. When prediction is perfect, all observations will lie on the diagonal. The percentage of observations on the diagonal is the percentage of correct classifications.

■ *Misclassification error-rate estimates*: A classification criterion can be evaluated by its performance in the classification of future observations. Two types of error-rate estimates are commonly used to evaluate the derived classification criterion based on posterior probability values estimated by the training sample. The posterior probability estimates provide good estimates of the error rate when the posterior probabilities are accurate. When a parametric classification criterion (linear or quadratic discriminant function) is derived from a nonnormal population, the resulting posterior probability error-rate estimators may not be appropriate. The overall classification error rate is estimated through a weighted average of the individual group level-specific error-rate estimates, where the prior probabilities are used as the weights.

■ *Apparent error rate*: When no independent validation data are available, the same dataset can be used both to define and to evaluate the classification criterion. The resulting error-count estimate has an optimistic bias and is called an *apparent error rate*.[12] To reduce the bias, you can split the data into two sets: one set for deriving the discriminant function and the other set for estimating the error rate. The error-count estimate is calculated by applying the classification criterion derived from the training sample to a test set and then counting the number of misclassified observations. The group-specific error-count estimate is the proportion of misclassified observations in the group. When the test set is independent of the training sample, the estimate is unbiased. However, it can have a large variance, especially if the test set is small. Such a split-sample method has the unfortunate effect of reducing the effective sample size.

■ *Cross-validation*: To reduce both the bias and the variance of the estimator, posterior probability estimates can be computed based on cross-validation. Cross-validation uses $n - 1$ out of n observations as a training set. It determines the classification criterion based on these $n - 1$ observations and then applies them to classify the one observation left out. This is done for each of the n training observations. The misclassification rate for each group is the proportion of sample observations in that group that are misclassified. This method achieves a nearly unbiased estimate but with a relatively large variance.[13]

▪ *Smoothed error rate*: To reduce the variance in an error-count estimate, smoothed error-rate estimates have been suggested.[14] Instead of summing terms that are either zero or one, as in the error-count estimator, the smoothed estimator uses the posterior probability scores, which are a continuum of values between zero and one in the terms that are summed. The resulting estimator has a smaller variance than the error-count estimate. Two types (unstratified and stratified) of smoothed posterior probability error-rate estimates are provided by the POSTERR option in the PROC DISCRIM statement. The stratified posterior probability error-rate estimates take into account the relative sizes of the groups, whereas the unstratified error-rate estimate does not.[5]

6.6 Applications of Discriminant Analysis

DA can be used to develop classification criteria for grouping bank customers, online shoppers, diabetes types, college admissions, etc. Given categorical (two or more levels) outcomes, DA can estimate the probability that a new cable TV customer will upgrade to digital cable or to cable modem. Other DA applications:

▪ Investigate differences between groups of college freshmen that drop out in one year, transfer to a new school, or complete the degree.
▪ Discriminate between types of cancer cells effectively.
▪ Select the important financial indicators for discriminating between fast-growing, stable, and diminishing stocks.
▪ Classify a new credit card customer into preexisting groups based on paying credit card bills.

It is very clear from these applications that DA can be a useful tool for decision-making endeavors.

6.7 Classification Tree Based on CHAID

Classification tree analysis is a segmentation technique designed to split a sample into two or more categories based on available attributes (e.g., gender, employment status, race, age). The results are often presented in the form of a tree. Classification tree analysis can be characterized as a hierarchical, highly flexible technique for predicting membership of cases in the classes of a categorical response using one or more predictor variables. The predictor variables used in developing a classification tree can be categorical, continuous, or any mix of the two types of predictors. The study and use of classification trees are widely used in many diverse fields such as medicine, business, biology, and social sciences. Classification tree analysis can sometimes be quite complex. However, graphical procedures can be developed to help simplify interpretation even for complex trees. The goal of a classification tree analysis is to obtain the most accurate prediction possible.

In a number of ways, classification trees are different from traditional statistical methods (DA, logistic regression) for predicting class membership on a categorical response. When stringent statistical assumptions of traditional methods are met, these methods may be preferable for classification problems. However, as an exploratory technique, or when traditional methods fail, classification trees are preferred. Classification methods employ a hierarchy of predictions, sometimes being applied to particular cases to sort the cases into predicted classes. Traditional methods use simultaneous techniques to make one—and only one—class membership prediction for each and every case. Tree-based methods have several attractive properties when compared to traditional methods, because they

- Provide a simple rule for classification or prediction of observations.
- Handle interactions among variables in a straightforward way.
- Can easily handle a large number of predictor variables.
- Do not require assumptions about the distribution of the data.

However, tree-based methods do not conform to the usual hypothesis-testing framework. For finding models that predict well, there is no substitute for a thorough understanding of the nature of the relationships between the predictor and response variables. A brief account of nonmathematical descriptions and applications of classification tree methods are discussed in this section. Additional details of decision tree methods are discussed elsewhere.[15,16]

6.7.1 Key Concepts and Terminology in Classification Tree Methods

Construction of classification trees: A decision tree partitions data into smaller segments called *terminal nodes* or *leaves*, which are homogeneous with respect to a target variable. Partitions are defined in terms of input variables, thereby defining a predictive relationship between the inputs and the target. This partitioning goes on until the subsets cannot be partitioned any further using one of many user-defined stopping criteria. By creating homogeneous groups, analysts can predict with greater certainty how individuals in each group will behave. Classification trees can be constructed based on splitting a single (univariate splits) ordinal-scale predictor variable after performing transformation that preserves the order of values on the ordinal variable. Thus, classification trees based on univariate splits can be computed without concern for whether a unit change on a continuous predictor represents a unit change on the dimension underlying the values on the predictor variable. In short, assumptions regarding the level of measurement of predictor variables are less stringent in classification tree analysis. Classification tree analyses are not limited to univariate splits on predictor variables. When continuous predictors are measured on at least an interval scale, linear combination splits, similar to the splits for linear DA, can be computed for classification trees. However, the linear combination splits computed for classification trees do differ in important ways from those splits computed for DA. In linear DA, the number of linear discriminant functions that can be extracted is the lesser of the number of predictor variables or the number of classes on the group variable minus one. The recursive approach implemented for the classification tree module does not face this limitation.

- *Chi-square automatic interaction detection method*: This is a classification tree method to study the relationship between a categorical response variable and a series of predictor variables. CHAID modeling selects a set of predictors and their interactions that optimally predict the group response membership. The developed classification tree (or data-partitioning tree) shows how major subsets are formed from the predictor variables by differentially predicting a criterion or response variable. The SAS TREEDISC macro uses CHAID-type analysis.[15] The decision tree in TREEDISC is constructed by partitioning the dataset into two or more subsets of observations based on the categories of one of the predictor variables. After the dataset is partitioned according to the chosen predictor variable, each subset is considered for further partitioning using the same criterion that was applied to the entire dataset. Each subset is partitioned without regard to any other subset. This process is repeated for each subset until some stopping criterion is met. This recursive partitioning forms a tree structure.[15]

- *Decision tree diagram*: It consists of a tree trunk that progressively splits into smaller and smaller branches. The *root* of the tree is the entire dataset. The subsets and sub-subsets form the *branches* of the tree. Subsets that meet a stopping criterion and thus are not partitioned are *leaves*, or the terminal node. The number of subsets in a partition can range from two up to the number of categories of the predictor variable. Note that the tree can be pictured in an orientation from *top-to-bottom* or *left-to-right* or *right-to-left* and that the results are identical. Different orientations of the same tree are sometimes useful to highlight different portions of the results.

- *Classification criterion*: The predictor variable used to form a partition is chosen to be the variable that is most significantly associated with the categorical response variable according to a *chi-squared* test of independence in a contingency table. The main stopping criterion used by the TREEDISC macro is the p-value from this chi-squared test. A small p-value indicates that the observed association between the predictor and the dependent variable is unlikely to have occurred solely as the result of sampling variability. If a predictor has more than two categories, then there may be a very large number of ways to partition the dataset based on the categories. A combinatorial search algorithm is used to find a partition that has a small p-value for the chi-squared test. The p-values for each chi-squared test are adjusted for the multiplicity of partitions. Predictors can be nominal, ordinal, or ordinal with a floating category.[15] For a nominal predictor, the categories are not ordered and therefore can be combined in any way to form a partition. For an ordinal predictor, the categories are ordered, and only categories that are adjacent in the order can be combined when forming a partition. A general issue that arises when applying tree classification is that the final trees can become very large. In practice, when the input data is complex and, for example, contains many different categories for classification problems, and many possible predictors for performing the classification, then the resulting trees can become very large and interpretation becomes difficult.

6.8 Applications of CHAID

■ In database marketing, decision trees can be used to segment groups of customers and develop customer profiles to help marketers produce targeted promotions that achieve higher response rates. Also, because the CHAID algorithm will often effectively yield many multiway frequency tables (e.g., when classifying a categorical response variable with many categories, based on categorical predictors with many classes), it has been particularly popular in marketing research, in the context of market segmentation studies.

■ A credit card issuer may use a decision tree to define a rule of the form: If the monthly mortgage-to-income ratio is less than 30%, the months posted late is less than 1, and salary is greater than $36,000, then issue a gold card.

6.9 Discriminant Analysis Using SAS Macro DISCRIM2

The DISCRIM2 macro is a powerful user-friendly enhanced SAS application for performing complete discriminant analysis. Options are available for obtaining various exploratory and diagnostic graphs and for performing different types of discriminant analyses. The SAS procedures STEPDISC and DISCRIM are the main tools used in the DISCRIM2 macro.[6-8] In addition to these SAS procedures, GPLOT, SGPLOT, and BOXPLOT procedures, and IML modules are also utilized in the DISCRIM2 macro. The enhanced features implemented in the DISCRIM2 macro are:

■ Data description table, including all numerical and categorical variable names and labels and the macro input values specified by the user in the last run of DISCRIM2 macro submission are produced.

- Variable selection using forward and stepwise selection and backward elimination methods are included.
- Exploratory bivariate plots for checking group discrimination in a simple scatter plot between two predictor variables are included.
- Plots for checking for multivariate normality and influential observations within each group level are available.
- Test statistics and *p*-values for testing equality in variance and covariance matrices within each group level are automatically produced.
- In the case of CDA, box plots of canonical discriminant functions by groups and biplot display of canonical discriminant function scores of observations and the structure loadings for the predictors are generated. When fitting DFA, box plots of the *i*th group level posterior probability scores by group levels are produced.
- Options are available for validating the discriminant model obtained from a training dataset using an independent validation dataset by comparing classification errors.
- Options for saving the output tables and graphics in WORD, HTML, PDF, and TXT formats are available.

Software requirements for using the DISCRIM2 macro:

- SAS/BASE, SAS/STAT, and SAS/GRAPH must be licensed and installed at your site. SAS/IML is required to check for multivariate normality.
- SAS version 9.13 and above is required for full utilization.

6.9.1 Steps Involved in Running the DISCRIM2 Macro

- Create a temporary SAS dataset containing at least one categorical group response (target) variables and many continuous and or ordinal predictor (input) variables.
- Open the DISCRIM2.SAS macro-call file into the SAS EDITOR Window. Instructions are given in Appendix 1 regarding downloading the macro-call and sample data files from this book's Web site. Click the RUN icon to submit the macro-call file DISCRIM2.SAS to open the MACRO–CALL window called DISCRIM2 (Figure 6.1).
- *Special note to SAS Enterprise Guide (EG) CODE window users*: Because the user-friendly SAS macro application included in this book uses SAS WINDOW/DISPLAY commands, and these commands are not compatible with SAS EG, open the traditional DISCRIM2 macro-call file included in the \dmsas2e\maccal\nodisplay\ into the SAS editor. Read the instructions given in Appendix 3 regarding using the traditional macro-call files in the SAS EG/SAS Learning Edition (LE) code window.
- Input the appropriate parameters in the macro-call window by following the instructions provided in the DISCRIM2 macro-help file (Appendix 2). Users can choose whether to include exploratory graphs and variable selections

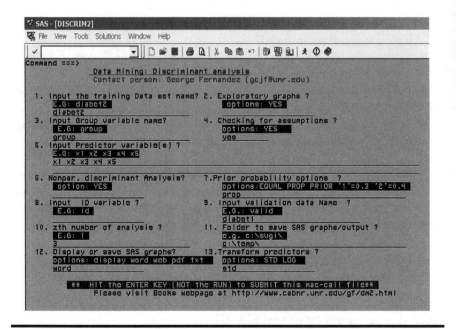

Figure 6.1 Screen copy of the DISCRIM2 macro-call window showing the macro-call parameters required for performing parametric discriminant analysis.

by stepwise methods or skip the exploratory methods. After inputting all the required macro parameters, check that the cursor is in the last input field, and then hit the ENTER key (not the RUN icon) to submit the macro.

■ Examine the LOG window (only in DISPLAY mode) for any macro execution errors. If you see any errors in the LOG window, activate the EDITOR window, resubmit the DISCRIM2.SAS macro-call file, check your macro input values, and correct any input errors.

■ If no errors are found in the LOG window, activate the EDITOR window, resubmit the DISCRIM2.SAS macro-call file, and change the macro input value from DISPLAY to any other desirable output format. The SAS output files from complete discriminant analysis and all the exploratory/descriptive graphs can be saved in user-specified formats in the user-specified folder.

6.10 Decision Tree Using SAS Macro CHAID2

The CHAID2 macro is a powerful SAS application for implementing classification models based on decision trees. Options are implemented for performing variable selection using SAS PROC STEPDISC, enhanced graphical plots to explore univariate association between predictor and the group response, for creating decision tree diagrams, classification plots, and validating the decision tree models using independent validation datasets. This CHAID2 macro is an enhanced version of SAS institute's TREEDESC[16] macro. The CHAID macro requires the SAS/IML product. To draw the tree, the SAS/OR product is required, and release 9.13 or later is recommended. In addition to these SAS procedures, GCHART and SGPLOT[18] procedures are also utilized in the CHAID2 macro to obtain classification charts.

The CHAID2 macro generates a SAS dataset, which describes a decision tree computed from a training dataset to predict a specified categorical response variable from one or more predictor variables. The tree can be listed, drawn, or used to generate code for a SAS DATA step to classify observations. Also, the classification results between the observed and the predicted levels for the training and the independent validation data can be viewed in a donut chart display.

The enhanced features of CHAID2 macro are:

- The CHAID2 macro has a user-friendly front end. SAS programming or SAS macro experience is not required to run the CHAID2 macro.
- Data description table, including all numerical and categorical variable names and labels and the macro input values specified by the user in the last run of CHAID2 macro submission.
- *New*: Variable selection using forward and stepwise selection and backward elimination methods.
- *New*: SAS SGPLOT graphics are included in data exploration.
- Options for creating donut charts illustrating the differences between the observed and the predicted group memberships are available.
- Options for validating the fitted CHAID model obtained from a training dataset using an independent validation dataset by comparing classification errors are available.
- Options for saving the output tables and graphics in WORD, HTML, PDF, and TXT formats are available.

Software requirements for using the CHAID2 macro are:

- SAS/BASE, SAS/IML, SAS/GRAPH, and optional SAS/OR must be licensed and installed at your site.
- SAS version 9.13 and above is required for full utilization.

6.10.1 Steps Involved in Running the CHAID2 Macro

- Create a SAS temporary dataset containing at least one categorical response (target) variable and many continuous and or categorical predictor (input) variables.
- Open the CHAID2.SAS macro-call file into the SAS EDITOR Window. Instructions are given in Appendix 1 regarding downloading the macro-call and sample data files from this book's Web site. Click the RUN icon to submit the macro-call file CHAID2.SAS to open the MACRO–CALL window called CHAID2 (Figure 6.14).

■ *Special note to SAS Enterprise Guide (EG) CODE window users*: Because the user-friendly SAS macro application included in this book uses SAS WINDOW/DISPLAY commands and these commands are not compatible with SAS EG, open the traditional CHAID2 macro-call file included in the \dmsas2e\maccal\nodisplay\ into the SAS editor. Read the instructions given in Appendix 3 regarding using the traditional macro-call files in the SAS EG/ SAS Learning Edition (LE) code window.

■ Input the appropriate parameters in the macro-call window by following the instructions provided in the CHAID2 macro-help file in Appendix 2. After inputting all the required macro parameters, check to see that the cursor is in the last input field, and then hit the ENTER key (not the RUN icon) to submit the macro.

■ Examine the LOG window (only in DISPLAY mode) for any macro execution errors. If you see any errors in the LOG window, activate the EDITOR window, resubmit the CHAID2.SAS macro-call file, check your macro input values, and correct if you see any input errors.

■ If no errors are found in the LOG window, activate the EDITOR window, resubmit the CHAID2.SAS macro-call file, and change the macro input value from DISPLAY to any other desirable output format. The SAS output files from complete CHAID modeling and diagnostic graphs can be saved in a user-specified format in the user-specified folder.

6.11 Case Study 1: Canonical Discriminant Analysis and Parametric Discriminant Function Analysis

The objective of this study is to discriminate between three clinical diabetes groups (normal, overt diabetic, and chemical diabetic) in a simulated multivariate normally distributed data using blood plasma and insulin measures. This simulated data satisfies the multivariate normality assumption and was generated using the group means and group variance–covariance estimates of real clinical diabetes data reported elsewhere.[11,17] This real diabetes data is used as the validation dataset to check the validity of the discriminant functions derived from the simulated data.

6.11.1 Study Objectives

1. *Data exploration using diagnostic plots*: Used to check for the discriminative potential of predictor variables, two at a time in simple scatter plots. In addition, these plots are also useful in examining the characteristics (scale of measurements, range of variability, extreme values) of predictor variables.

2. *Variable selection using stepwise selection methods*: Used to perform backward elimination and stepwise and forward selection methods to identify predictor variables that have significant discriminating potentials.

3. *Checking for any violations of discriminant analysis assumptions*: Used to perform statistical tests and graphical analysis to verify that no multivariate influential outliers are present and the data satisfies the multivariate normality assumption. The validity of canonical discriminant and parametric discriminant analyses results depends on satisfying the multivariate normality assumption.

4. *Parametric discriminant function analyses*: Used to perform canonical discriminant analysis to investigate the characteristics of significant discriminating predictor variables, perform parametric discriminant function analysis to develop classification functions and to assign observations into predefined group levels, and to measures the success of discrimination by comparing the classification error rates.

5. *Create and save "plotp," "stat," and "out2" datasets for future use*: These three temporary SAS datasets are created and saved in the work folder when you run the DISCRIM2 macro. The "plotp" dataset contains the observed predictor variables, group response value, canonical discriminant function scores, posterior probability scores, and new classification results. This canonical discriminant function and posterior probability scores for each observation in the dataset could be used as the base for developing the scorecards. The temporary SAS dataset called "stat" contains the canonical and discriminant function analysis parameter estimates. If you include an independent validation dataset, the classification results for the validation dataset are saved as a temporary SAS data called "out2," which could be used to develop scorecards for new members in the validation data.

6. *Validation*: Validate the derived discriminant functions by applying these classification criterions to an independent validation dataset (real diabetes dataset), and examine the success of classification.

6.11.2 Case Study 1: Parametric Discriminant Analysis

Dataset names	a. Training — Simulated dataset: Temporary SAS dataset Diabet1 located in the work folder
	b. Validation — Real dataset: Temporary SAS dataset Diabet2 located in the work folder
Group response (Group)	Group (three clinical diabetic groups: 1 — Normal; 2 — Overt diabetic; 3 — Chemical diabetic)
Predictor variables (*X*)	*X1*: Relative weight
	X2: Fasting plasma glucose level
	X3: Test plasma glucose
	X4: Plasma insulin during test
	X5: Steady-state plasma glucose level

Number of observations	a. Training data: 141
	b. Validation data 145
Source:	a. Training data: Simulated data that satisfies multivariate normality assumption
	b. Validation data: Real diabetes data[17]

Open the DISCRIM2.SAS macro-call file in the SAS EDITOR window, and click RUN icon to open the DISCRIM2 macro-call window (Figure 6.1). Input the appropriate macro-input values by following the suggestions given in the help file (Section 6.7.2).

Exploratory analysis/diagnostic plots: Input dataset name, group variable, predictor variable names, and prior probability option. Input YES in macro field #2 to perform data exploration and to create diagnostic plots. Submit the DISCRIM2 macro, and discriminant diagnostic plots and stepwise variable selection output will be produced.

Data exploration and checking: Examining the group level discrimination based on simple scatter plots between any two discrimination variables is the first step in data exploring. An example of simple two-dimensional scatter plots showing the discrimination of three diabetes groups is presented in Figure 6.2. These scatter plots are useful in examining the range of variation, presence of extreme observations, and the degree of linear associations between any two predictor variables. The scatter plot presented in Figure 6.2 revealed that a strong correlation existed between fasting plasma glucose level ($X2$) and test plasma glucose ($X3$). These two attributes appeared to discriminate the diabetes group 3 from the other two to a certain degree. Discrimination between the normal and the overt diabetes group is not very distinct.

Variable selection methods: Variable selection based on backward elimination, stepwise selection, and forward selections are performed automatically when you input YES to data exploration in the DISCRIM2 macro-call window. These variable selection methods are useful especially when there is a need for screening a large number of predictor variables. Information on the number of observations, number of group levels, discriminating variables, and the threshold significance level for eliminating nonsignificant predictor variables in the backward elimination method is presented in Table 6.1. About 50% of the cases in the training dataset appear to be in the normal group. Among the clinically diagnosed diabetes group, 25% of them are considered the overt type, and 23% are in the chemical diabetes group (Table 6.2).

The results of variable selection based on backward elimination are summarized in Table 6.1. Backward elimination starts with the full model, and the overall significance in discriminating between the diabetes groups is highly significant based on the *p*-value for Wilks lambda (<0.0001) and the *p*-value for the average

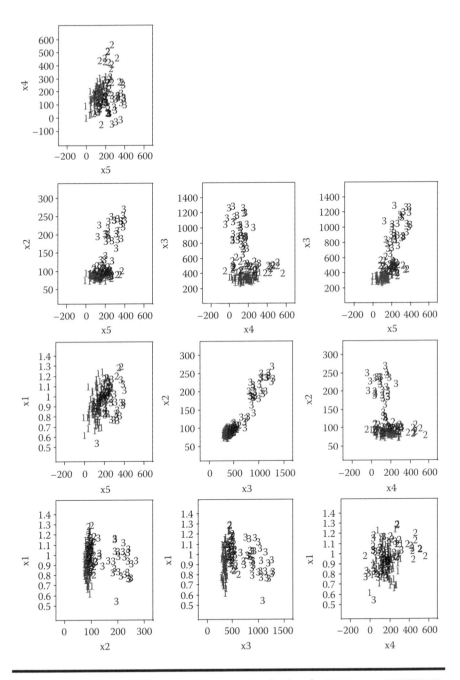

Figure 6.2 Bivariate exploratory plots generated using the SAS macro DISCRIM2: Group discrimination of three types of diabetic groups (data=diabetic1) in simple scatter plots.

Table 6.1   Data Exploration Using SAS Macro: DISCRIM2—Backward Elimination Summary

Step	Removed	Label	Partial R-square	F value	Pr > F	Wilks Lambda	Pr < Lambda	Average Squared Canonical Correlation (ASCC)	Pr > ASCC
0	None	None	.	.	.	0.080	<.0001	0.611	<.0001
1	X5	Steady state plasma glucose	0.0187	1.28	0.282	0.081	<.0001	0.605	<.0001

Note: Observations = 141; variables in the analysis = 5; class levels = 3; significance level to stay = 0.15.

Table 6.2   Data Exploration Using SAS Macro: DISCRIM2—Group Level Information

Group	Variable Name	Frequency	Weight	Proportion
1	_1	72	72	0.510
2	_2	36	36	0.255
3	_3	33	33	0.234

squared canonical correlations (<0.0001). In Step 1, the nonsignificant (p-value 0.28) steady-state plasma glucose ($X5$) is eliminated from the model and the resulting four-variable model is as good as the full model. The backward elimination is stopped here since no other variable can be dropped based on the p-value to stay (0.15) criterion. In the backward elimination method, once a variable is removed from the discriminant model, it cannot be reentered.

Information on number of observations, number of group levels, discriminating variables, and the threshold significance level for entering (0.15) and staying (0.15) in the stepwise selection method is presented in Table 6.3. The significance of predictor variables in discriminating between the three clinical diabetes groups is evaluated in a stepwise fashion. At each step, the significance of already entered predictor variables is evaluated based on the significance for staying (p-value: 0.15) criterion, and the significance of newly entering variables is evaluated based on the significance for entering (p-value: 0.15) criterion. The stepwise selection procedure stops when no variables can be removed or entered. The summary results of the stepwise selection method are presented in Table 6.3. The results of stepwise selection methods are in agreement with the backward elimination method since both methods choose variables $X1$–$X4$ as the significant predictors.

Information on number of observations, number of group levels, discriminating variables, and the threshold significance level for entering (p-value: 0.15) in the forward selection method is presented in Table 6.4. The significance of predictor variables in discriminating between the three clinical diabetes groups is evaluated one at a time. At each step, the significance of entering variables is evaluated based on the significance for entering (p-value: 0.15) criterion. The forward selection procedure stops when no variables can be entered by the entering p-value (0.15) criterion. In the forward selection method, once a variable is entered into the discriminant model, it cannot be removed. The summary results of the forward selection method are presented in Table 6.4. The results of the forward selection method are in agreement with both the backward elimination and the stepwise selection methods since all three methods choose variables $X1$–$X4$ as the significant predictors.

Parametric Discriminant analysis and checking for multivariate normality:

Table 6.3 Data Exploration Using SAS Macro: DISCRIM2—Variable Selection Using Stepwise Selection Method

Entered	Removed	Label	Partial R-Square	F Value	Pr > F	Wilks Lambda	Pr < Lambda	ASCC	Pr > ASCC
X3	—	Test plasma glucose	0.87	461.68	<0.0001	0.130	<0.0001	0.434	<0.0001
X2	—	Fasting plasma glucose	0.19	16.52	<0.0001	0.104	<0.0001	0.530	<0.0001
X1	—	Relative weight	0.16	13.82	<0.0001	0.087	<0.0001	0.584	<0.0001
X4	—	Plasma insulin during test	0.05	4.29	0.0156	0.081	<0.0001	0.605	<0.0001

Note: Observations = 141; variables in the analysis = 5; class levels = 3; significance level to enter = 0.15; significance level to stay = 0.15.

Table 6.4 Data Exploration Using SAS Macro: DISCRIM2—Forward Selection Summary

Step	Entered	Label	Partial R-Square	F Value	Pr > F	Wilks Lambda	Pr < Lambda	Average Squared Canonical Correlation (ASCC)	Pr > ASCC
1	x3	Test plasma glucose	0.8700	461.68	<0.0001	0.130	<0.0001	0.434	<0.0001
2	x2	Fasting plasma glucose	0.1943	16.52	<0.0001	0.104	<0.0001	0.530	<0.0001
3	x1	Relative weight	0.1689	13.82	<0.0001	0.087	<0.0001	0.584	<0.0001
4	x4	Plasma insulin during test	0.0598	4.29	00.0156	0.081	<0.0001	0.605	<0.0001

Note: Observations = 141; variables in the analysis = 5; class levels = 3; significance level to enter = 0.15.

Open the DISCRIM2 macro-call window, and input the dataset name, group variable, predictor variable names, and the prior probability option. Leave macro field #2 BLANK to perform CDA and parametric DFA. To perform the multivariate normality check, input YES in the macro field #4. Submit the DISCRIM2 macro, and you will get the multivariate normality check, CDA output and graphics, and parametric DFA output and graphics.

Checking for multivariate normality: The right choice for selecting parametric versus nonparametric discriminant analysis is dependent on the assumption of multivariate normality within each group level. The diabetes data within each clinical group is assumed to have a multivariate normal distribution. This multivariate normality assumption can be checked by estimating multivariate skewness, kurtosis, and testing for their significance levels. The quantile-quantile (Q-Q) plot of expected and observed distributions[9] of multiattributes can be used to graphically examine for multivariate normality. The estimated multivariate skewness and multivariate kurtosis (Figure 6.3) clearly support the hypothesis that within each group level, these four multiattributes have a joint multivariate normal distribution. A nonsignificant departure from the 45° angle reference line in the Q-Q plot (Figure 6.3) also supports this finding. Thus, parametric discriminant analysis can be considered to be the appropriate technique for discriminating between the three clinical groups based on these four attributes ($X1$ to $X4$).

Checking for the presence of multivariate outliers: Multivariate outliers can be detected in a plot between the differences of robust (Mahalanobis distance—chi-squared quantile) versus chi-squared quantile values.[9] Only two observations are identified as extreme multivariate outliers within a group since the differences between the robust *Mahalanobis* distance and the chi-squared quantile value are slightly larger than 2 and fall outside the critical region (Figure 6.4). The presence of only few borderline outliers can be expected since the training dataset used is a multivariate normally distributed simulated dataset.

6.11.2.1 Canonical Discriminant Analysis (CDA)

The main objective of CDA is to extract a set of linear combinations of the quantitative variables that best reveal the differences among the groups. The class level information, group frequency, and the prior probability values for training dataset diabet1 are presented in Table 6.5. The descriptive statistical measures, sample size mean, variance, standard deviation for the whole dataset (Table 6.6), and the within each group levels are presented in Table 6.7. Very large differences in the means and variances are observed for $X2$, $X3$, and $X4$ among the three levels of diabetes groups. Therefore, in CDA, it is customary to standardize the multiattributes so that the canonical variables have means that are equal to zero and the pooled within-class variances are equal to one.

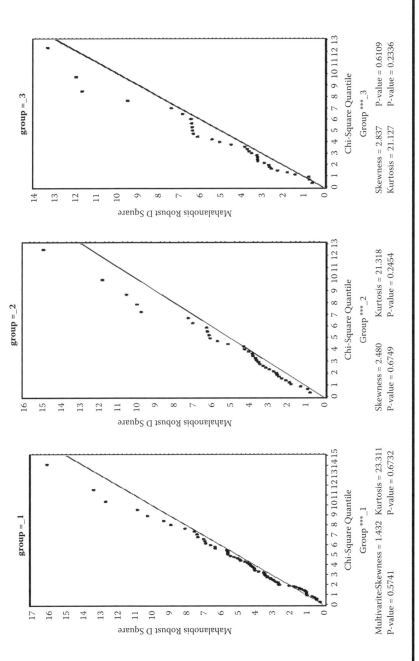

Figure 6.3 Checking for multivariate normality assumption for all three types of diabetic groups (data=diabet1) in a Q-Q plot using the SAS macro DISCRIM2.

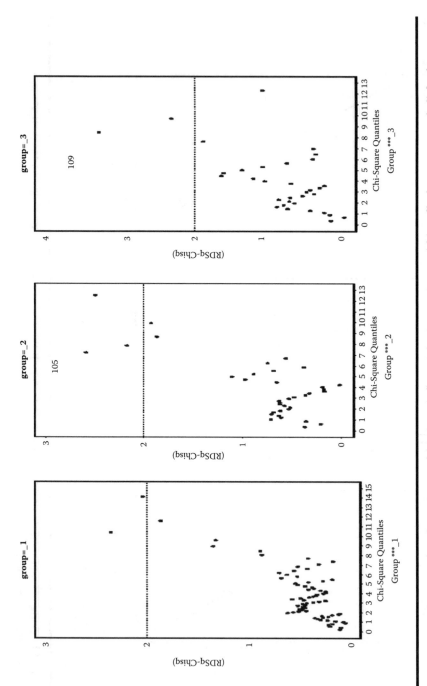

Figure 6.4 Diagnostic plot for detecting multivariate influential observations within all three types of diabetic groups (data=diabet1) generated using the SAS macro DISCRIM2.

Table 6.5   Canonical Discriminant Analysis Using SAS Macro: DISCRIM2—Group Level Information

Group	Variable Name	Frequency	Weight	Proportion	Prior Probability
1	_1	72	72	0.510	0.510
2	_2	36	36	0.255	0.255
3	_3	33	33	0.234	0.234

Note: Observations 141; degrees of freedom total 140.

The univariate ANOVA results indicate that highly significant group differences exist for all the predictor variables (Table 6.8). The total, between-group, and within-group variability in predictor variables are expressed in standard deviation terms. The R^2 statistic describes the amount of variability in each predictor variable accounted for by the group differences. The $(R^2/1-R^2)$ column expresses the ratio between accounted and unaccounted variation in the univariate ANOVA model. By comparing the R^2 and the $(R^2/1-R^2)$ statistics for each significant predictor variable, we can conclude that the test plasma glucose ($X3$) has the highest amount of significant discriminative potential, while the relative weight has the least amount of discriminative power in differentiating the three diabetes clinical groups. The relatively large average R^2 weighted by variances (Table 6.8) indicates that the four predictor variables have high discriminatory power in classifying the three clinical diabetes groups. This is further confirmed by the highly significant MANOVA test results by all four criteria (Table 6.9). If the MANOVA assumptions, multivariate normality, and equal variance–covariance are not, the validity of the MONOVA test

Table 6.6   Canonical Discriminant Analysis Using SAS Macro: DISCRIM2—Total Sample Statistics

Variable	Label	N	Sum	Mean	Variance	Standard Deviation
x1	Relative weight	141	135.191	0.958	0.02034	0.1426
x2	Fasting plasma glucose	141	16565	117.480	2531	50.3132
x3	Test plasma glucose	141	75170	533.123	70856	266.1886
x4	Plasma insulin during test	141	25895	183.651	12431	111.4925

Table 6.7   Canonical Discriminant Analysis Using SAS Macro: DISCRIM2—Within-Group Statistics

Variable	Label	N	Sum	Mean	Variance	Standard Deviation
			Group = 1			
x1	Relative weight	72	66.502	0.923	0.0193	0.1392
x2	Fasting plasma glucose	72	6487	90.094	58.5841	7.6540
x3	Test plasma glucose	72	25463	353.656	1038	32.2236
x4	Plasma insulin during test	72	12767	177.320	4736	68.8154
			Group = 2			
x1	Relative weight	36	37.527	1.042	0.0151	0.123
x2	Fasting plasma glucose	36	3551	98.645	80.3741	8.965
x3	Test plasma glucose	36	17749	493.027	2866	53.539
x4	Plasma insulin during test	36	9123	253.405	25895	160.920
			Group = 3			
x1	Relative weight	33	31.161	0.944	0.0185	0.136
x2	Fasting plasma glucose	33	6527	197.781	2121	46.050
x3	Test plasma glucose	33	31958	968.429	34867	186.727
x4	Plasma insulin during test	33	4005	121.370	5989	77.391

Table 6.8 Canonical Discriminant Analysis Using SAS Macro: DISCRIM2—Univariate Test Statistics

Variable	Label	Total Standard Deviation	Pooled Standard Deviation	Between Standard Deviation	R-Square	R-Square/ (1–RSq)	F Value	Pr > F
x1	Relative weight	0.142	0.134	0.060	0.121	0.139	9.60	0.0001
x2	Fasting plasma glucose	50.313	23.286	54.535	0.788	3.735	257.78	<0.0001
x3	Test plasma glucose	266.188	96.676	303.000	0.870	6.691	461.68	<0.0001
x4	Plasma insulin during test	111.492	101.945	57.060	0.175	0.213	14.72	<0.0001

Note: F statistics, numerator df = 2, denominator df = 138, R^2 weighted by variance = 0.767.

Table 6.9   Canonical Discriminant Analysis Using SAS Macro: DISCRIM2—Multivariate ANOVA Statistics

Statistic	Value	F Value	Numerator df	Denominator df	Pr > F
Wilks lambda	0.0818	84.21	8	270	<0.0001
Pillai's trace	1.2114	52.23	8	272	<0.0001
Hotelling–Lawley trace	7.6333	128.25	8	190.55	<0.0001
Roy's greatest root	7.1309	242.45	4	136	<0.0001

is questionable. Transforming the predictor variables into a log scale might alleviate the problems caused by the unequal variance–covariance between the groups.

In CDA, canonical variables that have the highest possible multiple correlations with the groups are extracted. The unstandardized coefficients used in computing the raw canonical variables are called the *canonical coefficients* or *canonical weights*. The standardized discriminant function coefficients indicate the partial contribution of each variable to the discriminant functions, controlling for other attributes entered in the equation. The total standardized discriminant functions given in Table 6.10 indicate that the predictor variable, test plasma glucose ($X3$), contributed significantly to the first canonical variable ($CAN1$). The fasting plasma glucose ($X2$) and test plasma glucose ($X3$) equally contributed in a negative way to the second canonical variable ($CAN2$).

These canonical variables are independent or orthogonal to each other; that is, their contributions to the discrimination between groups will not overlap. This maximal multiple correlation between the first canonical variable and the group variables are called the *first canonical correlation*. The second canonical correlation is obtained by finding the linear combination uncorrelated with the $CAN1$ that has the highest possible multiple correlations with the groups. In CDA, the process of

Table 6.10   Canonical Discriminant Analysis Using SAS Macro: DISCRIM2—Total Sample Standardized Canonical Coefficients

Variable	Label	CAN1	CAN2
X1	Relative weight	0.2516	0.5627
X2	Fasting plasma glucose	−0.4103	−2.499
X3	Test plasma glucose	3.227	2.4102
X4	Plasma insulin during test	0.0426	0.4699

extracting canonical variables is repeated until you extract the maximum number of canonical variables, which is equal to the number of groups minus one, or the number of variables in the analysis, whichever is smaller.

The correlation between the $CAN1$ and the clinical group is very high (>0.9), and about 87% of the variation in the first canonical variable can be attributed to the differences among the three clinical groups (Table 6.11). The first eigenvalue measures the variability in the $CAN1$ and accounts for 93% of the variability among the three group members in four predictor variables. The correlation between the $CAN2$ and the clinical group is moderate (0.5), and about 37% of the variation in the second canonical variable can be attributed to the differences among the three clinical groups (Table 6.12). The second eigenvalue measures the variability in the second canonical variable and accounts for the remaining 6% of the variability among the three group members in the four predictor variables. Both canonical variables are statistically highly significant based on the Wilks lambda test (Table 6.12). However, the statistical validity might be questionable if the multivariate normality or the equal variance–covariance assumptions are violated.

The total structure coefficients or loadings measure the simple correlations between the predictor variables and the canonical variable. These structure loadings are commonly used when interpreting the meaning of the canonical variable because the structure loadings appear to be more stable, and they allow for the interpretation of the canonical variable in a manner that is analogous to factor analysis. The structure loadings for the first two canonical variables are presented in Table 6.13. The first and the second canonical functions account for 93% and 6% of the variation in the discriminating variables (Table 6.11). Two variables, fasting plasma glucose and test plasma glucose, have very large positive loadings on CAN. The predictor variables relative weight ($X1$) and plasma insulin during test ($X4$) have moderate-size loadings on $CAN2$. Therefore, the $CAN1$ can be named as the *plasma glucose factor*. The standardized means of three clinical groups on $CAN1$ and $CAN2$ are presented in Table 6.14. The mean $CAN1$ and $CAN2$ for the normal group are relatively lower than the other two diabetic groups. The mean $CAN1$ for the chemical diabetic group is very high, and the overt group has a large mean for $CAN2$. Thus, in general, these two canonical variables successfully discriminate between the three diabetic groups.

For each observation in the training dataset, we can compute the standardized canonical variable scores. The box plot of the $CAN1$ score by group is very useful in visualizing the group differences based on the $CAN1$. The $CAN1$ score effectively discriminates between the three groups (Figure 6.5). The chemical diabetic (group3) has a relatively larger variation for $CAN1$ score than the other two groups.

These standardized canonical variable scores and the structure loadings can be used in two-dimensional biplots to aid visual interpretation of the group differences. Interrelationships among the four multiattributes and the discriminations of the three groups are presented in Figure 6.6. The first canonical variable that has the largest loadings on $X2$ and $X3$ discriminated between the normal (1), overt (2), and the chemical diabetic groups effectively. $CAN2$, which has a moderate size

Table 6.11 Canonical Discriminant Analysis Using SAS Macro: DISCRIM2—Canonical Correlations

| | Canonical Correlation | Adjusted Canonical Correlation | Approximate Standard Error | Squared Canonical Correlation | Eigenvalues of Inv(E)*H = CanRsq/(1–CanRsq) | | | |
					Eigenvalue	Difference	Proportion	Cumulative
1	0.936	0.934	0.010	0.877	7.130	6.628	0.934	0.934
2	0.578	0.572	0.056	0.334	0.502		0.0658	1.000

Table 6.12   Canonical Discriminant Analysis Using SAS Macro: DISCRIM2—Testing the Canonical Correlations in the Current Row and All that Follow Are Zero

	Likelihood Ratio	Approximate F Value	Numerator df	Denominator df	Pr > F
1	0.0818	84.21	8	270	<0.0001
2	0.6655	22.78	3	136	<0.0001

Table 6.13   Canonical Discriminant Analysis Using SAS Macro: DISCRIM2—Total Canonical Structure Loadings

Variable	Label	CAN1	CAN2
x1	Relative weight	0.049	0.599
x2	Fasting plasma glucose	0.926	−0.327
x3	Test plasma glucose	0.994	−0.096
x4	Plasma insulin during test	−0.229	0.623

Table 6.14   Canonical Discriminant Analysis Using SAS Macro: DISCRIM2—Group Means on Canonical Variables

Group	CAN1	CAN2
1	−2.017	−0.429
2	−0.158	1.196
3	4.573	−0.368

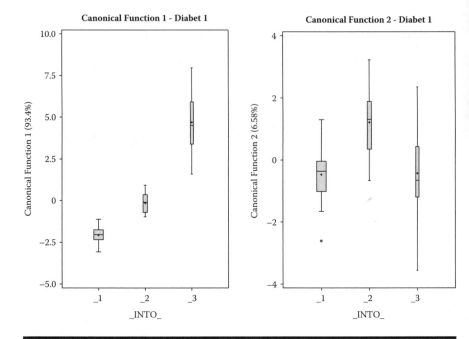

Figure 6.5   Box plot display of group discrimination using canonical discriminant functions 1 and 2 generated using the SAS macro DISCRIM2.

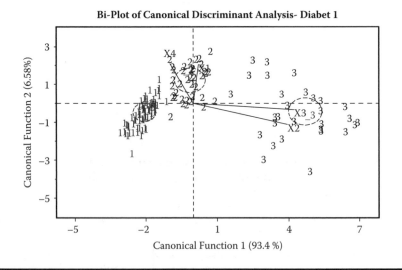

Figure 6.6   Biplot display of canonical discriminant functions and structure loadings generated using the SAS macro DISCRIM2.

loading on $X1$ and $X4$, discriminated between the normal (1) and the overt groups. However, $CAN2$ is not effective in separating the chemical diabetic group. The narrow angle between the $X2$ and $X3$ variable vector in the same direction indicates that the plasma glucose variables are positively highly correlated. The correlations between $X1$ and $X4$ are moderate in size and act in the opposite direction from the plasma glucose variables.

The two canonical variables extracted from the CDA effectively discriminated between the three clinical diabetic groups. The difference between the normal and the chemical group is distinct. The discrimination between the normal and the overt group is effective when both $CAN1$ and $CAN2$ are used simultaneously. Therefore, the CDA can be considered to be an effective descriptive tool in discriminating between groups based on continuous predictor variables. If the variance–covariance between the groups is assumed to be equal and the predictor variables have joint multivariate normal distributions within each group, then the group differences can be tested statistically for group differences.

Discriminant function analysis (DFA): It is commonly used for classifying observations to predefined groups based on the knowledge of their quantitative attributes. Because the distribution within each diabetes group is found to be multivariate normal (Figure 6.5), a parametric method can be used to develop a discriminant function using a measure of their generalized squared distance. The discriminant function, also known as a classification criterion, is estimated by measuring the generalized squared distance.[8] The Mahalanobis distance squares among the three group means are presented in Table 6.15. The generalized squared distance between the normal and the diabetes groups are farther apart than the distances between the two diabetes groups.

The classification criterion in DFA is based on the measure of squared distance and the prior probability estimates. Either the individual within-group covariance matrices (a quadratic function) or the pooled covariance matrix (a linear function) can be used in deriving the squared distance. Since the chi-square value for testing the heterogeneity of between variance–covariance is significant at the 0.1 level (Table 6.16), the within-covariance matrices are used in computing the quadratic discriminant function.

Table 6.15 Discriminant Function Analysis Using SAS Macro: DISCRIM2—Mahalanobis Distance Measures between Group Levels

From Group	1	2	3
1	0	12.70	16.67
2	19.02	0	7.21
3	493.631	131.23	0

Table 6.16 Parametric Discriminant Analysis Using SAS Macro: DISCRIM2—Chi-Square Test for Heterogeneity of Variance–Covariance among Groups

Chi-Square	DF	Pr > ChiSq
313.500033	20	<.0001

In computing the classification criterion, the prior probability proportional to the group frequency was used in this analysis. The posterior probability of an observation belonging to each class is estimated, and each observation is classified in the group from which it has the smallest generalized squared distance or larger posterior probability. The DISCRIM macro also outputs a table of the ith group posterior probability estimates for all observations in the training dataset. Table 6.17 provides a partial list of the ith group posterior probability estimates for some of the selected observations. These posterior probability values are very useful estimates since these

Table 6.17 Parametric Discriminant Function Analysis Using SAS Macro: DISCRIM2—Partial List of Posterior Probability Estimates (PPE) by Groups in Cross-Validation Using the Quadratic Discriminant Function

Obs	From Group	Classified into Group	Group 1 PPE	Group 2 PPE	Group 3 PPE
1	1	1	0.9997	0.0003	0.0000
2	1	1	1.0000	0.0000	0.0000
3	1	1	1.0000	0.0000	0.0000
4	1	1	0.9968	0.0031	0.0001
5	1	1	0.9997	0.0003	0.0000
6	1	1	0.9995	0.0005	0.0000
7	1	1	1.0000	0.0000	0.0000
		—[a]			
141	3	3	0.0000	0.0000	1.0000
142	3	3	0.0000	0.0000	1.0000
143	3	3	0.0000	0.0000	1.0000
144	3	3	0.0000	0.0000	1.0000
145	3	3	0.0000	0.0000	1.0000

[a] Partial list of posterior probability estimates.

Table 6.18   Quadratic Discriminant Function Analysis Based on Cross-Validation Using SAS Macro: DISCRIM2—Posterior Probability of Group Membership for Misclassified Cases

Observation	From Group	Classified into Group	Group 1 PPE	Group 2 PPE	Group 3 PPE
5	1	2*	0.1242	0.8731	0.0027
29	1	2*	0.0540	0.9392	0.0068
61	1	2*	0.1954	0.8008	0.0039
74	2	1*	0.9698	0.0297	0.0006
120	3	2*	0.0000	0.9991	0.0009

* Denote a new group classification.

estimates can be successfully used in developing scorecards and ranking the observations in the dataset.

The performance of a discriminant criterion in classification is evaluated by estimating the probabilities of misclassification or error-rates. These error-rate estimates include error-count estimates and error-rate estimates. When the training dataset is a valid SAS dataset, the error-rate can also be estimated by cross-validation. In cross-validation, $n - 1$ out of n observations in the training sample are used. The DFA estimates the discriminant functions based on these $n - 1$ observations and then applies them to classify the one observation left out. This kind of validation is called *cross-validation* and is performed for each of the n training observations. The misclassification rate for each group is estimated from the proportion of sample observations in that group that are misclassified.

Table 6.18 lists the misclassified observations based on the posterior probability estimates computed by the quadratic discriminant function by cross-validation. Three cases that belong to the normal group are classified into the overt group since their posterior probability estimates are larger than 0.8 for group2 (overt). Furthermore, one case that belongs to the overt group is classified into the normal group, and one case is switched from the chemical group to the overt group.

Classification results based on the quadratic DF and error rates based on cross-validation are presented in Table 6.19. The misclassification rates in groups 1, 2, and 3 are 3.5%, 2.7%, and 3.55%, respectively. The overall discrimination is quite satisfactory since the overall error rate is very low, at 3.5%. The overall error rate is estimated through a weighted average of the individual group-specific error-rate estimates, where the prior probabilities are used as the weights. The posterior probability estimates based on cross-validation reduces both the bias and the variance of the classification function. The resulting overall error estimates are intended to have both a low variance from using the posterior probability estimate and a low bias from cross-validation.

Table 6.19 Parametric Discriminant Function Analysis Using SAS Macro DISCRIM2—Classification Table and Error-Count Estimates by Groups in Cross-Validation Using Quadratic Discriminant Function

From Group	To Group 1	2	3	Total
1	69[a]	3	0	72
	95.83[b]	4.17	0.00	100.00
2	1	35	0	36
	2.78	97.22	0.00	100.00
3	0	1	32	33
	0.00	3.03	96.97	100.00
Total	70	39	32	141
	49.65	27.66	22.70	100.00

Error-Count Estimates for Group	1	2	3	Total
Error Rate	0.041	0.027	0.030	0.035
Prior probability	0.510	0.255	0.234	

[a] Frequency.
[b] Percentage.

A box plot display of the ith posterior probability estimates by group is a powerful graphical tool to summarize the classification results, to detect false positive and negatives, and to investigate the variation in the ith posterior probability estimates. Figure 6.7 illustrates the variation in the posterior probability estimates for all three group levels. The posterior probability estimates of a majority of the cases that belong to the normal group are larger than 0.9. Three observations (#5, #29, #61) are identified as false negatives. One observation, (#75), which belongs to the overt group, is identified as a false positive. The posterior probability estimates of a majority of the cases that belong to the overt group are larger than 0.85. One observation (#74) is identified as a false negative. Three observations, (#5, #29, #29), which belong to the normal group, and one observation, (#120), which belongs to the chemical group, are identified as false positives. The posterior probability estimates of a majority of the cases that belong to the chemical group, are larger than 0.95. One observation (#120) is identified as a false negative, while no observations are identified as false positives.

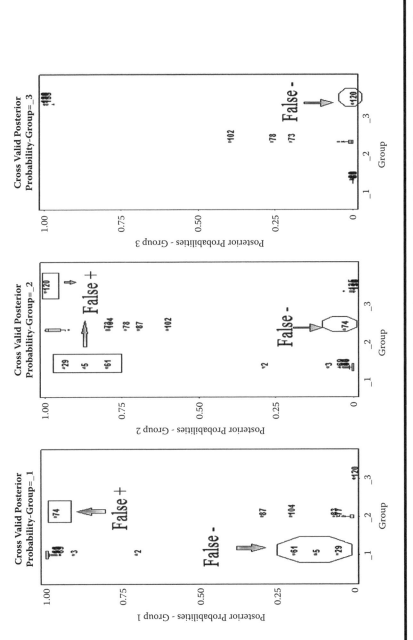

Figure 6.7 Box plot display of posterior probability estimates for all three group levels derived from parametric discriminant function analysis by cross-validation. This plot is generated using the SAS macro DISCRIM2.

Although the classification function by cross-validation achieves a nearly unbiased overall error estimate, it has a relatively large variance. To reduce the variance in an error-count estimate, a smoothed error-rate estimate is suggested.[14] Instead of summing values that are either zero or one as in the error-count estimation, the smoothed estimator uses a continuum of values between zero and one in the terms that are summed. The resulting estimator has a smaller variance than the error-count estimate. The posterior probability error-rate estimates are smoothed error-rate estimates. The posterior probability error-rate estimates for each group are based on the posterior probabilities of the observations classified into that same group. The posterior probability estimates provide good estimates of the error-rate when the posterior probabilities are accurate. When a parametric classification criterion (linear or quadratic discriminant function) is derived from a nonnormal population, the resulting posterior probability error-rate estimators may not be appropriate.

The smoothed posterior probability error-rate estimates based on the cross-validation quadratic DF are presented in Table 6.20. The overall error rate for stratified

Table 6.20   Parametric Discriminant Function Analysis Using SAS Macro DISCRIM2—Classification Table and Posterior Probability Error-Rate Estimates by Groups in Cross-Validation Using Quadratic Discriminant Functions

	To Group			
From Group	*1*	*2*	*3*	
1	69[a]	3	0	
	0.990[b]	0.871	0.00	
2	1	35	0	
	0.969	0.947	0.00	
3	0	1	32	
	0.00	0.999	0.999	
Total	70	39	32	
	0.989	0.942	0.999	

Posterior Probability Error-Rate Estimates for Group				
Estimate	*1*	*2*	*3*	*Total*
Stratified	0.037	−0.021	0.030	0.020
Unstratified	0.037	−0.021	0.030	0.020
Priors	0.510	0.255	0.234	

[a] Frequency.
[b] Percentage.

and unstratified estimates is equal since the prior probability option proportional to group frequency is used. The overall discrimination is quite satisfactory since the overall error rate using the smoothed posterior probability error rate is very low, at 2.1%.

The classification function derived from the training dataset can be validated by classifying the observations in an independent validation dataset. If the classification error rate obtained for the validation data is small and similar to the classification error rate for the training data, then we can conclude that the derived classification function has good discriminative potential. Classification results for the validation dataset based on the quadratic discriminant function are presented in Table 6.21. The misclassification rates in groups 1, 2, and 3 are 1.3%, 5.5%, and 9.0%, respectively. The overall discrimination in the validation dataset is quite good since the weighted error rate is very low, at 4.2%. A total of five observations in the validation dataset are misclassified, and Table 6.22 lists these five misclassified observations. For example, observation number 75 is classified into overt from the normal diabetes group. Because the training and validation datasets gave

Table 6.21 Parametric Discriminant Function Analysis Using SAS Macro DISCRIM2—Classification Table and Error-Count Estimates by Groups for Validation Data Using Quadratic Discriminant Functions

From Group	To Group			Total
	1	*2*	*3*	*Total*
1	75[a]	1	0	76
	98.68[b]	1.32	0.00	100.00
2	2	34	0	36
	5.56	94.44	0.00	100.00
3	0	3	30	33
	0.00	9.09	90.91	100.00
Total	77	38	30	145
	53.10	26.21	20.69	100.00

Error-Count Estimates for Group				
	1	*2*	*3*	*Total*
Rate	0.013	0.055	0.090	0.042
Priors	0.510	0.255	0.234	

[a] Frequency
[b] Percentage

Table 6.22 Parametric Discriminant Function Analysis Using SAS Macro DISCRIM2—Misclassified Observations in Validation Data Using Quadratic Discriminant Functions

Observation	X1	X2	X3	X4	From Group	Into Group
75	1.11	93	393	490	1	2
83	1.08	94	426	213	2	1
110	0.94	88	423	212	2	1
131	1.07	124	538	460	3	2
134	0.81	123	557	130	3	2
136	1.01	120	636	314	3	2

comparable classification results based on parametric quadratic functions, we can conclude that blood plasma and insulin measures are quite effective in grouping the clinical diabetes.

6.12 Case Study 2: Nonparametric Discriminant Function Analysis

When the assumption regarding multivariate normality within each group level is not met, nonparametric discriminant function analysis (DFA) is considered to be a suitable classification technique. The diabetes dataset (diabet2)[11,17] used in Section 6.11 for validation is used as the training dataset to derive the classification function using blood plasma and insulin measures. Because the multivariate normality assumption was not met for this diabetes dataset, this is a suitable candidate for performing nonparametric DFA. The simulated diabetes dataset (diabet1) used in Section 6.11 for training is used as the validation dataset.

6.12.1 Study Objectives

To discriminate between three clinical diabetic groups (normal, overt diabetic, and chemical diabetic) using blood plasma and insulin measures.

1. *Checking for any violations of discriminant analysis assumptions*: Perform statistical tests and graphical analysis to detect multivariate influential outliers and departure from multivariate normality.
2. *Nonparametric discriminant analyses*: Because this dataset significantly violates the multivariate normality assumption, you can perform nonparametric

discriminant function analyses based on the nearest-neighbor and kernel density methods. It develops classification functions based on nonparametric posterior probability density estimates and assigns observations into predefined group levels and measures the success of discrimination by comparing the classification error rates.

3. *Saving "plotp" and "out2" datasets for future use*: Running the DISCRIM2 macro creates these two temporary SAS datasets and saves them in the work folder. The "plotp" dataset contains the observed predictor variables, group response value, posterior probability scores, and new classification results. This posterior probability score for each observation in the dataset can be used as the base for developing the scorecards and ranking the patients. If you include an independent validation dataset, the classification results for the validation dataset are saved in a temporary SAS dataset called "out2," which can be used to develop scorecards for new patients.

4. *Validation*: This step validates the derived discriminant functions obtained from the training data by applying these classification criteria to the independent simulated dataset and verifying the success of classification.

6.12.2 Data Descriptions

Dataset names	a. Training: SAS dataset diabet2[11,17] located in the SAS work folder
	b. Validation: SAS dataset diabet1 (simulated) located in the SAS work folder
Group response (Group)	Group (three clinical diabetic groups: 1—normal; 2—overt diabetic; 3—chemical diabetic)
Predictor variables (X)	$X1$: Relative weight
	$X2$: Fasting plasma glucose level
	$X3$: Test plasma glucose
	$X4$: Plasma insulin during test
	$X5$: Steady-state plasma glucose level
Number of observations	Training data (diabet2): 145
	Validation data (diabet1): 141
Source	c. Training data: real data.[11,17]
	d. Validation data: simulated data

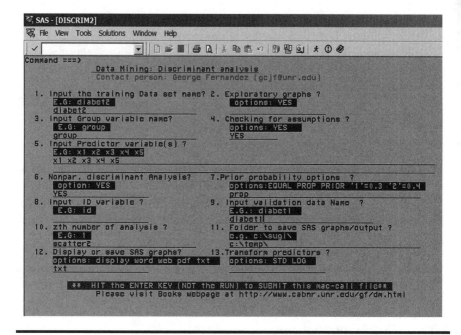

Figure 6.8   Screen copy of DISCRIM2 macro-call window showing the macro-call parameters required for performing nonparametric discriminant analysis.

Open the DISCRIM2.SAS macro-call file in the SAS EDITOR window, and click RUN to open the DISCRIM2 macro-call window (Figure 6.8). Input the appropriate macro-input values by following the suggestions given in the help file (Appendix 2).

Exploratory analysis/diagnostic plots: Input dataset name, group variable, predictor variable names, and the prior probability option. Input YES in macro field #2 to perform data exploration and create diagnostic plots. Submit the DISCRIM2 macro and discriminant diagnostic plots, and automatic variable selection output will be produced.

Data exploration and checking: A simple two-dimensional scatter plot matrix showing the discrimination of three diabetes groups is presented in Figure 6.9. These scatter plots are useful in examining the range of variation in the predictor variables and the degree of correlations between any two predictor variables. The scatter plot presented in Figure 6.9 revealed that a strong correlation existed between fasting plasma glucose level ($X2$) and test plasma glucose ($X3$). These two attributes appeared to discriminate diabetes group 3 from the other two groups to a certain degree. Discrimination between the normal and the overt diabetes group is not very distinct. The details of variable selection results are not discussed here since these results are similar to Case Study 1 in this chapter.

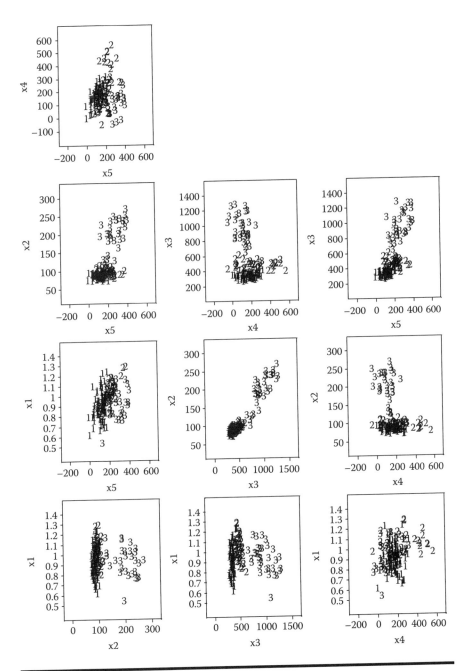

Figure 6.9 Bivariate exploratory plots generated using the SAS macro DISCRIM2: Group discrimination of three types of diabetic groups (data=diabet2) in simple scatter plots.

Discriminant analysis and checking for multivariate normality: Open the DISCRIM2.SAS macro-call file in the SAS EDITOR window, and click RUN to open the DISCRIM2 macro-call window (Figure 6.8). Input the appropriate macro-input values by following the suggestions given in the help file (Appendix 2). Input the dataset name, group variable, predictor variable names, and the prior probability option. Leave macro field #2 BLANK, and input YES in option #6 to perform nonparametric DFA. Also input YES to perform a multivariate normality check in macro field #4. Submit the DISCRIM2 macro, and you will get the multivariate normality check and the nonparametric DFA output and graphics.

Checking for multivariate normality: This multivariate normality assumption can be checked by estimating multivariate skewness, kurtosis, and testing for their significance levels. The quantile-quantile (Q-Q) plot of expected and observed distributions[9] of multiattribute residuals can be used to graphically examine multivariate normality for each response group levels. The estimated multivariate skewness and multivariate kurtosis (Figure 6.10) clearly support the hypothesis that these five multiattributes do not have a joint multivariate normal distribution. A significant departure from the 45° angle reference line in the Q-Q plot (Figure 6.10) also supports this finding. Thus, nonparametric discriminant analysis must be considered to be the appropriate technique for discriminating between the three clinical groups based on these five attributes ($X1$ to $X5$).

Checking for the presence of multivariate outliers: Multivariate outliers can be detected in a plot between the differences of robust (*Mahalanobis distance–chi-squared quantile*) versus chi-squared quantile value.[9] Eight observations are identified as influential observations (Table 6.23) because the difference between robust Mahalanobis distance and chi-squared quantile values is larger than 2 and falls outside the critical region (Figure 6.11).

When the distribution within each group is assumed to not have multivariate normal distribution, nonparametric DFA methods can be used to estimate the group-specific densities. Nonparametric discriminant methods are based on nonparametric estimates of group-specific probability densities. Either a kernel method or the k-nearest-neighbor method can be used to generate a nonparametric density estimate for each group level and to produce a classification criterion.

The group-level information and the prior probability estimate used in performing the nonparametric DFA are given in Table 6.24. By default, the DISCRIM2 macro performs three (k = 2, 3, and 4) nearest-neighbor (NN) and one kernel density (KD) (unequal bandwidth kernel density) nonparametric DFA. We can compare the classification summary and the misclassification rates of these four different nonparametric DFA methods and can pick one that gives the smallest classification error in the cross-validation.

Among the three NN-DFA (k = 2, 3, 4), classification results based on the second NN nonparametric DFA gave the smallest classification error. The classification summary and the error rates for NN (k = 2) are presented in Table 6.25. When

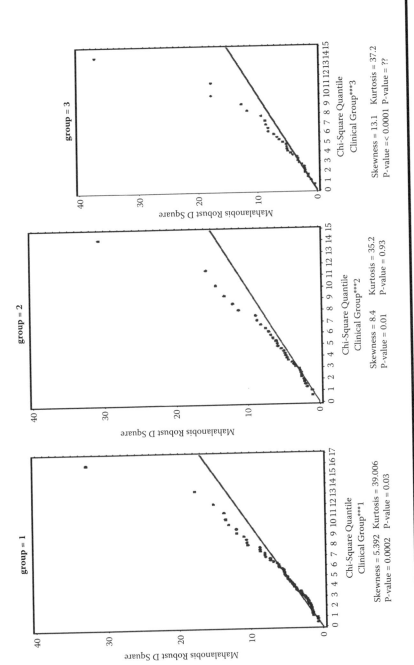

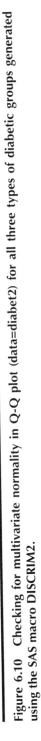

Figure 6.10 Checking for multivariate normality in Q-Q plot (data=diabet2) for all three types of diabetic groups generated using the SAS macro DISCRIM2.

Table 6.23 Detecting Multivariate Outliers and Influential Observations with SAS Macro DISCRIM2

Observation ID	Robust Distance Squared (RDSQ)	Chi-Square	Difference (RDSQ–Chi-Square)
82	29.218	17.629	11.588
86	23.420	15.004	8.415
69	20.861	12.920	7.941
131	21.087	13.755	7.332
111	15.461	12.289	3.172
26	14.725	11.779	2.945
76	14.099	11.352	2.747
31	13.564	10.982	2.582

the k-nearest-neighbor method is used, the Mahalanobis distances are estimated based on the pooled covariance matrix. Classification results based on NN ($k = 2$) and error rates based on cross-validation are presented in Table 6.25. The misclassification rates in group levels 1, 2, and 3 are 1.3%, 0%, and 12.0%, respectively. The overall discrimination is quite satisfactory since the overall error rate is very low at 3.45%. The posterior probability estimates based on cross-validation reduces both the bias and the variance of classification function. The resulting overall error estimates are intended to have both low variance from using the posterior probability estimate and a low bias from cross-validation.

Figure 6.12 illustrates the variation in the posterior probability estimates for the three diabetic group levels. The posterior probability estimates of a majority of the cases that belong to the normal group are larger than 0.95. One observation (#69) is identified as a false negative, while no other observation is identified as a false positive. A small amount of intragroup variation for the posterior probability estimates was observed. A relatively large variability for the posterior probability estimates is observed for the second overt diabetes group and ranges from 0.5 to 1. No observation is identified as a false negative. However, five observations, one belonging to the normal group and 4 observations belonging to the chemical group, are identified as false positives. The posterior probability estimates for a majority of the cases that belong to the chemical group are larger than 0.95. One observation is identified as a false negative, but no observations are identified as false positives.

The DISCRIM2 macro also output a table of the *i*th group posterior probability estimates for all observations in the training dataset. Table 6.26 provides a partial list of the *i*th group posterior probability estimates for some of the selected

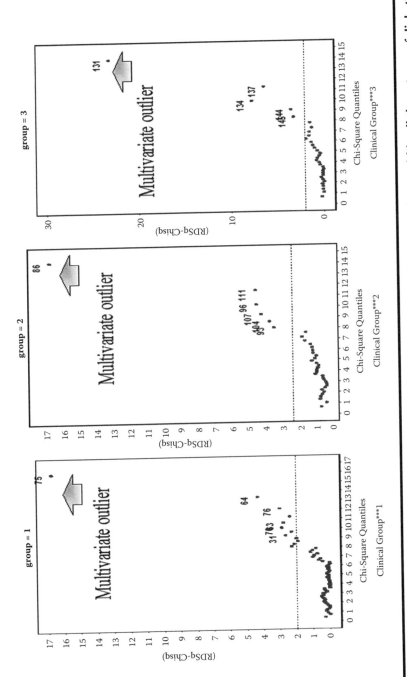

Figure 6.11   Diagnostic plot for detecting multivariate influential observations (data=diabet2) within all three types of diabetic groups generated using the SAS macro DISCRIM2.

Table 6.24 Nonparametric Discriminant Function Analysis Using SAS Macro DISCRIM2—Class-Level Information

Group	Group Level Name	Frequency	Weight	Proportion	Prior Probability
1	_1	76	76	0.524	0.524
2	_2	36	36	0.248	0.248
3	_3	33	33	0.227	0.227

observations in the table. These posterior probability values are very useful estimates since they can be successfully used in developing scorecards and ranking the observations in the dataset.

Smoothed posterior probability error rate: The posterior probability error-rate estimates for each group are based on the posterior probabilities of the observations

Table 6.25 Nearest-Neighbor ($k = 2$) Nonparametric Discriminant Function Analysis Using SAS Macro DISCRIM2: Classification Summary Using Cross-Validation

From Group	To Group			Total
	1	2	3	
1	75[a]	1	0	76
	98.68[b]	1.32	0.00	100.00
2	0	36	0	36
	0.00	100.00	0.00	100.00
3	0	4	29	33
	0.00	12.12	87.88	100.00
Total	75	41	29	145
	51.72	28.28	20.00	100.00

Error-Count Estimates for Group				
	1	2	3	Total
Rate	0.013	0.000	0.121	0.034
Priors	0.524	0.248	0.227	

[a] Number of observations.
[b] Percent.

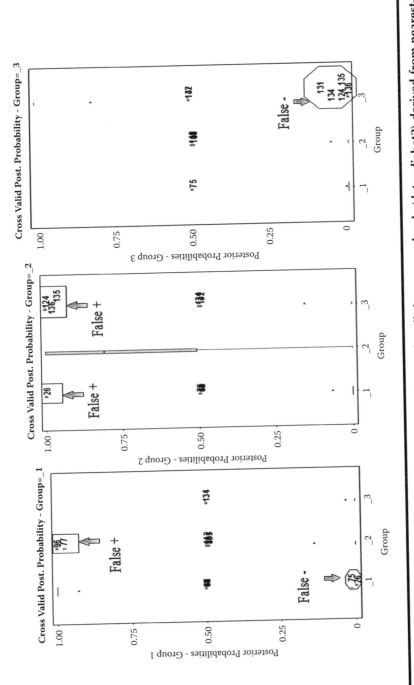

Figure 6.12 Box plot display of posterior probability estimates for all three group levels (data=diabet2) derived from nearest-neighbor (*k* = 2) nonparametric discriminant function analysis by cross-validation. This plot is generated using the SAS macro DISCRIM2.

Table 6.26 Nearest-Neighbor ($k = 2$) Nonparametric Discriminant Function Analysis Using SAS Macro DISCRIM2: Partial List of Posterior Probability Estimates by Group Levels in Cross-Validation

		Posterior Probability of Membership in Group			
Obs	From Group	Classified into Group	1	2	3
1	1	1	0.9999	0.0001	0.0000
2	1	2*	0.1223	0.8777	0.0001
3	1	1	0.7947	0.2053	0.0000
4	1	1	0.9018	0.0982	0.0000
5	1	2*	0.4356	0.5643	0.0001
6	1	1	0.8738	0.1262	0.0000
7	1	1	0.9762	0.0238	0.0000
8	1	1	0.9082	0.0918	0.0000
		Partial List of Posterior Probability Estimates			
137	3	1*	0.9401	0.0448	0.0151
138	3	3	0.0000	0.3121	0.6879
139	3	3	0.0000	0.0047	0.9953
140	3	3	0.0000	0.0000	1.0000
141	3	3	0.0000	0.0011	0.9988

classified into that same group level. The posterior probability estimates provide good estimates of the error rate when the posterior probabilities are accurate. The smoothed posterior probability error-rate estimates based on the cross-validation quadratic DF are presented in Table 6.27. The overall error rate for stratified and unstratified estimates is equal since group proportion was used as the prior probability estimate. The overall discrimination is quite satisfactory since the overall error rate using the smoothed posterior probability error rate is relatively low, at 6.8%.

If the classification error rate obtained for the validation data is small and similar to the classification error rate for the training data, then we can conclude that the derived classification function has good discriminative potential. Classification results for the validation dataset based on NN ($k = 2$) classification functions are presented in Table 6.28. The misclassification rates in group levels 1, 2, and 3 are 4.1%, 25%, and 15.1%, respectively.

Table 6.27 Nearest-Neighbor (*k* = 2) Nonparametric Discriminant Function Analysis Using SAS Macro DISCRIM2: Classification Summary and Smoothed Posterior Probability Error-Rate in Cross-Validation

From Group	*To Group* 1	2	3	
1	75 0.960	1 1.000	0 0.00	
2	0 0.00	36 0.835	0 0.00	
3	0 0.00	4 1.000	29 0.966	
Total	75 0.960	41 0.855	29 0.966	

Posterior Probability Error-Rate Estimates for Group				
Estimate	*1*	*2*	*3*	*Total*
Stratified	0.052	0.025	0.151	0.068
Unstratified	0.052	0.025	0.151	0.068
Priors	0.524	0.248	0.227	

The overall discrimination in the validation dataset (diabet1) is moderately good since the weighted error rate is 11.2%. A total of 17 observations in the validation dataset are misclassified. Table 6.29 shows a partial list of probability density estimates and the classification information for all the observations in the validation dataset. The misclassification error rate estimated for the validation dataset is relatively higher than that obtained from the training data. We can conclude that the classification criterion derived using NN (*k* = 2) performed poorly in validating the independent validation dataset. The presence of multivariate influential observations in the training dataset might be one of the contributing factors for this poor performance in validation. Using larger *k* values in NN DFA might do a better job of classifying the validation dataset.

DISCRIM2 also performs nonparametric discriminant analysis based on nonparametric kernel density (KD) estimates with unequal bandwidth. The kernel method in the DISCRIM2 macro uses normal kernels in the density estimation. In the KD method, the Mahalanobis distances based on either the individual within-group covariance matrices or the pooled covariance matrix can be used.

Table 6.28 Classification Summary and Error-Rate for Validation Dataset Using k = 2 Nearest-Neighbor Method Using SAS Macro DISCRIM2

From Group	Into Group			Total
	1	*2*	*3*	*Total*
1	69	3	0	72
	95.83	4.17	0.00	100.00
2	9	27	0	36
	25.00	75.00	0.00	100.00
3	2	3	28	33
	6.06	9.09	84.85	100.00
Total	80	33	28	141
	56.74	23.40	19.86	100.00
Error-Count Estimates for Group				
	1	*2*	*3*	*Total Error Rate*
Error-rate	0.0417	0.2500	0.1515	0.1184
Prior Probability	0.5241	0.2483	0.2276	

Table 6.29 Partial List of Classification Results in Validation Dataset by Nearest-Neighbor (k = 2) Method Using SAS Macro DISCRIM2

Observation Number	*x1*	*x2*	*x3*	*x4*	From Group	Into Group
1	0.81	80	356	124	1	1
2	0.95	97	289	117	1	1
3	0.94	105	319	143	1	1
4	1.04	90	356	199	1	1
	—a					
142	0.91	180	923	77	3	3
143	0.9	213	1025	29	3	3
144	1.11	328	1246	124	3	3
145	0.74	346	1568	15	3	3

a Partial list.

Table 6.30 Unequal Bandwidth Kernel Density Discriminant Function Analysis Using SAS Macro DISCRIM2: Classification Summary Using Cross-Validation Results

Number of Observations and Percentage Classified into Group				
From Group	1	2	3	Total
1	70	5	1	76
	92.11	6.58	1.32	100.00
2	2	30	4	36
	5.56	83.33	11.11	100.00
3	0	3	30	33
	0.00	9.09	90.91	100.00
Total	72	38	35	145
	49.66	26.21	24.14	100.00
Error-Count Estimates for Group				
	1	2	3	Total Error
Error-rate	0.078	0.166	0.090	0.103
Prior probability	0.524	0.248	0.227	

The classification of observations in the training data is based on the estimated group-specific densities from the training dataset. From these estimated densities, the posterior probabilities of group membership are estimated.

The classification summary using KD (normal, unequal bandwidth) nonparametric DFA and the error rates using cross-validation are presented in Table 6.30. The misclassification rates in groups 1, 2, and 3 are 7.8%, 16.6%, and 9.0%, respectively. Thus, an overall success rate of correct discrimination is about 90% since the overall error rate is about 10.3% higher than the overall error rate (3.4%) for the $k = 2$ NN method.

Figure 6.13 illustrates the variation in the posterior probability estimates for all three groups. The posterior probability estimates for a majority of the cases that belong to the normal group are larger than 0.95. Seven observations belonging to the normal group level are identified as false negatives. Two observations belonging to the overt group are identified as false positive. Very small amounts of intragroup variation for the normal group posterior probability estimates were observed. A relatively large variability for the posterior probability estimates is observed for the second overt diabetes group level and ranged from 0.75 to 1. Six observations belonging to the overt group are identified as false negatives.

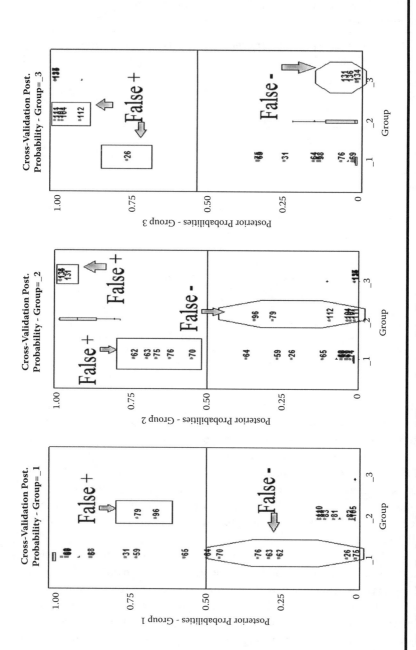

Figure 6.13 Box plot display of posterior probability estimates for all three group levels (data=diabet2) derived from nonparametric unequal bandwidth kernel density discriminant function analysis by cross-validation. This plot is generated using the SAS macro DISCRIM2.

Five observations belong to the normal group, and two observations belonging to the chemical group are identified as false positives. The posterior probability estimates for a majority of the cases that belong to the chemical group are larger than 0.95. Two observations from the chemical group are identified as false negatives, while four observations belonging to the overt group are identified as false positives.

The posterior probability error-rate estimates for each group are based on the posterior probabilities of the observations classified into that same group. The smoothed posterior probability error-rate estimates based on cross-validation DFA are presented in Table 6.31. The overall error-rate for stratified and unstratified estimates are equal since group proportion was used as the prior probability estimate. The overall discrimination is quite satisfactory since the overall error-rate using the smoothed posterior probability error rate is relatively low, at 4.3%.

If the classification error-rate obtained for the validation data is small and similar to the classification error-rate for the training data, then we can conclude that

Table 6.31   Unequal Bandwidth Kernel Density Discriminant Function Analysis Using SAS Macro DISCRIM2: Classification Summary and Smoothed Posterior Probability Error-Rates Based on Cross-Validation

	To Group			
From Group	*1*	*2*	*3*	
1	70	5	1	
	0.970	0.658	0.759	
2	2	30	4	
	0.690	0.946	0.969	
3	0	3	30	
	0.00	0.996	0.999	
Total	72	38	35	
	0.963	0.912	0.989	
Posterior Probability Error-Rate Estimates for group				
Estimate	*1*	*2*	*3*	*Total*
Stratified	0.087	0.036	−0.049	0.043
Unstratified	0.087	0.036	−0.049	0.043
Prior probability	0.524	0.248	0.227	

Table 6.32   Classification Summary and Error-Rate for Validation Dataset (Diabet1) Using Unequal Bandwidth Kernel Density Discriminant Function Analysis with SAS Macro DISCRIM2

Number of Observations and Percentage Classified into Group				
From Group	1	2	3	Total
1	69	3	0	72
	95.83	4.17	0.00	100.00
2	9	27	0	36
	25.00	75.00	0.00	100.00
3	2	3	28	33
	6.06	9.09	84.85	100.00
Total	80	33	28	141
	56.74	23.40	19.86	100.00
Priors	0.52414	0.24828	0.22759	
Error-Count Estimates for Group				
	1	2	3	Total Error-Rate
Error-rate	0.041	0.250	0.151	0.118
Prior probability	0.524	0.248	0.227	

the derived classification function has good discriminative potential. Classification results for the validation dataset based on KD (normal, unequal bandwidth) non-parametric DFA classification functions are presented in Table 6.32. The mis-classification rates in three group levels 1, 2, and 3 are 4.1%, 25%, and 15.1%, respectively.

The overall discrimination in the validation dataset is moderately good since the weighted error-rate is 11.8%. A total of 17 observations in the validation dataset are misclassified. Table 6.33 shows a partial list of probability density estimates and the classification information for all the observations in the validation dataset. The misclassification error-rate estimated for the validation dataset is higher than the error-rate obtained from the training data. We can conclude that the classification criterion derived using KD (normal and unequal bandwidth) performed poorly in validating the independent validation data-set. The presence of multivariate influential observations in the training dataset might be one of the contributing factors to this poor performance in validation.

Table 6.33 Misclassified Observations in Validation (Diabet1) Dataset by the Unequal Bandwidth Kernel Density Discriminant Function Analysis Using SAS Macro DISCRIM2

Observation ID	x1	x2	x3	x4	x5	From Group	Into Group
2	1.0017	76.127	378.69	296.34	206.07	1	2
29	1.2519	94.588	439.34	260.81	187	1	2
61	1.1095	101.87	423.92	324.94	207.82	1	2
80	0.9916	98.293	445.43	−43.83	162.16	2	1
82	0.8679	92.868	498.67	59.006	113.38	2	1
83	0.9162	87.208	450.96	125.4	155.03	2	1
93	0.8956	112.02	580.41	40.301	90.217	2	1
98	0.9194	89.872	459.24	137.31	103.32	2	1
102	0.9492	117.89	510.63	233.84	122.3	2	1
104	1.0744	93.793	435.95	110.52	187.27	2	1
105	1.1961	114.51	492.52	44.762	232.42	2	1
108	0.9343	106.97	508.96	185	87.276	2	1
120	1.0439	103.84	547.63	322.49	259.23	3	2
121	0.9947	203.72	816.96	176.81	201.23	3	1
126	0.9436	119.83	732.16	162.04	173.31	3	2
135	1.026	133.56	665.97	149.9	193.46	3	2
137	0.9905	228.35	849.61	92.44	145.71	3	1

Using other types of density options might do a better job of classifying the validation dataset.

6.13 Case Study 3: Classification Tree Using CHAID

CHAID analysis is a powerful classification tree method suitable for classifying observations into predetermined groups based on easy-to-follow decision rules. Unlike discriminant analysis, both categorical (nominal) and continuous variables can be used as predictors. Multivariate normality or between-group equal variance–covariance assumptions are not required to run the CHAID analysis. In Case Study 3, the features of CHAID analysis are revealed by fitting a classification model using the diabetes dataset described in Section 6.12. By comparing the classification summary and misclassification error-rates, the similarities and differences between the CHAID and DFA are presented here. The diabetes dataset[11,17] used in Section 6.12 is used as the training dataset to derive the decision tree using blood plasma and insulin measures. The simulated diabetes dataset used in Section 6.12 is used as the validation dataset.

6.13.1 Study Objectives

To discriminate between three clinical diabetic groups (normal, overt diabetic, and chemical diabetic) using blood plasma and insulin measures by developing a decision tree model.

> *Develop classification tree functions*: Develop classification functions based on CHAID analysis and assign observations into predefined group levels. Measure the success of classification by comparing the classification error rates.
>
> *Construct the decision tree*: Construct easy-to-follow decision trees using the decision rules generated in the CHAID analysis.
>
> *Validation*: Validates the derived classification functions obtained from the training data (diabet2) by applying these classification criteria to the independent validation dataset (diabet1) and verifying the success of classification.

6.13.2 Data Descriptions

Dataset names	a. Training: SAS dataset diabet2[11,17]
	b. Validation: SAS dataset diabet1 (simulated)
Group response *(Group) and predictor Variables (X)*	Group (three clinical diabetic groups: 1—normal; 2—overt diabetic; 3—chemical diabetic)
	Continuous predictors:
	$X1$: Relative weight
	$X4$: Plasma insulin during test
	$X5$: Steady-state plasma glucose level
	Nominal predictors:
	Fastplgp: Fasting plasma glucose values (*L:* <100; *M* = 100–200; *H* ≥ 200)

	Tstplgp1: Test plasma glucose values (L: <600; M = 600–1200; $H \geq$ 1200)
Number of observations	a. Training data: 145
	b. Validation data 141
Source:	a. Training data: Real data[11,17]
	b. Validation data; simulated data

Open the CHAID2.SAS macro-call file in the SAS EDITOR window, and click RUN to open the CHAID2 macro-call window (Figure 6.14). Input the appropriate macro-input values by following the suggestions given in the help file (Appendix 2). Input the dataset name, group response variable, and nominal and ordinal predictor variable names. Also, input the validation dataset name if you would like to validate the classification tree. Submit the CHAID2 macro and stepwise variable selection summary, exploratory graphics, classification summary plots, and the decision tree diagrams will be produced.

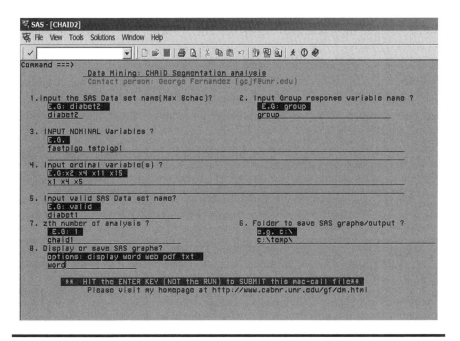

Figure 6.14   Screen copy of CHAID2 macro-call window showing the macro-call parameters required for performing chi-squared automatic interaction detection analysis.

The CHAID analysis macro included in the first edition of the book generates relatively small amounts of output and graphics. Therefore, in the revised CHAID2 macro, new exploratory analyses such as stepwise variable selection, exploratory graphical analysis between the response group variable, and all nominal and the ordinal predictor variables are included. However, for a through data exploration and preliminary analysis, fitting classification models using LOGISTIC2 and DISCRIM2 models before trying the CHAID analysis using the CHAID2 macro is highly recommended.

The results of the preliminary stepwise selection using the SAS STEPDISC procedure are presented in Table 6.34. All five variables (both two nominal and three ordinal) included in the analysis were identified as potentially significant predictors that can be used in the classification of the three diabetes groups. The CHAID2 macro converts all nominal variables to dummy variables before running the stepwise variable selection. In addition, exploratory graphical analysis plots are also automatically generated between all nominal and ordinal variables with the response group variables. The donut and the frequency plots generated between the two nominal variables and the response group variable clearly indicated the classification potential of these two variables (Figures 6.15 and 6.16). For ordinal or continuous variables, the box plot display is used to show the classification potential. Figure 6.17 clearly shows the strong classification potential of the X_5: steady-state plasma glucose level variable. These exploratory analyses can be successfully used in eliminating potentially nonsignificant predictor variables before starting the CHAID analysis.

If the CHAID2 macro runs successfully, the classification summary for the training data will be displayed in a donut chart. Classification results based on the CHAID analysis are presented in Figure 6.18. The misclassification rates in groups 1, 2, and 3 are 8%, 28%, and 3%, respectively. The overall error-rate is not computed directly. However, an overall error-rate can be computed by an average error-rate weighted by the group frequency. In case of a validation dataset, relatively more classification errors were observed (Figure 6.19). The misclassification rates in groups 1, 2, and 3 are 4%, 44%, and 3%, respectively. By comparing the classification errors obtained from the CHAID2 and the DISCRIM2 macros, we can conclude that the discriminatory power of the DISCRIM2 macro is superior to the CHAID2 macro in correctly classifying the three diabetes groups.

The CHAID2 macro generates a decision tree and produces a graphics file using the NETDRAW procedure available in the SAS OR module. Therefore, to generate the decision tree, the SAS OR module should be installed in your computer. The decision tree generated by the CHAID2 macro is given in Figure 6.20. When the decision tree graphics are too large to fit within a page, SAS splits the graphics file and outputs them in multiple pages. These split graphics files could be joined back into a single graphic. When the graphics files are joined, the resolution of the graphics file could be reduced and, as a result, the interpretation of the decision tree can become unclear.

Table 6.34 Stepwise Variable Selection Using PROC STEPDISC for Excluding Nonsignificant Variables in the Exploratory Analysis of CHLID Using SAS Macro CHAID2

Step	Number In	Entered	Removed	Label	Partial R-Square	F Value	Pr > F	Wilks' Lambda	Pr < Lambda	Average Squared Canonical Correlation	Pr > ASCC
1	1	C5		TSTPLGP1 L	0.8857	550.04	<0.0001	0.1143	<0.0001	0.44283788	<0.0001
2	2	X5		Steady-state plasma glucose	0.3024	30.56	<0.0001	0.0797	<0.0001	0.55771944	<0.0001
3	3	C3		FASTPLGP m	0.1562	12.95	<0.0001	0.067	<0.0001	0.60692243	<0.0001
4	4	X4		Plasma insulin during test	0.0448	3.26	0.0415	0.064	<0.0001	0.62199512	<0.0001
5	5	X1		Relative weight	0.0359	2.57	0.0801	0.061	<0.0001	0.62778313	<0.0001

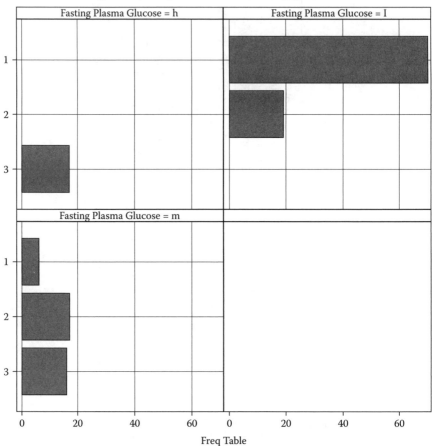

Exploratory Analysis of Nominal Data Fastplgp by GROUP

Freq Table

Figure 6.15 Exploratory frequency chart showing the diabetes group classification by nominal variable fasting plasma glucose. This plot is generated using the SAS macro CHAID2.

One way to solve this problem is to generate the decision tree manually using the decision rule generated by the CHAID2 macro. The SAS OR module is not required to generate the decision rules of classification, and a copy of the decision rule is presented in Table 6.35. Based on these rules, a decision tree can be constructed manually using any suitable drawing program. The decision tree generated by the Corel Presentation DRAW program is presented in Figure 6.21.

In the decision tree construction, first, all the observations in the training dataset are grouped together in the tree trunk node. In the first step, the nominal

Exploratory Analysis of Nominal Data tstplgp1 by GROUP

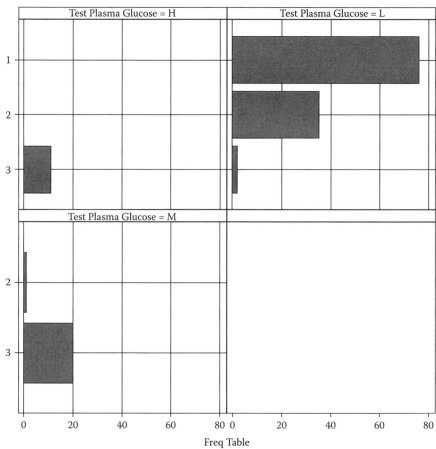

Figure 6.16 Exploratory frequency chart showing the diabetes group classification by nominal variable test plasma glucose. This plot is generated using the SAS macro CHAID2.

variable test plasma glucose (TSTPLGP1) is identified as the most significant predictor in separating observations in the three clinical groups. In the first split, all except one chemical diabetes case is correctly identified based on the following decision rule:

Chemical group (Terminal/leaf node): TSTPLGP1 >= 600 (with one misclassification)

Normal and overt: TSTPLGP1 < 600 (with one misclassification)

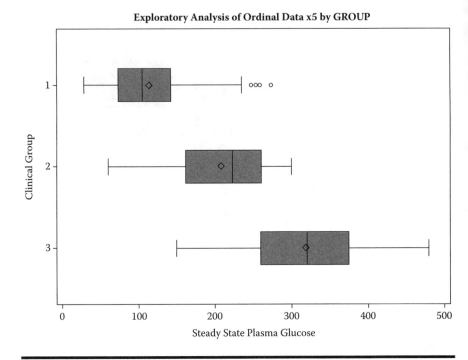

Figure 6.17 Exploratory box plot chart showing the diabetes group classification by ordinal variable steady-state plasma glucose. This plot is generated using the SAS macro CHAID2.

In the second split, the steady-state plasma glucose level is selected as the important predictor to discriminate the normal group from the overt group by the following decision rule:

Normal group: $X5 < 170$ (69 out of 72 normal
group plus 9 overt cases misclassified as the normal)

overt: $X5 > 170$ (26 out of 35 overt group plus 4 normal cases misclassified as overt)

The subsequent splits are statistically not significant. Thus, we could stop at this step and interpret the decision tree. Therefore, using two predictor variables and easy-to-follow decision rules, 126 out of the 141 cases can be correctly classified. Thus, the CHAID2 SAS macro provides a simple but very valuable classification tool to classify the categorical response with acceptable predictive accuracy.

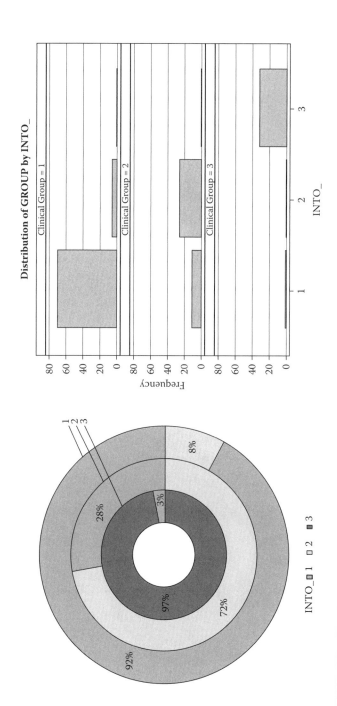

Figure 6.18 Donut chart showing the training data classification summary display generated using the SAS macro CHAID2.

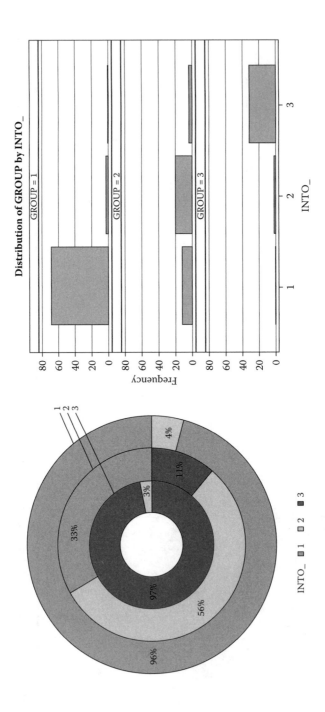

Figure 6.19 Donut chart showing the validation data classification summary display generated using the SAS macro CHAID2.

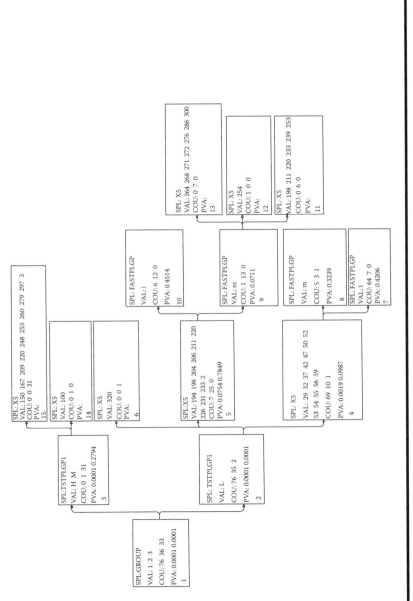

Figure 6.20 Decision tree diagram generated by the SAS macro CHAID2 using the net draw procedure available in the SAS/OR module.

Table 6.35   Decision Rules Generated by the CHAID Analysis Using the SAS Macro CHAID2: CHAID Analysis of Dependent Variable (DV) GROUP

GROUP value(s): 1 2 3
DV counts: 76 36 33 Best p-value(s): 0.0001 0.0001
TSTPLGP1 value(s): L
DV counts: 76 35 2 Best p-value(s): 0.0001 0.0001
X5 value(s):
29 32 37 42 47 50 52 53 54 55 56 59 60 66 68 71 73 74 76 78 80 83 85 90
91 93 94 96 98 99 102 103 105 106 108 109 111 117 118 119 122 124 128 132 13
5 etc.
DV counts: 69 10 1 Best p-value(s): 0.0019 0.0987
FASTPLGP value(s): M
DV counts: 8 4 0 Best p-value(s): 0.3636 0.3636
FASTPLGP value(s): L
DV counts: 60 5 0 Best p-value(s): 0.2136 0.4416
X5 value(s):
179.86824278 187.00215516 187.27177317 188.26867046 196.45058388 198.862023
11 206.06732876 206.21903124 207.82210852 208.4789315 210.51382161 218.196
15889 223.67582281 225.12568308 etc.
DV counts: 4 19 0 Best p-value(s): 0.3146 0.4017
X5 value(s): 259.22566232
DV counts: 0 0 1
X5 value(s):
263.35227522 271.4201007 272.46053531 274.01080629 328.18269653 351.9001982
4 374.56331699
DV counts: 0 7 0
TSTPLGP1 value(s): H M
DV counts: 0 1 32 Best p-value(s): 0.0001 0.0001
X5 value(s): 128.43457924 145.70590001
DV counts: 0 0 2
X5 value(s): 147.23281792
DV counts: 0 1 0
X5 value(s):
160.54684235 173.30593671 193.46150047 201.22972662 206.06366955 211.140313
45 211.70428831 216.74283674 231.27195626 234.62979942 237.96478751 240.75
302441 245.17 271.21569933 etc.
DV counts: 0 0 30

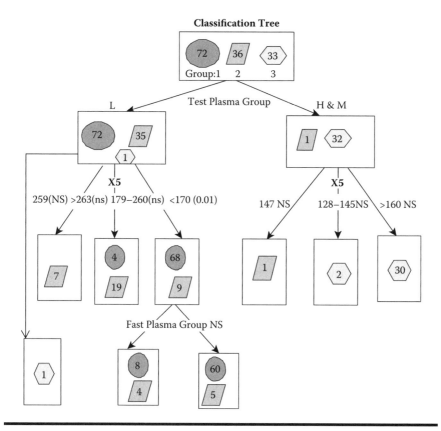

Figure 6.21 Decision tree diagram generated manually using the decision tree information generated using the SAS macro CHAID2.

6.14 Summary

The methods of performing supervised classification models, for grouping categorical group response variables, and using the user-friendly SAS macro applications are covered in this chapter. Graphical methods to perform diagnostic and exploratory analysis, classification and discrimination, decision tree analysis, model assessment, and validation are presented using a clinical diabetes dataset. Steps involved in using the user-friendly SAS macro applications DISCRIM2 for performing parametric and nonparametric discriminant analysis and CHAID2 for performing CHAID analysis and generating decision trees are presented here.

References

1. Sharma, S., *Applied Multivariate Techniques*, John Wiley & Sons, New York, 1996, chap. 8,9.
2. Johnson, R. A. and Wichern, D. W., *Applied Multivariate Statistical Analysis*, 5th ed., Prentice Hall, NJ, 2002, chap. 11.
3. Breiman, L., Friedman, J. H., Olshen, R. A., and Stone, C. J., *Classification and Regression Trees*, Wadsworth Int. Group, Belmont, California, 1984.
4. SAS Institute Inc., *SAS/STAT 9.1 Users Guide*, Cary, NC, 2004.
5. Khattree, R. and Naik, D. N., *Multivariate Data Reduction and Discrimination with SAS Software*, 1st ed. SAS Institute, Inc., Cary, NC, 2000, chap. 5.
6. SAS institute Inc., The STEPDISC Procedure, An overview (last accessed 07/21/09).
7. SAS institute Inc., The CANDISC Procedure, An overview (last accessed 07/21/09).
8. SAS institute Inc., *The DISCRIM Procedure*, An overview (last accessed 05/21/09).
9. Khattree, R. and Naik, D. N., *Applied Multivariate Statistics with SAS Software*, Cary, NC, SAS Institute Inc., 1995, chap. 1.
10. Gabriel, K. R., Bi-plot display of multivariate matrices for inspection of data and diagnosis. In V. Barnett (Ed.). *Interpreting Multivariate Data*. London: John Wiley & Sons, 1981.
11. SAS Institute Inc., *SAS Systems for Statistical Graphics*, 1st ed., Cary, NC, 1991, chap. 9.
12. Lachenbruch, P. A. and Mickey, M. A., Estimation of error rates in discriminant analysis, *Technometrics*. 10: 1–10, 1968.
13. Hora, S. C. and Wilcox, J. B., Estimation of error rates in several population discriminant analyses. *Journal of Marketing Research*, 19: 57–61, 1982.
14. Glick, N. Additive estimators for probabilities of correct classification. *Pattern Recognition*, 10:211–222, 1 1978.
15. Berry, M. J. A. and Linoff, G. S. *Data Mining Techniques: For Marketing, Sales, and Customer Support*. John Wiley & Sons, New York, 1997, chap. 12.
16. SAS Institute Inc., The TREEDISC macro for CHAID analysis, at http://www.stat. lsu.edu/faculty/moser/exst7037/treedisc.html (last accessed 3/10/10).
17. Reaven, G. M. and Miller, R. G. An attempt to define the nature of chemical diabetes using a multidimensional analysis. *Diabetologia*, 16: 17–24, 1979.
18. SAS Institute Inc., SAS/GRAPH 9.2: Statistical Graphics Procedures Guide: The SGPLOT Procedure at http://support.sas.com/documentation/cdl/en/grstatproc/61948/ HTML/default/sgplot-chap.htm (last accessed 05/21/09).

Chapter 7

Advanced Analytics and Other SAS Data Mining Resources

7.1 Introduction

Advanced analytics plays a major role in this fast-changing corporate finance world, where enterprise goals change abruptly. During these uncertain times, decision makers in the corporate world count on their analysts to adopt advanced analytic tools using their own data, enabling more effective and proactive decision making that drives superior enterprise performance. The successes of an organization's business strategy depend on utilizing the right information at the right time. There has never been a greater need for proactive, evidence-based decisions and agile strategies. As information is collected and enriched into actionable business intelligence, the challenge becomes making this intelligence readily available to the right people, in the appropriate form.

Neural net (NN) applications and market basket association (MBA) analysis are two of the predictive analytic methods that help to make proactive, forward-looking decisions in data mining applications. NN, or parallel-distributed processing as it is sometimes called, is an information-processing paradigm that closely resembles the densely interconnected parallel structure of the mammalian brain. NN techniques include collections of prediction and classification models that emulate biological nervous systems and draw on the analogies of adaptive biological learning. MBA is a computer algorithm that examines many transactions in order to determine which items are most frequently purchased together and provides this valuable information to the retail store management for better marketing.

The purpose of this chapter is to gently introduce the concept of these advanced analytics and provide some additional information regarding the capabilities of the SAS software for performing these analyses. Refer to the following publications for additional information neural net applications and MBA analysis: References 1, 2, and 3.

7.2 Artificial Neural Network Methods

The recent explosion of artificial neural net (NN) technology has led data miners to explore a variety of computer engineering applications that are not based on traditional statistical theory. Borrowing the concept from the human brain, neural systems fit models by learning in repeated trials to achieve the best prediction. In other words, NN *learns* from examples. The network is composed of a large number of highly interconnected processing elements (neurons) working in parallel to solve a specific problem. In NN systems, the input, output, and the intermediate variables act as nodes that are interconnected by weighted network paths of a network diagram. The input layer contains a unit for each input layer. The output layer represents the target. The hidden layer contains hidden units (neurons) that are the intermediate transformed inputs. The connections in the network path represent the unknown parameter coefficients that are estimated by fitting the model to the data.[2,7]

Many NN applications use the supervised learning approach. For supervised learning, training data that includes both the input and the target variables must be provided. After successful training, you can use test data to the NN (that is, input data without the target value), and the NN will compute an output value that approximates the response. If trained successfully, NNs may exhibit generalization beyond the training data and predict correct results for new cases in the validation dataset. However, for successful training, big training data and lengthy computer training time are essential.

NN modeling can be used for both prediction and classification. NN models enable the construction of train and validate multiplayer feed-forward network models for modeling large data and complex interactions with many predictor variables. They usually contain more parameters than a typical statistical model, and the results are not easily interpreted and no explicit rationale is given for the prediction. All variables are treated as numeric, and all nominal variables are coded as binary. Categorical variables must be encoded into numbers before being given to the network. Relatively more training time is needed to fit the NN models.

The NN models are considered flexible multivariate function estimators. Technically speaking, they are multistage parametric nonlinear regression models and classification models. The most common type of NN model used for supervised prediction is the *multilayer perceptron*, that is, the feed-forward NN that uses sigmoid hyperbolic functions.[9] For mathematical aspects of NN, see References 3 and 13. For a worked-out example of fitting an NN model using the SAS enterprise miner, see Reference 14.

There is considerable overlap between NN and statistics fields.[12] Feed-forward nets with no hidden layer (including functional-link NNs and higher-order NNs) are basically generalized linear models. Probabilistic NNs are identical to kernel discriminant analysis.[8] Kohonen nets for adaptive vector analysis are very similar to *k*-means cluster analysis.[8] Hebbian learning is closely related to principal component analysis.[9] It is sometimes claimed that neural networks, unlike statistical models, require no distributional assumptions. In fact, neural networks involve exactly the same sort of distributional assumptions as statistical models,[2] but statisticians study the consequences and importance of these assumptions, while many neural networkers ignore them. There are many methods in statistical literature that can be used for flexible nonlinear modeling. These methods include polynomial regression, *k*-nearest-neighbor regression, kernel regression, and discriminant analysis.

7.3 Market Basket Association Analysis

The objective of market basket association analysis (MBA) is to find out what products and services customers purchase together. Knowing what products people purchase as a group can be very helpful to any business. A retail store could use this information to display products frequently sold together in the same aisle. A World Wide Web-based Internet merchant could use MBA to determine the layout of Web page catalog forms. Banks and telephone companies could use the MBA results to determine what new products to offer their existing customers. Once the association rule that customers who buy one product is likely to buy another is known, it is possible for the company to market the products together, or to make the purchasers of one product the target prospects for another. This is the purpose of MBA—to improve the effectiveness of marketing and sales tactics using customer data already available to the company. For a nontechnical account of MBA and its applications, see Reference 12. For a mathematical discussion of association rules used in MBA, see Reference 14. For an example of performing MBA analysis using the SAS Enterprise Miner, see Reference 13.

The strength of MBA is that customers' sales data can provide valuable information regarding what products consumers would logically buy together. This is a good example of data-driven marketing. There are several advantages to MBA over other types of data mining. First of all, it is undirected. It is not necessary to choose a product that you want to focus on in order to run a basket analysis. Instead, all products are considered, and the data mining software reveals which products are most important to the analysis. In addition, the results of basket analysis are clear, simple, and understandable association rules that can be utilized immediately for better business advantage.

7.3.1 Benefits of MBA

Impulse buying: Knowing which products sell together can be very useful to any business. The most obvious effect is the increase in sales that a retail store can achieve by reorganizing its products so that things that sell together are found together. This facilitates impulse buying and helps ensure that customers who would buy a product do not forget to buy it on account of not having seen it.

Customer satisfaction: In addition, MBA has the side effect of improving customer satisfaction. Once they have found one of the items they want, the customer does not have to search the store for the other item they want to buy. Their other purchases are already located conveniently close together. Internet merchants get the same benefit by conveniently organizing their Web site so that items that sell together are found together.

Actionable: Unlike most promotions, advertising based on MBA findings is almost sure to pay off: the convenience store has the data to back it up before even beginning the advertising. This is an example of the best kind of MBA result.

Product bundling: For companies that do not have a physical store, such as mail-order companies, Internet businesses, and catalog merchants, MBA can be more useful for developing promotions rather than reorganizing product placement. By offering promotions such that the buyers of one item get discounts on another item they have been found likely to buy, sales of both items may be increased.

In addition, basket analysis can be useful for direct marketers for reducing the number of mailings or calls that need to be made. By calling only customers who have shown themselves likely to want a product, the cost of marketing can be reduced, while the response rate is increased.

Stock inventory: It can be useful for operations purposes to know which products sell together in order to stock inventory. Running out of one item can affect sales of associated items; perhaps the reorder point of a product should be based on the inventory levels of several products, rather than just one.

7.3.2 Limitations of Market Basket Association Analysis

Though useful and productive, MBA does have a few limitations. It is necessary to have a large number of real transactions to get meaningful data, but the data's accuracy is compromised if all of the products do not occur with similar frequency. Second, MBA can sometimes present results that are actually due to the success of previous marketing campaigns. Third, association rules sometimes generated by MBA; that is, the trivial and the inexplicable may not always useful. A trivial rule is one that would be obvious to anyone with some familiarity with the industry at hand. Finally, MBA occasionally produces inexplicable rules. These rules are not only obvious but also do not lend them selves to immediate marketing use. An inexplicable rule is not necessarily useless, but its business value is not obvious and it does not lend itself to immediate use for cross-selling.[12]

7.4 SAS Software: The Leader in Data Mining

SAS Institute,[15] the industry leader in analytical and decision-support solutions, offers a comprehensive data mining solution that allows one to explore large quantities of data and discover relationships and patterns that lead to proactive decision making. SAS software provides the industry's most powerful easy-to-use, metadata-driven warehouse management and ETL capabilities with the added value of integrated data quality assessment and monitoring to ensure that the consolidated information is consistent and accurate. Data mining is very effective when it is a part of an integrated enterprise knowledge delivery strategy. SAS/Ware house administrator, a component of the SAS data warehousing solution, OLAP, NN, and MBA analysis are integrated seamlessly with the SAS Enterprise Minor software.[16,17] Enterprise Miner functions well as an advanced data mining and business intelligence solution. SAS Enterprise Miner streamlines the data mining process to create highly accurate predictive and descriptive models based on analysis of vast amounts of data from across the enterprise.

Another feature available in EM is its text mining for analyzing sales notes, e-mail text, customer and prospect Web documents for additional business intelligence applications, and online blogs. Other areas for exploration are trend and sequence analysis in large invoice transaction data and hierarchical clustering for text mining and semiautomated customer profile modeling. Integrating text-based information with structured data enriches your predictive modeling capabilities and provides new stores of insightful information for driving your business and research initiatives forward. Text mining supports a wide variety of applications such as categorizing huge collections of call center data, Web text and blog analysis, finding patterns in customer feedback or employee surveys, detecting emerging product issues, analyzing competitive intelligence reports or patent databases, and classifying reports and company information by topic or business issue.

Some of the main strengths of SAS Enterprise Miner are the general ease of use in designing and building data mining process flow diagrams with its drag-and-drop node capability, ease in setting up validation and test sets, and the results of those cross-validations. Enterprise Miner has very advanced powerful algorithms for data mining such as ensemble models and memory-based reasoning. Many other vendors do not have these algorithms, and we have found them very useful. Also, when building clustering or neural network models, the user does not have to recode categorical or nominal variables; it is done automatically. Data manipulation or data preparation is easily accomplished within the Enterprise Miner process flow environment.

7.5 Summary

The SAS institute, the industry leader in analytical and decision support solutions, offers the powerful software called *Enterprise Miner* to implement complete data mining solutions. The SAS data mining solution provides business technologists and quantitative experts the necessary tools to obtain the enterprise knowledge for helping their organizations achieve a competitive advantage. SAS macros for performing these advanced analytic methods are not include in this book since Enterprise Miner software is required to perform these analysis. However, the 13 user-friendly applications used in the book are very valuable resources to SAS Enterprise Miner software users.

References

1. Ripley, B. D., *Pattern Recognition and Neural Networks*, Cambridge University Press, Cambridge, U.K., 1996.
2. Bishop, C. M., *Neural Networks for Pattern Recognition*, Oxford University Press, Oxford, 1995.
3. Berry, M. J. A. and Linoff, G. S., *Data Mining Techniques: For Marketing, Sales, and Customer Support*, John Wiley & Sons, New York, 1997, chap. 8.
4. Hair, J. E, Anderson, R. E., Tatham, R. L., and Black, W. C., *Multivariate Data Analysis*, 5th ed., Prentice Hall, NJ, 1998, chap. 12.
5. Berry, M. J. A. and Linoff, G. S., *Data Mining Techniques: For Marketing, Sales, and Customer Support*. John Wiley & Sons, New York, 1997, chap. 15.
6. SAS Institute Inc., Online Analytical Processing (OLAP) at http://www.sas.com/technologies/biolap/exploration.html (last accessed 3/10/10).
7. Sarle, W. S., Ed., *Neural Network FAQ, part 1 of 7: Introduction, periodic posting to the Usenet newsgroup comp.ai.neural-nets*, URL: ftp://ftp.sas.com/pub/neural/FAQ.html (last accessed 3/10/10).
8. SAS Institute Inc., *Neural Network Modeling Course Notes*, Cary, NC, 2000.
9. Hastie, T., Tibshirani, R., and Friedman, J., *The Elements of Statistical Learning—Data Mining, Inference, and Prediction*, Springer series in Statistics, New York, 2001, chap. 11.
10. Johnson, R. A. and Wichern, D. W., *Applied Multivariate Statistical Analysis*, 5th ed., Prentice Hall, NJ, 2002, chap. 11.
11. Sarle, W. S., Neural networks and statistical models, Proceedings of the Nineteenth Annual SAS Users Group International Conference, SAS Institute, Cary, NC, pp. 1538–1550. 1994.
12 Berry, M. J. A. and Linoff, G. S., *Data Mining Techniques: For Marketing, Sales, and Customer Support*, John Wiley & Sons, New York, 1997, chap. 8.
13. Hastie, T., Tibshirani, R., and Friedman, J., *The Elements of Statistical Learning—Data Mining, Inference, and Prediction*. Springer series in Statistics, New York, 2001, chap. 14.
14. SAS Institute Inc., *Data Mining Using Enterprise Miner Software: A Case Study Approach*, 1st ed., Cary, NC, 2000.
15 SAS Institute Inc., The Power to Know at http://www.sas.com.
16. SAS Institute Inc., The Enterprise Miner at http://www.sas.com/technologies/analytics/datamining/miner/index.html (last accessed 3/10/10).
17. SAS Institute Inc., SAS Enterprise Miner Product Fact Sheet at http://www.sas.com/resources/factsheet/sas-enterprise-miner-factsheet.pdf.

Appendix I: Instruction for Using the SAS Macros

A.1 Prerequisites for Using the SAS Macros

- Read all the instructions given in this appendix first.
- SAS software requirements:
  - SAS/BASE, SAS/STAT, and SAS/GRAPH must be licensed and installed at your site. SAS/IML is required to run the CHAID2 macro and check for multivariate normality in FACTOR2, DISJCLS2, and DISCRIM2 macros. SAS/ACCESS (PC-file types) is required to convert PC files (Excel, Access, Dbase, etc.) to SAS datasets in EXCLSAS2 macro. SAS/ QC is required to produce control charts in UNIVAR2 macro.
  - SAS version 9.0 and above is required for full utilization.
- Internet requirements:
  - Working Internet connection is required for initial downloading the compiled macro files, macro-call files, and the sample datasets from the book's Web site.
- System requirements:
  - SAS system for Microsoft Windows (XP, Vista [Business, Ultimate]) is required to run these macros.
- SAS experience: No experience in SAS macros or SAS graphics is needed to run these macros. However, a working knowledge in SAS for Windows and creating temporary and permanent SAS datasets are helpful.

A.2 Instructions for Downloading Macro-Call Files

- Visit the book's Web page at http://www.cabnr.unr.edu/gf/dm.html
- Click the download link and you will be directed to the password-protected compiled macro, macro-call download page to download the zip file.

383

- Input the following username (lowercase only) and the password to go to the download page.
  - Username: gcjf
- Password: 41552mannar321
      Click the download link and download the zipped file called dm2e.zip (If you are using SAS in a 32 bit Windows environment), and save it in a folder in your PC. However, if you are using SAS 9.2 in a 64 bit Windows Vista, then download dm2164.zip file.
- Unzip the dm2e.zip file in your PC using any unzip program, and install it in the C:\ drive. (Note: If you install it in other dives (D:\, E:\), then you need to change the path in the maccall files.) You will find three folders, mcatalog, maccall, and sasdata, and one text file called README. TXT in the zipped file. The sasdata folder contains the Excel data files and the permanent SAS data files used in the book. In the maccall folder, you will find 13 macro-call files corresponding to the 13 data mining macros described in the book (use these file in traditional SAS display manager mode) and a folder called NODISPLAY. Because the "blue window-" based macro-call files are not compatible with SAS Enterprise Guide (EG) Code window, I have created traditional non-window-based separate macro-call files and saved them in the NODISPLAY folder. Use these files in SAS EG or with SAS Learning Edition 4.1. (See Appendix 3 for more details.)
- The mcatalog folder contains the compiled SAS macro files sasmacr.sas7bcat and a folder NODISPLAY. You will find another file, sasmacr.sas7bcat, which is compatible with SAS EG nonwindow mode. Do not change or modify the contents of these compiled macro files.
- SASDATA folder contains the permanent SAS data files and Excel files used in the book.
- Read the README.TXT file for the version number and any update information.
- Visit the book's Web page at least once a month for any news about the update information.

A.3 Instructions for Running the SAS Macros

- It is highly recommended that the reader run these macros using the sample data included in the sasdata folder before using these macros on actual data.
- Create a temporary SAS dataset using one of the sample permanent dataset. For example, to create a temporary dataset called Train from the permanent

dataset Sales saved in the c:\dmsas2e\sasdata\ folder, type the following statements in the SAS program editor window:

```
LIBNAME GF 'c:\dmsas2e\sasdata\';
/*Assign a libname GF to the sasdata folder containing
  the sample data files*/
DATA train;
     SET GF.sales;
RUN;
```

After finished entering, click the RUN button to create a temporary dataset called Train from the permanent dataset Sales saved in the folder called sasdata.

■ For example, to run multiple linear regression, click the editor window, open the REGDIAG2.SAS into the editor (do not make any changes to the macro-call file), and click the RUN icon to open the CYAN color macro-call window REGDIAG2. Following the instructions given in the specific help file for the REGDIAG2 macro (Appendix 2), input the necessary macro-input values. When your cursor blinks at the last macro field, hit the ENTER key (not the RUN icon) to execute the macro. To check for any macro execution errors in the LOG window, always run with the DISPLAY option first. Ignore any warnings related to FONT substitution, since these font specifications are system specific. If you do not have any macro-execution errors, then you can save the output/graphics by changing DISPLAY to the desired file formats (WORD, WEB, PDF, and TXT).

■ Read the specific chapter and macro-help files for specific details.

Appendix II: Data Mining SAS Macro Help Files

A.1 Help File for SAS Macro EXLSAS2

1. Macro-call parameters:	Options/Explanations (allowed file type):
Input PC file type? (required parameter)	• *Excel* — (XLS) files Office 2003 and before
Descriptions & Explanation: Include the PC file type you would like to import.	• *Lotus* — (WK4) files • *dBase* — (III & IV) files • *Access* — (mdb) files 97 and 2003 files • *Tab* — (TAB) tab-delimited files • *CSV* — (CSV) Comma-delimited files
2. Macro-call parameters:	Options/Explanations:
Input folder name containing the PC file? (required parameter)	Possible values: • *e:* — *"e" drive*
Descriptions & Explanation: Input the location (path) of folder name containing the PC file.	• *c:\excel* — folder name "Excel" in the C drive (Be sure to include the backslash at the end of folder name.)

3. Macro-call parameters:	Options/Explanations:
Input PC file names? (required statement)	Examples:
Descriptions & Explanation:	• *BASEBALL fraud (multiple file example)*
List the names of PC files (without the file extension) being imported.	• *CRIME*
	• *customer99*
	Use only short single-word file names. The same file name will be used for naming the imported SAS dataset. If you list multiple PC files, all of them can be imported in one operation.
4. Macro-call parameters:	**Options/Explanations:**
Optional LIBNAME?	Examples:
Descriptions & Explanation:	• *SASUSER*
To save the imported PC file as a permanent SAS dataset, input the preassigned library (LIBNAME) name.	• *MYLIB*
	Save the imported PC files as permanent SAS data in the folder library called SASUSER or MYLIB.
	The predefined LIBNAME will tell SAS in which folder to save the permanent dataset. If this field is left blank, a temporary data file will be created and saved in the work folder.
5. Macro-call parameters:	**Options/Explanations:**
Optional SAS data step statements?	Examples:
(New option in the second edition)	*Rename x1=y2 x4=y3; Logx2 = log(x2)*
Descriptions & Explanation:	This is a valid SAS data step statement instructing SAS to rename X1 and X4 to Y2 and Y3, respectively, and perform a log transformation and convert X2 to Log x2.
Input any valid SAS data step statements to modify a variable name, assign a label, and perform a data transformation.	
6. Macro-call parameters:	**Options/Explanations:**
Folder to save SAS output? (optional statement)	Possible values:
	c:\output—folder name OUTPUT in the c drive
	s:\george—folder name "George" in the network drive S

Descriptions & Explanation:	Be sure you include the backslash at the end of the folder name. The same imported SAS dataset name will be assigned to the output file. If this field is left blank, the output file will be saved in the default folder.
To save the SAS output files created by the macro in a specific folder, input the full path of the folder.	

7. Macro-call parameters:	**Options/Explanations:**
Display SAS output in the output window or save SAS output to a file?	Possible values:
(required statement)	• *DISPLAY*: Output will be displayed in the OUTPUT window. System messages will be displayed in LOG window.
Descriptions & Explanation:	
Option for displaying all output files in the OUTPUT window or save as a specific format in a folder specified in option 5.	• *WORD*: Output will be saved in the user-specified folder and viewed in the RESULTS VIEWER window as a single RTF format (version 8.2 and later).
	• *WEB*: Output will be saved in the user-specified folder and viewed in the RESULTS VIEWER window as a single HTML (version 8.2 and later) file.
	• *PDF*: Output will be saved in the user-specified folder and viewed in the RESULTS VIEWER window as a single PDF (version 8.2 and later) file.
	• *TXT*: Output will be saved as a TXT file in the user-specified folder in all SAS versions. No output will be displayed in the OUTPUT window.
	NOTE: All system messages will be deleted from the LOG window at the end of macro execution if you do not select DISPLAY as the macro input in #6.

A.2 Help File for SAS Macro RANSPLIT2

1. Macro-call parameters:	Options/Explanations:
Input the temporary SAS dataset name? (required parameter)	• *Fraud* (temporary SAS data called "fraud")
Descriptions & Explanation:	• *cars93*
Input the temporary SAS dataset name from which you would like to draw training and validation samples.	
2. Macro-call parameters:	**Options/Examples:**
Input continuous response variables names? (optional parameter)	Possible values:
	• netsales
Descriptions & Explanation:	• fraud
To assess the similarities between the training and validation samples, the distribution aspects of the specified response variables are compared by box plots.	
3. Macro-call parameters:	**Options/Explanations:**
Input the sample size (n) in train data. (required statement)	Examples:
	• 800
Descriptions & Explanation:	• 600
Input the desired sample size number for the "Training" data. Usually, 40% of the database is selected.	
4. Macro-call parameters:	**Options/Examples:**
Observation number in validation data? (optional parameter)	Examples:
	• 600
Descriptions & Explanation:	• 400
Input the desired sample size number for the VALIDATION sample. Usually, 30% of the database is selected for validation sample.	The leftover observations in the database after the TRAINING and VALIDATION samples are selected will be included in the TEST sample. If you leave this field blank, all of the leftover observations in the database after the TRAINING sample is selected will be included in the VALIDATION set.

5. Macro-call parameters:	Options/Explanations:
Folder to save SAS output? (optional statement)	Possible values:
Descriptions & Explanation:	*c:\output* —folder name OUTPUT in the C drive
To save the SAS output files created by the macro in a specific folder, input the full path of the folder.	*s:\george* —folder name "George" in the network drive S
	Be sure you include the backslash at the end of the folder name. The same imported SAS dataset name will be assigned to the output file. If this field is left blank, the output file will be saved in the default folder.

6. Macro-call parameters:	Options/Explanations:
Folder to save SAS graphics files? (optional statement)	Possible values:
Descriptions & Explanation:	*c:\output* —folder name OUTPUT in the C drive
To save the SAS graphics files in EMF format (specify output format as TXT) or JPEG format (specify output format as WEB) in the folder specified here. If the graphics folder field (macro input number 6) is left blank, the graphics file will be saved in the default folder.	*s:\george* —folder name "George" in the network drive S
	Be sure you include the backslash at the end of the folder name.

7. Macro-call parameters:	Options/Explanations:
Display SAS output in the output window or save SAS output to a file? (required statement)	Possible values:
	• *DISPLAY*: Output will be displayed in the OUTPUT window. System messages will be displayed in LOG window.
	• *WORD*: Output will be saved in the user-specified folder and viewed in the RESULTS VIEWER window as a single RTF format (version 8.2 and later).
	• *WEB*: Output will be saved in the user-specified folder and viewed in the RESULTS VIEWER window as a single HTML (version 8.2 and later) file.

Descriptions & Explanation:	• **PDF**: Output will be saved in the user-specified folder and viewed in the RESULTS VIEWER window as a single PDF (version 8.2 and later) file.
Option for displaying all output files in the OUTPUT window or saving in a specific format in a folder specified in option 5.	
	• **TXT**: Output will be saved as a TXT file in the user-specified folder in all SAS versions. No output will be displayed in the OUTPUT window.
	NOTE: All System messages will be deleted from the LOG window at the end of macro execution if you do not select DISPLAY as the macro input in #7.

8. Macro-call parameters:	**Options/Explanations:**
A counter value: zth number of run? (required statement)	• 1
	• A
Descriptions & Explanation:	• A1
SAS output files created by the RANSPLIT2 will be saved by forming a file name from the original SAS dataset name and the counter number provided in the macro input field 8.	Numbers 1 to 10 and any letters are valid.
	For example, if the original SAS dataset name is "fraud" and the counter number included is 1, the SAS output files will be saved as "fraud1.*" in the user-specified folder. By changing the counter numbers, the users can avoid replacing the previous SAS output files with the new outputs.

9. Macro-call parameters:	**Options/Explanations:**
Optional LIBNAME for saving permanent SAS data?	Examples:
	• *SASUSER*: Save the permanent SAS data in the library called SASUSER.
Descriptions & Explanation:	The predefined LIBNAME will tell SAS in which folder to save the permanent datasets. If this field is left blank, temporary WORK data files will be created for all samples.
To save the TRAINING, VALIDATION, and TEST datasets as permanent SAS datasets, input the preassigned library (LIBNAME) name.	

10. Macro-call parameters:	Options/Explanations:
Input optional seed number.	Examples:
Descriptions & Explanation:	• 5678905
To impose a similar split in the repeated runs, input a 5–7-digit random seed number.	• 4688943
	If you leave this field blank, a random number seed will be obtained by current time, and every time you run with blank filled, new splits will be generated.
11. Macro-call parameters:	**Options/Explanations:**
Optional stratification variables.	Options:
Descriptions & Explanation:	• Manager
To generate stratified random samples, input the stratification variable here, and input the stratification method in macro input #12.	• C3
	If you leave this field blank, only a simple random sample will be generated.
12. Macro-call parameters:	**Options/Explanations:**
Optional stratified sampling method.	Options:
Descriptions & Explanation:	• SRS—simple random sampling
After inputting the stratification variable in #11, input the stratification method in macro input field 12.	• SYS—systematic random sampling
	• URS—unrestricted random sampling, which is selection with equal probability
	If you leave this field blank but input a stratification variable in macro input #11, a stratified random sampling using SRS will be generated.

A.3 Help File for SAS Macro FREQ2

1. Macro-call parameters:	Options/Explanations:
Input the temporary SAS dataset name? (required parameter)	• *Fraud* (temporary SAS data called "fraud")
Descriptions & Explanation:	• *cars93*
Input the temporary SAS dataset name, on which you would like to perform data exploration of categorical variables.	

2. Macro-call parameters:	Options/Examples:
Input Response categorical variable names? (required parameter)	Possible values: • mpg (name of a target categorical variable)
Descriptions & Explanation:	• Q1 Q2 Q4 (multiple variables are allowed)
Input name of the categorical variables you would like to treat as the output variables in a two or three-way analysis.	However, for creating one-way tables and charts, input the categorical variable names and leave the macro field #3 and #4 blank.
	Note: Two input rows are included to add very long string of variables.
	X1-X10 syntax is allowed. However, statistics will be computed for X1 through X10. But graphics will be produced only for X1 and X10.

3. Macro-call parameters:	*Options*/Explanations:
Input GROUP variable name? (optional statement)	Examples: • C3
Descriptions & Explanation:	• C5
Input the name of the first-level categorical variable for a two-way analysis.	

4. Macro-call parameters:	Options/Examples:
Input BLOCK variable (second level) name? (optional statement)	Examples: • *B2* • *mgr*

Descriptions & Explanation: Input the name of the second-level categorical variable for a three-way analysis.	
5. Macro-call parameters: Input plot type (frequency or percentage) options? (required statement)	**Options/Explanations:** Possible values: • **Percent:** report percentages • **Freq:** report frequencies
Descriptions & Explanation: The type of frequency/percentage statistics desired in the charts.	• **Cpercent:** report cumulative percentages • **Cfreq:** report cumulative frequencies
6. Macro-call parameters: Type of patterns used in bar charts? (required statement)	**Options/Explanations:** Possible values: • **Midpoint:** Changes patterns when the midpoint value changes. If the GROUP= option is specified, the respective midpoint patterns are repeated for each group report percentages.
Descriptions & Explanation: Selecting the pattern specifications in different types of bar charts.	• **Group:** Changes patterns when the group variable changes. All bars within each group use the same pattern, but a different pattern is used for each group. • **Subgroup:** Changes patterns when the value of the subgroup variable changes.
7. Macro-call parameters: **Input** Color options? (required statement)	**Options/Explanations:** Examples: **Color:** preassigned colors used in charts
Descriptions & Explanation: Input whether color or black-and-white charts are required.	**Gray:** preassigned gray used in shades

8. Macro-call parameters:	Options/Explanations:
A counter value: zth number of run? (required statement)	• 1 • A
Descriptions & Explanation:	• A1
SAS output files created by the FREQ2 will be saved by forming a file name from the original SAS dataset name and the counter number provided in the macro input field 8.	Numbers 1 to 10 and any letters are valid. For example, if the original SAS dataset name is "fraud" and counter number included is 1, the SAS output files will be saved as "fraud1.*" in the user-specified folder. By changing the counter numbers, the users can avoid replacing the previous SAS output files with the new outputs.
9. Macro-call parameters:	**Options/Explanations:**
Folder to save SAS output? (optional statement)	Possible values: *c:\output* —folder name OUTPUT in the C drive
Descriptions & Explanation:	*s:\george* —folder name "George" in the network drive S
To save the SAS output files created by the macro in a specific folder, input the full path of the folder.	Be sure you include the backslash at the end of the folder name. The same imported SAS dataset name will be assigned to the output file. If this field is left blank, the output file will be saved in the default folder.
10. Macro-call parameters:	**Options/Explanations:**
Folder to save SAS graphics files? (optional statement)	Possible values: *c:\output* —folder name OUTPUT in the C drive
Descriptions & Explanation:	*s:\george* —folder name "George" in the network drive S
To save the SAS graphics files in EMF format (specify output format as TXT) or JPEG format (specify output format as WEB) in the folder specified here. If the graphics folder field (macro input number 10) is left blank, the graphics file will be saved in the default folder.	Be sure you include the backslash at the end of the folder name.

11. Macro-call parameters:	**Options/Explanations:**
Display SAS output in the output window or save SAS output to a file? (required statement)	Possible values:
Descriptions & Explanation:	• *DISPLAY*: Output will be displayed in the OUTPUT window. System messages will be displayed in the LOG window.
Option for displaying all output files in the OUTPUT window or saving in a specific format in a folder specified in option 5.	• *WORD*: Output will be saved in the user-specified folder and viewed in the RESULTS VIEWER window as a single RTF format (version 8.2 and later).
	• *WEB*: Output will be saved in the user-specified folder and viewed in the RESULTS VIEWER window as a single HTML (version 8.2 and later) file.
	• *PDF*: Output will be saved in the user-specified folder and viewed in the RESULTS VIEWER window as a single PDF (version 8:2 and later) file.
	• *TXT*: Output will be saved as a TXT file in the user-specified folder in all SAS versions. No output will be displayed in the OUTPUT window.
	NOTE: All system messages will be deleted from the LOG window at the end of macro execution if you do not select DISPLAY as the macro input in #7.
12. Macro-call parameters:	**Options/Explanations:**
Optional SAS SURVEYFREQ options	Examples:
Descriptions & Explanation:	• **Strata gender; weight wt**
To estimate population estimates and 95% confidence intervals for survey data and check if survey weights are available.	This statement identifies the stratification variable and instruct SAS SURVEYFREQ to use the survey weights and estimate adjusted percentages and frequencies.

A.4 Help File for SAS Macro UNIVAR2

1. Macro-call parameters: Input the temporary SAS dataset name? (required parameter)	**Options/Explanations:** • *Fraud* (temporary SAS data called "fraud")
Descriptions & Explanation: Input the temporary SAS dataset name, on which you would like to perform data exploration of categorical variables.	• *cars93*
2. Macro-call parameters: Input continuous response variable names? (required parameter)	**Options/Examples:** Possible values: • Y2 (name of a continuous variable) • Y2 Y4 (multiple variables are allowed)
Descriptions & Explanation: Input continuous variable names you would like to explore. If you include multiple variable names, exploratory analysis will be performed and a new dataset excluding the outliers will be created for each variable specified.	For example, if you input two continuous variables Y1 and Y2 from a temporary SAS data cars93 and leave the group variable field (#3) blank, separate exploratory analysis will be performed on these two variables. Also, two new SAS datasets, cars931 and cars932, will be created after excluding the outliers. Note: Two input rows are included to add very long string of variables. X1-X10 syntax is NOT allowed.
3. Macro-call parameters: Input GROUP variable name? (optional statement)	***Options/*Explanations:** Examples: • C3
Descriptions & Explanation: If you would like to perform data exploration by a group variable, specify the group variable name in this field. If you include group variable names, exploratory analysis will be performed and separate datasets excluding the outliers will be created for each variable specified and each level within the group variable.	• C5 For example, if you input one continuous variable Y2 from a temporary SAS data cars93 and input a group variable name such as b2 (origin with two levels) in field #3, separate exploratory analyses will be performed for each level within a group variable. Also, two new SAS datasets, cars9311 and cars9312, will be created after excluding the outliers.

4. Macro-call parameters:	Options/Examples:
Input confidence levels? (required statement)	Examples: • 95
Descriptions & Explanation:	• 90
Input the level of confidence you would like to assign to your parameter estimates.	

5. Macro-call parameters:	Options/Examples:
Input ID variable?	• Car
(optional statement)	• id
Descriptions & Explanation:	• model
Input the name of the variable you would like to treat as the ID.	If you leave this field blank, a character variable will be created from the observational number and will be treated as the ID variable.

6. Macro-call parameters:	Options/Explanations:
A counter value: zth number of run? (required statement)	• 1
	• A
Descriptions & Explanation:	• A1
SAS output files created by the UNIVAR2 will be saved by forming a file name from the original SAS dataset name and the counter number provided in the macro input field 6.	Numbers 1 to 10 and any letters are valid. For example, if the original SAS dataset name is "fraud" and the counter number included is 1, the SAS output files will be saved as "fraud1.*" in the user-specified folder. By changing the counter numbers, the users can avoid replacing the previous SAS output files with the new outputs.

7. Macro-call parameters:	Options/Explanations:
Folder to save SAS output? (optional statement)	Possible values: *c:\output* — folder name OUTPUT in the C drive
Descriptions & Explanation:	*s:\george* — folder name "George" in the network drive S
To save the SAS output files created by the macro in a specific folder, input the full path of the folder.	Be sure you include the backslash at the end of the folder name. The same imported SAS dataset name will be assigned to the output file. If this field is left blank, the output file will be saved in the default folder.

8. Macro-call parameters:	Options/Explanations:
Folder to save SAS graphics files? (optional statement)	Possible values:
	c:\output — folder name OUTPUT in the C drive
Descriptions & Explanation:	*s:\george* — folder name "George" in the network drive S
To save the SAS graphics files in EMF format (specify output format as TXT) or JPEG format (specify output format as WEB) in the folder specified here.	Be sure you include the backslash at the end of the folder name.
	If the graphics folder field (macro input number 8) is left blank, the graphics file will be saved in the default folder.

9. Macro-call parameters:	Options/Explanations:
Display SAS output in the output window or save SAS output to a file? (required statement)	Possible values:
	• *DISPLAY*: Output will be displayed in the OUTPUT window. System messages will be displayed in LOG window.
Descriptions & Explanation:	• *WORD*: Output will be saved in the user-specified folder and viewed in the RESULTS VIEWER window as a single RTF format (version 8.2 and later).
Option for displaying all output files in the OUTPUT window or saving in a specific format in a folder specified in option 5.	• *WEB*: Output will be saved in the user-specified folder and viewed in the RESULTS VIEWER window as a single HTML (version 8.2 and later) file.
	• *PDF*: Output will be saved in the user-specified folder and viewed in the RESULTS VIEWER window as a single PDF (version 8.2 and later) file.
	• *TXT*: Output will be saved as a TXT file in the user-specified folder in all SAS versions. No output will be displayed in the OUTPUT window.
	Note: All system messages will be deleted from the LOG window at the end of macro execution if you do not select DISPLAY as the macro input in #9.

10. Macro-call parameters:	Options/Explanations:
Optional SAS SURVEYMEAN options.	Examples: • **Strata gender; weight wt**
Descriptions & Explanation: To estimate population estimates and 95% confidence intervals for survey data and if survey weights are available.	This statement identifies the stratification variable and instructs SAS SURVEYMEAN to use the survey weights and estimate adjusted means and confidence intervals.

A.5 Help File for SAS Macro FACTOR2

1. Macro-call parameters:	Options/Explanations:
Input the temporary SAS dataset name? (required parameter)	• *Fraud* (temporary SAS data called "fraud")
Descriptions & Explanation: Input the temporary SAS dataset name for, on which PCA or EFA will be performed using the raw data.	• *cars93* If only the correlation matrix of the raw data is available and PCA or EFA is to be performed on this correlation matrix, create SAS special correlation matrix data. See an example (Figure 4.15) of the special SAS correlation dataset.
2. Macro-call parameter: Exploratory graphs? (optional parameter)	**Options/Examples** • **Yes:** Only simple descriptive statistics, correlation matrix, and scatter plot matrix of all variables are produced. PCA/EFA is not performed.
Descriptions & Explanation: This macro-call parameter is used to select the type of analysis (exploratory graphics and descriptive statistics or PCA/EFA).	• **Blank:** Only PCA or EFA is performed, depending on the macro input in #6. If the macro input field is left blank, no descriptive statistics, correlation matrix, or scatter plot matrix are produced.

3. Macro-call parameters:	**Options/Examples:**
Input continuous multivariate variable names? (required parameter for PCA or EFA on raw data)	• Y2 Y4 X4 X8 X11 X15 (name of a continuous multiattributes). Acceptable format for both exploratory analysis (Macro input 2 = Yes) and for PCA /EFA (Macro input 2 = blank).
If this field is left blank, the macro is expected to perform PCA or FACTOR analysis on a special correlation matrix. The dataset specified in #1 should be in the form of a special correlation matrix data.	• X2-X15 is allowed. Acceptable format for only for PCA/EFA (Macro input 2 = blank).
Descriptions & Explanation:	If you do not have the raw data, but only have the correlation matrix, and you would like to perform PCA or EFA analysis on this correlation matrix, then leave this field blank. However, make sure the SAS dataset type specified in macro #1 is a special correlation matrix.
Input continuous multiattribute names from the SAS dataset that you would like to include in the PCA/EFA.	
4. Macro-call parameters:	**Options/Explanations:**
Check for multivariate normality and to detect influential outliers when performing PCA/ EFA (optional statement).	• **Yes:** Statistical estimates for multivariate skewness, multivariate kurtosis, and their statistical significance are produced. In addition Q-Q plot for checking multivariate normality and multivariate outlier detection plots are also produced.
Descriptions & Explanation:	
If you would like to check for multivariate normality and check for the presence of any extreme multivariate outliers/influential data, input YES. If you leave this field blank, this step will be omitted.	• **Blank:** If the macro input field is left blank, no statistical estimates for checking for multivariate normality or detecting outliers are performed.
5. Macro-call parameters:	**Options/Examples:**
Input initial estimate of number of PC or factors to be extracted? (required options)	• 1
	• 2
	• 3
Descriptions & Explanation:	• 4
Input the number of principal components or factors to be extracted.	The allowed number should be between 1 to the total number of multiattributes.

6. Macro-call parameters: Input the factor analysis method? (required option) **Descriptions & Explanation:** Different factor analysis methods are available in SAS PROC FACTOR. But to perform PCA, input the factor analysis method "*P.*" This macro will use the default prior communality estimate "1." To perform factor analysis, select any one of the factor analysis methods other than "*P*" and select a different PRIOR = option in macro input 9.	**Options/Explanations:** Possible values: • **P**—PCA analysis with the default prior communality estimate "1." The screen plot analysis also includes a parallel analysis plot. • **PRINIT**—Iterative PCA with the default prior communality estimate SMC. The screen plot analysis also includes a parallel analysis plot. • **ML**—Maximum Likelihood Factor Analysis with the default prior communality estimate SMC. The screen plot analysis does not include parallel analysis plot. (For other EFA methods, refer to the PROC FACTOR section in SAS online manual [16])
7. Macro-call parameters: Input the factor rotation method? (required option) **Descriptions & Explanation:** Different factor rotation methods are available in SAS PROC FACTOR. But to perform PCA, input the factor rotation method "None."	**Options/Explanations:** • **None**—default • **V**—Varimax, orthogonal rotation • **P**—Promax, oblique rotation (For other rotation methods, refer to the PROC FACTOR section in the SAS online manual [16])
8. Macro-call parameters: Input ID variable? (optional statement) **Descriptions & Explanation:** Input the name of the variable you would like to treat as the ID.	**Options/Examples:** • Car • id • model If you leave this field blank, a character variable will be created from the observational number and will be treated as the ID variable.
9. Macro-call parameters: Input any optional PROC FACTOR statement options. (optional statement)	**Options/Explanations:** Possible values: • **COV**—running PCA or EFA using a covariance matrix. (default option uses a correlation matrix)

Descriptions & Explanation: By specifying different PROC FACTOR statement options, many variations of FACTOR analysis can be performed (new feature in the second edition of the book).	• **%STR(priors=smc)** — specifying prior communality option SMC. • **%STR(cover=0.45 alpha =0.1)** — With ML factor analysis, you can test whether the factor loadings are significantly different from a user-specified value such as 0.45 and estimate 90% CI for the loadings. (Refer to the PROC FACTOR section in SAS online manual [16] for this SAS 9.2 enhancement feature).
10. Macro-call parameters: Folder to save SAS output? (optional statement)	**Options/Explanations:** Possible values: *c:\output* — folder name OUTPUT in the C drive
Descriptions & Explanation: To save the SAS output files created by the macro in a specific folder, input the full path of the folder.	*s:\george* — folder name "George" in the network drive S Be sure you include the backslash at the end of the folder name. The same imported SAS dataset name will be assigned to the output file. If this field is left blank, the output file will be saved in the default folder.
11. Macro-call parameters: Folder to save SAS graphics files? (optional statement)	**Options/Explanations:** Possible values: *c:\output* — folder name OUTPUT in the C drive
Descriptions & Explanation: To save the SAS graphics files in EMF format (specify output format as TXT) or JPEG format (specify output format as WEB) in the folder specified here. If the graphics folder field (macro input number 11) is left blank, the graphics file will be saved in the default folder.	*s:\george* — folder name "George" in the network drive S Be sure you include the backslash at the end of the folder name.
12. Macro-call parameters: A counter value: zth number of run? (required statement)	**Options/Explanations:** • 1 • A • A1

Descriptions & Explanation: SAS output files created by the FACTOR2 will be saved by forming a file name from the original SAS dataset name and the counter number provided in the macro input field 12.	Numbers 1 to 10 and any letters are valid. For example, if the original SAS dataset name is "fraud" and the counter number included is 1, the SAS output files will be saved as "fraud1.*" in the user-specified folder. By changing the counter numbers, the users can avoid replacing the previous SAS output files with the new outputs.
13. Macro-call parameters: Display SAS output in the output window or save SAS output to a file? (required statement) **Descriptions & Explanation:** Option for displaying all output files in the OUTPUT window or saving in a specific format in a folder specified in option 5.	**Options/Explanations:** Possible values: • *DISPLAY*: Output will be displayed in the OUTPUT window. System messages will be displayed in the LOG window. • *WORD*: Output will be saved in the user-specified folder and viewed in the RESULTS VIEWER window as a single RTF format (version 8.2 and later). • *WEB*: Output will be saved in the user-specified folder and viewed in the RESULTS VIEWER window as a single HTML (version 8.2 and later) file. • *PDF*: Output will be saved in the user-specified folder and viewed in the RESULTS VIEWER window as a single PDF (version 8.2 and later) file. • *TXT*: Output will be saved as a TXT file in the user-specified folder in all SAS versions. No output will be displayed in the OUTPUT window. Note: All system messages will be deleted from the LOG window at the end of macro execution if you do not select DISPLAY as the macro input in #13.

A.6 Help File for SAS Macro DISJCLS2

1. Macro-call parameters:	Options/Explanations:
Input the temporary SAS dataset name? (required parameter)	• *Fraud* (temporary SAS data called "fraud")
Descriptions & Explanation:	• *cars93*
Input the temporary SAS dataset name on which to perform a disjoint cluster analysis.	It should be in the form of coordinate data (rows = cases and columns = variables).

2. Macro-call parameter:	Options/Examples
Exploratory disjoint cluster analysis? (optional parameter)	• **Yes:**
Descriptions & Explanation:	a) The results of disjoint cluster analysis are displayed in a simple scatter plot matrix if the number of multiattributes is less than 8.
To display the results of cluster groupings in a simple two-variable scatter plot display; to verify the optimum cluster number by CCC, pseudo F-statistic (PSF), and pseudo t2 (PST2) statistics; and to select variables using the backward elimination method in stepwise discriminant analysis.	b) Plots of CCC, PSF, and PST2 against cluster numbers ranging from 1 to 20 for verifying the optimum cluster number are produced.
	c) Results of backward elimination variable selection in stepwise discriminant analysis are performed.
	d) Verification of cluster groupings by canonical discriminant analysis or checking for multivariate normality is not performed if you select YES in this field.
	e) Blank: Disjoint cluster analysis and the cluster differences are verified by canonical discriminant analysis.

3. Macro-call parameters:	*Examples:*
Input the number of disjoint clusters. (required options)	• 3
	• 10
Descriptions & Explanation:	• 15
Input the number of disjoint clusters you would like to extract?	

4. Macro-call parameters:	Options/Explanations:
Check for multivariate normality assumptions. (optional statement)	• **Yes:** Statistical estimates for multivariate skewness, multivariate kurtosis, and their statistical significance are produced for each cluster. In addition, Q-Q plot for checking multivariate normality and multivariate outlier detection plots are also produced for each cluster.
Descriptions & Explanation:	
If you would like to check for multivariate normality and check for the presence of any extreme multivariate outliers/influential data, input YES. If you leave this field blank, this step will be omitted. Multivariate normality is a requirement for canonical discriminant analysis.	
	• **Blank:** If the macro input field is left blank, no statistical estimates for checking for multivariate normality or detecting outliers are performed.
	However, this is not a requirement for DCA. If you leave this field blank, this step will be omitted.

5. Macro-call parameters:	Options/Examples:
Input continuous multiattribute variable names? (required parameter for performing disjoint cluster analysis on coordinate data)	• Y2 Y4 X4 X8 X11 X15 (name of a continuous multiattribute). Acceptable format for both exploratory analysis when (Macro input 2 = Yes) and for disjoint cluster analysis (Macro input 2 = blank).
Descriptions & Explanation:	
Input continuous multiattribute names from the SAS dataset that you would like to include in the disjoint cluster analysis.	
	• X2-X15 format is also allowed. This is an acceptable format for only disjoint cluster analysis (Macro input 2 = blank).
	New feature: Input all continuous variables in the input line 1. Any binary or ordinal variables can be included in the input line 2.

6. Macro-call parameters:	Options/Examples:
Input any optional PROC CLUSTER options? (optional statement)	• Method=ave
	• Method=cen
Descriptions & Explanation:	• Model=ward
Input any valid PROC CLUSTER options.	If you leave this field blank, the default cluster analysis method will be used.

7. Macro-call parameters:	Options/Examples:
Input ID variable? (optional statement)	• Car • id
Descriptions & Explanation:	• model
Input the name of the variable you would like to treat as the ID.	If you leave this field blank, a character variable will be created from the observational number and will be treated as the ID variable.
8. Macro-call parameters:	Options/Explanations:
Input any optional variable standardization method. (optional statement)	Possible values: • **PRIN**—Disjoint cluster analysis will be performed using all principal component scores.
Descriptions & Explanation:	• **STD**—Disjoint cluster analysis will be performed using standardized scores (mean = 0; STD = 1).
To standardize the continuous variables specified in macro input #5, select the standardization method.	• **IQR**—Disjoint cluster analysis will be performed using standardized scores (using IQR as the standardization option in PROC STD). • **MAD**—Disjoint cluster analysis will be performed using standardized scores (using Mean Absolute Differences as the standardization option in PROC STD).
9. Macro-call parameters:	Options/Explanations:
Folder to save SAS output? (optional statement)	Possible values: *c:\output* —folder name OUTPUT in the C drive
Descriptions & Explanation:	*s:\george* —folder name "George" in the network drive S
To save the SAS output files created by the macro in a specific folder, input the full path of the folder.	Be sure you include the backslash at the end of the folder name. The same imported SAS dataset name will be assigned to the output file. If this field is left blank, the output file will be saved in the default folder.

10. Macro-call parameters:	Options/Explanations:
Folder to save SAS graphics files? (optional statement)	Possible values:
	c:\output —folder name OUTPUT in the C drive
Descriptions & Explanation:	
To save the SAS graphics files in EMF format (specify output format as 'TXT') or JPEG format (specify output format as 'WEB') in the folder specified here. If the graphics folder field (macro input number 10) is left blank, the graphics file will be saved in the default folder.	*s:\george* —folder name "George" in the network drive S
	Be sure you include the backslash at the end of the folder name.

11. Macro-call parameters:	Options/Explanations:
A counter value: zth number of run? (required statement)	• 1
	• A
Descriptions & Explanation:	• A1
SAS output files created by the DISJCLS2 will be saved by forming a file name from the original SAS dataset name and the counter number provided in the macro input field 11.	Numbers 1 to 10 and any letters are valid.
	For example, if the original SAS dataset name is "fraud" and the counter number included is 1, the SAS output files will be saved as "fraud1.*" in the user-specified folder. By changing the counter numbers, the users can avoid replacing the previous SAS output files with the new outputs.

12. Macro-call parameters:	Options/Explanations:
Display SAS output in the output window or save SAS output to a file? (required statement)	Possible values:
	• *DISPLAY*: Output will be displayed in the OUTPUT window. System massages will be displayed in the LOG window.
	• *WORD*: Output will be saved in the user-specified folder and viewed in the RESULTS VIEWER window (single RTF format and later). (v

10. Macro-call parameters: Folder to save SAS graphics files? (optional statement) **Descriptions & Explanation:** To save the SAS graphics files in EMF format (specify output format as 'TXT') or JPEG format (specify output format as 'WEB') in the folder specified here. If the graphics folder field (macro input number 10) is left blank, the graphics file will be saved in the default folder.	**Options/Explanations:** Possible values: *c:\output* —folder name OUTPUT in the C drive *s:\george* —folder name "George" in the network drive S Be sure you include the backslash at the end of the folder name.
11. Macro-call parameters: A counter value: *z*th number of run? (required statement) **Descriptions & Explanation:** SAS output files created by the DISJCLS2 will be saved by forming a file name from the original SAS dataset name and the counter number provided in the macro input field 11.	**Options/Explanations:** • 1 • A • A1 Numbers 1 to 10 and any letters are valid. For example, if the original SAS dataset name is "fraud" and the counter number included is 1, the SAS output files will be saved as "fraud1.*" in the user-specified folder. By changing the counter numbers, the users can avoid replacing the previous SAS output files with the new outputs.
12. Macro-call parameters: Display SAS output in the output window or save SAS output to a file? (required statement)	**Options/Explanations:** Possible values: • *DISPLAY*: Output will be displayed in the OUTPUT window. System massages will be displayed in the LOG window. • *WORD*: Output will be saved in the user-specified folder and viewed in the RESULTS VIEWER window as a single RTF format (version 8.2 and later).

Descriptions & Explanation:	• **WEB**: Output will be saved in the user-specified folder and viewed in the RESULTS VIEWER window as a single HTML (version 8.2 and later) file.
Option for displaying all output files in the OUTPUT window or saving in a specific format in a folder specified in option 10.	• **PDF**: Output will be saved in the user-specified folder and viewed in the RESULTS VIEWER window as a single PDF (version 8.2 and later) file.
	• **TXT**: Output will be saved as a TXT file in the user-specified folder in all SAS versions. No output will be displayed in the OUTPUT window.
	Note: All system messages will be deleted from the LOG window at the end of macro execution if you do not select DISPLAY as the macro input in #12.

A.7 Help File for SAS Macro—REGDIAG2

1. Macro-call parameters:	Options/Explanations:
Input the temporary SAS dataset name? (required parameter)	• *Fraud* (temporary SAS data called "fraud")
Descriptions & Explanation:	• *cars93*
Input the temporary SAS dataset name on which to perform a multiple linear regression analysis.	It should be in the form of coordinate data (rows = cases and columns = variables).
2. Macro-call parameter:	Options/Examples
Input single continuous response variable name? (required parameter)	• Y (name of a continuous response)
Descriptions & Explanation:	
Input the continuous response variable name from the SAS dataset that you would like to model as the target (dependent) variable.	

3. Macro-call parameters:	**Options/Examples:**
Input group or categorical variable names? (optional statement)	For example, month manager:
Descriptions & Explanation:	If you leave this field blank, the REGDIAG2 macro will use PROC REG for regression modeling and fit the regression model using the continuous variables specified in the macro input field #5.
To include categorical variables from the SAS dataset as predictors in regression modeling, input the names of these variables. The REGDIAG2 macro will use PROC GLM for regression modeling and use the categorical variable names in the GLM CLASS statement.	
4. Macro-call parameters:	**Options/Explanations:**
Input the alpha level? (required parameter)	• 0.05
	• 0.01
Descriptions & Explanation:	• 0.10
Input the alpha level for computing the confidence interval estimates for parameter estimates.	However, this is not a requirement for DCA. If you leave this field blank, this step will be omitted.
5. Macro-call parameters:	**Options/Examples:**
Input the predictor variables? (optional statement)	• Y2 Y4 X4 X8 X11 X15 (name of a continuous predictor)
Descriptions & Explanation:	• X2-X15 format is also allowed. This is an acceptable format for only when the exploratory graph field is left blank.
Input the continuous predictor variable names.	
If the macro input field #3 is left blank, PROC REG is used to fit the MLR modeling using these variables as predictors. The checking of significant quadratic and cross-product effects for all continuous predictor variables will be performed using PROC RSREG. Model selection based on all possible combinations of predictor variables using PROC REG (MAXR2) will be performed using the continuous variables listed in this macro input. If the regression diagnostic plot option in the macro input field #14 is selected as YES, regression diagnostic plots for each variable will be generated.	If the macro input field #3 is not blank and categorical variable names are specified,
	PROC GLM is used to fit the MLR model using the model terms specified in macro input #6. If the regression diagnostics option in the macro input field #14 is selected as YES, partial plots are also generated. Model selection based on AICC and SBC criteria will be performed and the best candidate model will be identified by converting all specified categorical levels to dummy variables using SAS PROC GLMMOD automatically.

6. Macro-call parameters:	**Options/Examples:**
Input Model terms? (required parameter)	**X1 X2 X1X2 X1SQ** (MLR using PROC REG: X1 & X2—linear predictors, X1X2—interaction term, X1SQ—quadratic term for X1)
Descriptions & Explanation:	**X1 SOURCE X1*SOURCE** (MLR with indicator variable SOURCE using PROC GLM: X1—linear predictor, SOURCE—indicator variable, X1*SOURCE—interaction term between X1 and SOURCE)
This macro input field is equivalent to the right side of the equal sign in the PROC REG and PROC GLM model statement. When fitting MLR with continuous variables using PROC REG, input names of the continuous variables, quadratic, and cross product terms. Note that quadratic and cross-product terms must be created in the SAS dataset before specifying this in this macro field. However, to fit an MLR with indicator variables using PROC GLM, input continuous predictor variable names, the categorical variable names specified in input #3, and any possible quadratic and interaction terms.	(For details about specifying model statements, refer to SAS online manuals on PROC REG and PROC GLM.)
7. Macro-call parameters:	**Options/Examples:**
Input optional model options? (optional statement)	**Optional REG options:**
	• **INFLUENCE:** Additional influential statistics
Descriptions & Explanation:	• **COLINOINT:** Multicollinearity diagnostic test statistics
Input any optional SAS PROC GLM (when CLASS input #3 is not blank) or PROC REG model options (when CLASS input #3 is blank). Depending on the type of model fitted (GLM or REG) can add these model options for additional statistics.	Optional GLM options: • SS1 • NOINT **NOINT:** No intercept model (For details about specifying model options, refer to SAS online manuals on PROC REG and PROC GLM.)
8. Macro-call parameters:	**Options/Examples:**
Input ID variable. (optional statement)	• **ID** • **NUM**

Descriptions & Explanation: If a unique ID variable can be used to identify each record in the SAS data, input that variable name here. This will be used as the ID variable so that any influential outlier observations can be identified. If no ID variable is available in the dataset, leave this field blank. This macro can create an ID variable based on the observation number from the database.	
9. Macro-call parameters: Adjust for extreme influential observations? (optional parameter)	**Options/Explanations:** • **Yes:** Extreme outliers will be excluded from the analysis.
Descriptions & Explanation: This macro will fit the regression model after excluding extreme observations, if you input YES to this option. Any DIFFTS > 1.5 or RSTUDENT > 4 value in the initial MLR will be treated as an outlier and will be excluded from the analysis. A printout of all excluded observations is also produced.	• **Blank:** All observations in the dataset will be used.
10. Macro-call parameters: Input validation dataset name? (optional parameter)	**Options/Explanations:** • Any valid temporary SAS dataset name.
Descriptions & Explanation: To validate the regression model obtained from the training dataset by using an independent validation dataset, input the name of the SAS temporary validation dataset. This macro fits same regression model to the validation data and both predicted values and residuals can be compared visually.	

11. Macro-call parameters:	Options/Explanations:
Display SAS output in the output window or save SAS output to a file? (required statement)	Possible values:

Descriptions & Explanation:

Option for displaying all output files in the OUTPUT window or saving in a specific format in a folder specified in option 12.

Options/Explanations continued:

- *DISPLAY*: Output will be displayed in the OUTPUT window. System messages will be displayed in the LOG window.

- *WORD*: Output will be saved in the user-specified folder and viewed in the RESULTS VIEWER window as a single RTF format (version 8.2 and later).

- *WEB*: Output will be saved in the user-specified folder and viewed in the RESULTS VIEWER window as a single HTML (version 8.2 and later) file.

- *PDF*: Output will be saved in the user-specified folder and viewed in the RESULTS VIEWER window as a single PDF (version 8.2 and later) file.

- *TXT*: Output will be saved as a TXT file in the user-specified folder in all SAS versions. No output will be displayed in the OUTPUT window.

Note: All system messages will be deleted from the LOG window at the end of macro execution if you do not select DISPLAY as the macro input in #11.

12. Macro-call parameters:	Options/example:
Input folder name to save SAS graphs and output files. (optional statement)	c:\output\ —folder name OUTPUT in the C drive

Descriptions & Explanation:

To save the SAS output files created by the macro in a specific folder, input the full path of the folder.

Be sure you include the backslash at the end of the folder name. The same SAS dataset name will be assigned to the output file. If this field is left blank, the output file will be saved in the default SAS folder.

13. Macro-call parameters: A counter value: *z*th number of analysis. (required statement)	**Options/Explanations:** • 1 • 1reg
Descriptions & Explanation: SAS output files created by the REGDIAG2 will be saved by forming a file name from the original SAS dataset name and the counter number provided in the macro input field 13.	• A1 Numbers 1 to 10 and any letters are valid. For example, if the original SAS dataset name is "fraud" and the counter number included is 1, the SAS output files will be saved as "fraud1.*" in the user-specified folder. By changing the counter numbers, the users can avoid replacing the previous SAS output files with the new outputs.
14. Macro-call parameters: Regression diagnostic plots. (optional parameter)	**Options/Explanations:** • **Yes:** Diagnostic plots are produced for each predictor variable.
Descriptions & Explanation: If YES is input, the macro will produce regression diagnostic plots (augmented partial residuals, partial leverage plots, VIF diagnostic plots, and detecting significant interaction) for each continuous predictor variable and categorical variable. If you leave this macro field blank, no diagnostic plots are produced.	• **Blank:** No diagnostic plots are produced.
15. Macro-call parameters: Optional SAS SURVEYREG options.	**Options/Explanations:** Examples:
Descriptions & Explanation: To estimate population regression model estimates and 95% confidence intervals for survey data and if survey weights are available.	• **Strata gender; weight wt** This statement identifies the stratification variable and instructs SAS SURVEYREG to use the survey weights and estimate adjusted regression coefficients.

A.8 Help File for SAS Macro—LIFT2

1. Macro-call parameters:	Options/Explanations:
Input the temporary SAS dataset name? (required parameter)	• *Fraud* (temporary SAS data called "fraud")
Descriptions & Explanation: Input the temporary SAS dataset name on which to perform a multiple linear regression analysis.	• *cars93* It should be in the form of coordinate data (rows = cases and columns = variables).
2. Macro-call parameters: Input group or categorical variables names? (optional statement)	**Options/Examples:** • **month manager:** (Regression modeling using PROC GLM or LOGISTIC)
Descriptions & Explanation: To include categorical variables from the SAS dataset as predictors in regression modeling, input the names of these variables. Use this option with SAS procedures GLM and LOGISTIC (Version 8 and later).	• **Blank:** (If the macro input field is left blank, PROC REG or LOGISTIC)
3. Macro-call parameter: Input single continuous response variable name? (required parameter)	**Options/Examples** • Y (name of a continuous response—PROC REG or GLM) • Y (name of a binary response—PROC LOGISTIC)
Descriptions & Explanation: Input the continuous response variable name from the SAS dataset that you would like to model as the target (dependent) variable.	
4. Macro-call parameters: Input variable name of interest? (required options)	**Options/Examples:** • X1 • Manager
Descriptions & Explanation: Input the name of the predictor variable to control in the IF-THEN analysis. The variable of interest can be continuous (PROC REG, GLM, AND LOGISTIC) or a binary categorical variable (GLM and LOGISTIC).	

5. Macro-call parameters: Input the SAS PROC name? (required parameter)	**Options/Explanations:** • **REG** (MLR with continuous response and continuous predictor variables)
Descriptions & Explanation: Input the alpha level for computing the confidence interval estimates for parameter estimates.	• **GLM** (MLR with continuous response and continuous and categorical predictor variables) • **LOGISTIC**: (Logistic regression with binary response and continuous and categorical predictor variables) (For details about specifying model statements, refer to SAS online manuals on PROC REG, PROC GLM, and PROC LOGISTIC.)
6. Macro-call parameters: Input the LIFT variable fixed value? (required statement)	**Options/Examples:** • **50000** (Fixed value for a continuous variable X1; applicable in REG, GLM, and LOGISTIC)
Descriptions & Explanation: Input the fixed value for the variable of interest.	• **D** (Fixed categorical level when the variable of interest is categorical; Valid only in GLM and LOGISTIC)
7. Macro-call parameters: Input Model terms? (required parameter)	**Options/Examples:** • **X1X2 X3 X2X3 X2SQ** (MLR using PROC REG: X1, X2, and X3 are linear predictors; X2X3 is the interaction term; and X2SQ is the quadratic term for X2.)
Descriptions & Explanation: This macro input is equivalent to the right side of the equal sign in the PROC REG, PROC GLM, or PROC LOGISTIC model statement. In case of fitting MLR with continuous variables using PROC REG, input names of the continuous variables, quadratic, and cross-product terms. Note that quadratic and cross-product terms should be created in the SAS dataset before specifying this in the macro field; however, to fit a regression model with indicator variables using PROC GLM/LOGISTIC (Version 8.0	• **X1 SOURCE X1*SOURCE** (regression with indicator variable SOURCE using PROC GLM/ LOGISTIC: X1 is the linear predictor, SOURCE is the indicator variable, and X1*SOURCE is the interaction term between X1 and SOURCE).

and later), input continuous predictor variable names, categorical variable names you specified in the macro input #3, and any possible quadratic and interaction terms.	
8. Macro-call parameters: Any other optional statements. (optional statement)	**Options/Examples:** • **LSMEANS source /pdiff ; Contrast "source a vs b" source –1 1 0**
Descriptions & Explanation: Input any other optional statements associated with the SAS procedures; see the example in the next column.	(Only valid in GLM and LOGISTIC. Note that if you are using more than one statement, use ";" at the end of the first statement.) • **Test x1 –x2 = 0; Restrict intercept = 0** (In PROC REG only). Note if you are using more than one statement, use ";" at the end of the first statement. (For details about specifying other PROC statements, refer to SAS online manuals on PROC REG PROC GLM and PROC LOGISTIC.)
9. Macro-call parameters: Input ID variable. (optional statement)	**Options/Examples:** • **ID** • **NUM**
Descriptions & Explanation: If a unique ID variable can be used to identify each record in the SAS data, input that variable name here. This will be used as the ID variable so that any influential outlier observations can be identified.	If no ID variable is available in the dataset, leave this field blank. This macro can create an ID variable based on the observation number from the database.
10. Macro-call parameters: Input weight variable. (optional parameter)	**Options/Explanations:** • **E.G: _Wt_:** (Name of the weight variable.) • **Blank:** (1 will be used as the weights.)
Descriptions & Explanation: To perform a weighted regression analysis (weights to adjust for influential observations or heteroscedasticity) and if the weight	

variable exists in your dataset, input the name of this weight variable. If you leave this parameter blank, a weight of 1 will be used for all observations.	
11. Macro-call parameters: Adjust for extreme influential observations? (optional parameter)	**Options/Explanations:** • **Yes:** Extreme outliers will be excluded from the analysis. • **Blank:** All observations in the dataset will be used.
Descriptions & Explanation: This macro will fit the specified model after excluding extreme observations.	
12. Macro-call parameters: Input folder name to save SAS graphs and output files. (optional statement)	**Options/example:** *c:\output* —folder name OUTPUT in the C drive
Descriptions & Explanation: To save the SAS output files created by the macro in a specific folder, input the full path of the folder.	Be sure you include the backslash at the end of the folder name. The same SAS dataset name will be assigned to the output file. If this field is left blank, the output file will be saved in the default SAS folder.
13. Macro-call parameters: A counter value: zth number of analysis. (required statement)	**Options/Explanations:** • 1 • 1reg • A1
Descriptions & Explanation: SAS output files created by the LIFT2 will be saved by forming a file name from the original SAS dataset name and the counter number provided in the macro input field 13.	Numbers 1 to 10 and any letters are valid. For example, if the original SAS dataset name is "fraud" and the counter number included is 1, the SAS output files will be saved as "fraud1.*" in the user-specified folder. By changing the counter numbers, the users can avoid replacing the previous SAS output files with the new outputs.

14. Macro-call parameters:	Options/Explanations:
Display SAS output in the output window or save SAS output to a file? (required statement)	Possible values:
Descriptions & Explanation:	• *DISPLAY*: Output will be displayed in the OUTPUT window. System messages will be displayed in the LOG window.
Option for displaying all output files in the OUTPUT window or saving in a specific format in a folder specified in option 12.	• *WORD*: Output will be saved in the user-specified folder and viewed in the RESULTS VIEWER window as a single RTF format (version 8.2 and later).
	• *WEB*: Output will be saved in the user-specified folder and viewed in the RESULTS VIEWER window as a single HTML (version 8.2 and later) file.
	• *PDF*: Output will be saved in the user-specified folder and viewed in the RESULTS VIEWER window as a single PDF (version 8.2 and later) file.
	• *TXT*: Output will be saved as a TXT file in the user-specified folder in all SAS versions. No output will be displayed in the OUTPUT window.
	Note: All system messages will be deleted from the LOG window at the end of macro execution if you do not select DISPLAY as the macro input in 14.

A.9 Help File for SAS Macro RSCORE2

1. Macro-call parameters:	Options/Explanations:
Input name of the new scoring SAS dataset name. (required parameter)	• *Fraud2* (temporary SAS data called "fraud")
Descriptions & Explanation:	• *cars932*
Input the temporary SAS dataset name on which to perform scoring using the established regression model.	The data format should be in the form of coordinate data (rows = cases and columns = variables).

2. Macro-call parameters:	Options/Examples:
Input the SAS data name containing the regression model estimates. (required parameter)	• *Regest1*—Temporary SAS dataset
Descriptions & Explanation:	
Input the name of the temporary SAS dataset containing the regression model estimates derived by running the REGDIAG2 macro.	
3. Macro-call parameter:	**Options/Examples**
Input single continuous response variable name? (optional parameter)	• Y (name of a continuous response)
Descriptions & Explanation:	• If a response value is not available, leave this field blank.
If a response (dependent) variable is available in the scoring dataset, input the name of the response variable. The RSCORE2 macro can also estimate the residual and compute the R^2 for prediction.	
4. Macro-call parameters:	**Options/Examples:**
Input the model terms included in the original model. (required options)	• **X1 X2 X3 X2X3 X2SQ**
Descriptions & Explanation:	(X1, X2, and X3 are linear predictors; X2X3 is the interaction term; and X2SQ is the quadratic term for X2)
Input the regression model used to develop the original regression estimates. This must be identical to the PROC REG model statement specified when estimating the regression model using REGDIAG2 macro.	
5. Macro-call parameters:	**Options/Examples:**
Input ID variable name. (optional statement)	• **ID**
	• **NUM**
Descriptions & Explanation:	If no ID variable is available in the dataset, leave this field blank. This macro can create an ID variable based on the observation number from the database.
If a unique ID variable can be used to identify each record in the SAS data, input that variable name here. This will be used as the ID variable so that any influential outlier observations can be identified.	

6. Macro-call parameters:	Options/Example:
Input folder name to save SAS graphs and output files.	*c:\output* —folder name OUTPUT in the C drive
(optional statement)	Be sure you include the backslash at the end of the folder name. The same SAS dataset name will be assigned to the output file. If this field is left blank, the output file will be saved in the default SAS folder.
Descriptions & Explanation:	
To save the SAS output files created by the macro in a specific folder, input the full path of the folder.	

7. Macro-call parameters:	Options/Explanations:
A counter value: zth number of analysis. (required statement)	• 1
	• 1rcore2
Descriptions & Explanation:	• A1
SAS output files created by RSCORE2 will be saved by forming a file name from the original SAS dataset name and the counter number provided in this macro input field 7.	Numbers 1 to 10 and any letters are valid.
	For example, if the original SAS dataset name is "fraud" and the counter number included is 1, the SAS output files will be saved as "fraud1.*" in the user-specified folder. By changing the counter numbers, the users can avoid replacing the previous SAS output files with the new outputs.

8. Macro-call parameters:	Options/Explanations:
Display SAS output in the output window or save SAS output to a file? (required statement)	Possible values:
	DISPLAY: Output will be displayed in the OUTPUT window. System messages will be displayed in the LOG window.
	WORD: Output will be saved in the user-specified folder and viewed in the RESULTS VIEWER window as a single RTF format (version 8.2 and later).
	WEB: Output will be saved in the user-specified folder and viewed in the RESULTS VIEWER window as a single HTML (version 8.2 and later) file.

Descriptions & Explanation:	**PDF**: Output will be saved in the user-specified folder and viewed in the RESULTS VIEWER window as a single PDF (version 8.2 and later) file.
Option for displaying all output files in the OUTPUT window or save as a specific format in a folder specified in option 6.	**TXT**: Output will be saved as a TXT file in the user-specified folder in all SAS versions. No output will be displayed in the OUTPUT window.
	Note: All system messages will be deleted from the LOG window at the end of macro execution if you do not select DISPLAY as the macro input in 14.

A.10 Help File for SAS Macro—LOGIST2

1. Macro-call parameters:	Options/Explanations:
Input the temporary SAS dataset name? (required parameter)	• *Fraud* (temporary SAS data called "fraud")
Descriptions & Explanation:	• *cars93*
Input the temporary SAS dataset name on which to perform a logistic regression analysis with binary or ordinal response.	It should be in the form of coordinate data (rows = cases and columns = variables).

2. Macro-call parameter:	Options/Examples
Input binary response variable name? (required parameter)	Y (name of a binary (0 or 1)/ ordinal response (1 to 4)
Descriptions & Explanation:	
Input the numeric binary/ordinal response variable name from the SAS dataset that you would like to model as the target (dependent) variable.	

3. Macro-call parameters:	Options/Examples:
Input group or categorical variable names? (optional statement)	For example: • month manager • Blank

Descriptions & Explanation:	If you leave this field blank, the
To include categorical variables from the SAS dataset as predictors in logistic regression modeling, input the names of these variables. The LOGIST2 macro will use PROC LOGISTIC for logistic regression modeling and use the categorical variable names in the CLASS statement.	LOGIST2 macro will fit the model without the categorical variable. The class variable coding used in this LOGIST2 macro would be the GLM like reference coding.

4. Macro-call parameters:	**Options/Examples:**
Input continuous predictor variable names? (optional statement)	*EG*: X1 X2 X3
Descriptions & Explanation:	This diagnostic plot is useful in checking for nonlinearity, the significance of the parameter estimate, and the presence of multicollinearity among the predictor variable. If you also include categorical variables in #3, the logistic parameter estimates are adjusted for these categorical variables. No interaction terms between categorical and the continuous predictors are included in the diagnostic plots when computing the delta logit values. Leave this field blank if you have only categorical variables in the model statement.
For each continuous predictor variable you specify, an overlaid partial delta logit–simple logit plot and significant two-factor interaction plots will be automatically produced if you input YES to exploratory plots.	

5. Macro-call parameters:	**Options/Examples:**
Input model terms? (required option)	**X1 X2 X1*X2 X1*X1** (Continuous predictor variables including linear predictors X1 & X2; interaction term X1*X2; and the quadratic term for X1, X1*X1)
Descriptions & Explanation:	
This macro input field is equivalent to the right side of the equal sign in the PROC LOGISTIC model statement. You can include main effects of categorical variables, linear effects of continuous variables, and any interactions among these effects.	**X1 SOURCE X1*SOURCE** (A logistic regression with categorical variables SOURCE, linear predictor X1, and an interaction term between X1 and SOURCE, X1*SOURCE)
	(For details about specifying model statements, refer to SAS online manuals on PROC LOGISTIC.)

6. Macro-call parameters:	Options/Examples:
Exploratory analysis? (optional statement)	**YES**: Perform exploratory analysis **Blank**: Skip exploratory analysis
Descriptions & Explanation: To perform exploratory analysis, input YES in this field. Partial delta logit plots and all possible variable selection within best candidate subset selection method for continuous predictors or predicted probability plots for categorical variable model will be generated.	If you leave this field blank, this macro skips the exploratory analysis step and goes directly to the logistic regression step.
7. Macro-call parameters:	Options/Examples:
Overdispersion correction? (required statement)	• **NONE**: No overdispersion adjustment • **DEVIANCE**: Overdispersion adjustment by DEVIANCE factor • **PEARSON**: Overdispersion adjustment by PEARSON factor
Descriptions & Explanation: To not adjust for overdispersion, input NONE. However, if the test for overdispersion indicates that a high degree of overdispersion exists and you want to adjust for it by using either the DEVIANCE or PEARSON method, then input either DEVIANCE or PEARSON.	
8. Macro-call parameters:	Options/Examples:
Customized odds ratios/parameter test? (optional statement)	• **Units x1 = 0.5 – 0.5** (To obtain customized odds ratio estimate for X1 predictor when X1 is increased or decreased by 0.5 unit.)
Descriptions & Explanation: Input appropriate statements (UNITS and/or TEST) to obtain customized odds ratio estimates (UNITS option) and/or test the parameter estimates equal for specific values (TEST option).	• **Test x1 = 0.05** (To test the hypothesis that the X1 parameter estimate is equal to 0.5.) • **Units x1 = 0.5 – 0.5; Test x1 = 0.05** (To obtain both customized odds ratio and perform parameter test.) Note the ";" after the first statement. If you specify more than one statement, include a ";" at the end of all statements except the last statement.

9. Macro-call parameters:	Options/Explanations:
Input validation dataset name? (optional parameter)	• VALID1 (Temporary SAS dataset: valid (SAS dataset name)
Descriptions & Explanation:	This macro estimates predicted probabilities for the validation dataset using the model estimates derived from the training data. The success of this prediction could be verified by checking the deviance residual and the Brier scores for the validation dataset.
If you would like to validate the logistic regression model obtained from a training dataset by using an independent validation dataset, input the name of the SAS validation dataset.	

10. Macro-call parameters:	Options/Explanations:
Input ID variable name. (optional statement)	For example: **ID NUM** If no ID variable is available in the dataset, you can leave this field blank. This macro can create an ID variable based on the observation number from the database.
Descriptions & Explanation:	
If you have a unique ID variable that can be used to identify each record in the database, input that variable name here. This will be used as the ID variable so that any outlier/ influential observations can be detected.	

11. Macro-call parameters:	Options/Explanations:
Input a counter value: zth number of analysis. (required statement)	• 1 • 1reg • A1
Descriptions & Explanation:	Numbers 1 to 10 and any letters are valid.
SAS output files created by the LOGIST2 will be saved by forming a file name from the original SAS dataset name and the counter number provided in the macro input field 11.	For example, if the original SAS dataset name is "fraud" and the counter number included is 1, the SAS output files will be saved as "fraud1.*" in the user-specified folder. By changing the counter numbers, the users can avoid replacing the previous SAS output files with the new outputs.

12. Macro-call parameters:	Options/Explanations:
Display SAS output in the output window or save SAS output to a file? (required statement)	Possible values:
	DISPLAY: Output will be displayed in the OUTPUT window. System messages will be displayed in the LOG window.
Descriptions & Explanation:	
Option for displaying all output files in the OUTPUT window or save as a specific format in a folder specified in option 13.	*WORD*: Output will be saved in the user-specified folder and viewed in the RESULTS VIEWER window as a single RTF format (version 8.2 and later).
	WEB: Output will be saved in the user-specified folder and viewed in the RESULTS VIEWER window as a single HTML (version 8.2 and later) file.
	PDF: Output will be saved in the user-specified folder and viewed in the RESULTS VIEWER window as a single PDF (version 8.2 and later) file.
	TXT: Output will be saved as a TXT file in the user-specified folder in all SAS versions. No output will be displayed in the OUTPUT window.
	Note: All system messages will be deleted from the LOG window at the end of macro execution if you do not select DISPLAY as the macro input in #11.
13. Macro-call parameters:	**Options/example:**
Input folder name to save SAS graphs and output files. (optional statement)	*c:\output*—folder name 'OUTPUT' in the c drive
Descriptions & Explanation:	Be sure you include the backslash at the end of the folder name. The same SAS dataset name will be assigned to the output file. If this field is left blank, the output file will be saved in the default SAS folder.
To save the SAS output files created by the macro in a specific folder, input the full path of the folder.	

14. Macro-call parameters:	Options/Explanations:
Adjust for extreme influential observations? (optional parameter)	• **Yes:** Extreme outliers will be excluded from the analysis.
Descriptions & Explanation:	• **Blank:** All observations in the dataset will be used.
If you input YES to this option, the macro will fit the logistic regression model after excluding extreme observations (delta deviance > 4.0). An output of all excluded observations is also produced.	

15. Macro-call parameters:	Options/Explanations:
Input cutoff *p*-value? (required options)	• 0.45
Descriptions & Explanation	• 0.5
Input the cutoff *p*-value for classifying the predicted probability as event and nonevent.	• 0.55
	• 0.60

15. Macro-call parameters:	Options/Explanations:
Optional SAS SURVEYLOGISTIC options.	Examples:
	• **Strata gender; weight wt**
Descriptions & Explanation:	This statement identifies the stratification variable and instructs SAS SURVEYLOGISTIC to use the survey weights and estimate adjusted logit coefficients.
To estimate population logistic regression model estimates and 95% confidence intervals for survey data and if survey weights are available.	

A.11 Help File for SAS Macro LSCORE2

1. Macro-call parameters:	Options/Explanations:
Input the name of the new scoring SAS dataset name. (required parameter)	• *Fraud2* (temporary SAS dataset called "fraud")
	• *cars932*
Descriptions & Explanation:	The data format should be in the form of coordinate data (rows = cases and columns = variables).
Input the temporary SAS dataset name on which to perform scoring using the established logistic regression model estimates.	

2. Macro-call parameters:	Options/Examples:
Input optional categorical variables. (optional statement)	• **Month manager**: categorical variables
Descriptions & Explanation:	• **Blank**: If the macro input field is left blank, no categorical variables were used in the original model building.
If categorical variables are in the new "score" dataset and categorical variables were used in the original logistic regression as predictors, input the names of these variables.	

3. Macro-call parameters:	Options/Examples:
Input the optional binary response variable name. (optional parameter)	RESP2 — Optional binary response variable
Descriptions & Explanation:	
If a binary response variable is available in your NEW dataset, input the name of the response variable. The LSCORE2 macro can also estimate the residual and investigate the model fit graphically. If a binary response value is not available, leave this field blank.	

4. Macro-call parameters:	Options/Examples:
Input the model terms included in the original model. (required options)	• **X1 X2 X3 X2X3 X2SQ**
Descriptions & Explanation:	(X1, X2, and X3 are linear predictors; X2X3 is the interaction term; and X2SQ is the quadratic term for X2)
Input the regression model used to develop the original logistic regression estimates. This must be identical to the PROC LOGISTIC model statement specified when estimating the regression model using LOGIST2 macro.	

5. Macro-call parameters:	Options/Examples:
Input ID variable name. (optional statement)	• **ID**
	• **NUM**
Descriptions & Explanation:	If no ID variable is available in the dataset, leave this field blank. This macro can create an ID variable based on the observation number from the database.
If a unique ID variable can be used to identify each record in the SAS data, input that variable name here. This will be used as the ID variable so that any influential outlier observations can be identified.	

6. Macro-call parameters:	Options/Example:
Input folder name to save SAS graphs and output files. (optional statement)	*c:\output* — folder name OUTPUT in the C drive
Descriptions & Explanation: To save the SAS output files created by the macro in a specific folder, input the full path of the folder.	Be sure you include the backslash at the end of the folder name. The same SAS dataset name will be assigned to the output file. If this field is left blank, the output file will be saved in the default SAS folder.
7. Macro-call parameters:	Options/Explanations:
A counter value: *z*th number of analysis. (required statement)	• 1 • 1rcore2 • A1
Descriptions & Explanation: SAS output files created by the LSCORE2 will be saved by forming a file name from the original SAS dataset name and the counter number provided in this macro input field 7.	Numbers 1 to 10 and any letters are valid. For example, if the original SAS dataset name is "fraud" and the counter number included is 1, the SAS output files will be saved as "fraud1.*" in the user-specified folder. By changing the counter numbers, the users can avoid replacing the previous SAS output files with the new outputs.
8. Macro-call parameters:	Options/Explanations:
Display SAS output in the output window or save SAS output to a file? (required statement)	Possible values: • *DISPLAY*: Output will be displayed in the OUTPUT window. System messages will be displayed in the LOG window.

Descriptions & Explanation:	
Option for displaying all output files in the OUTPUT window or saving in a specific format in a folder specified in option 6.	• **WORD**: Output will be saved in the user-specified folder and viewed in the RESULTS VIEWER window as a single RTF format (version 8.2 and later).
	• **WEB**: Output will be saved in the user-specified folder and viewed in the RESULTS VIEWER window as a single HTML (version 8.2 and later) file.
	• **PDF**: Output will be saved in the user-specified folder and viewed in the RESULTS VIEWER window as a single PDF (version 8.2 and later) file.
	• **TXT**: Output will be saved as a TXT file in the user-specified folder in all SAS versions. No output will be displayed in the OUTPUT window.
	Note: All system messages will be deleted from the LOG window at the end of macro execution if you do not select DISPLAY as the macro input in 14.

A.12 Help File for SAS Macro DISCRIM2

1. Macro-call parameters:	Options/Explanations:
Input the temporary SAS dataset name? (required parameter)	• *Fraud* (temporary SAS dataset called "fraud")
Descriptions & Explanation:	• *Diabet2*
Input the temporary SAS dataset name on which to perform a discriminant analysis.	It should be in the form of coordinate data (rows = cases and columns = variables).

2. Macro-call parameter:	Options/Examples
Exploratory discriminant analysis? (optional parameter)	• **Yes:** Only the scatter plot matrix of all predictor variables by group response is produced. Variable selection by forward selection, backward elimination, and stepwise selection methods are performed. Discriminant analysis (CDA or DFA) is not performed.
Descriptions & Explanation:	
This macro-call parameter is used to select the type of analysis between exploratory graphics analysis and variable selection or CDA and DFA.	• **Blank:** If the macro input field is left blank, exploratory analysis and variable selection are not performed. Only CDA and parametric or nonparametric DFA are performed.
3. Macro-call parameters:	**Examples:**
Input categorical group response variable name? (required parameter)	• group
	(name of a categorical response)
Descriptions & Explanation:	
Input the categorical group response name from the SAS dataset that you would like to model as the target variable.	
4. Macro-call parameters:	**Options/Explanations:**
Check for multivariate normality assumptions. (optional statement)	• **Yes:** Statistical estimates for multivariate skewness, multivariate kurtosis, and their statistical significance are produced. In addition, Q-Q plots for checking multivariate normality and multivariate outlier detection plots are also produced.
Descriptions & Explanation:	
If you would like to check for multivariate normality and check for the presence of any extreme multivariate outliers/influential observations, input YES. If you leave this field blank, this step will be omitted.	• **Blank:** If the macro input field is left blank, no statistical estimates for checking for multivariate normality or detecting outliers are performed.

5. Macro-call parameters:	Options/Examples:
Input numeric multiattribute variable names? (required parameter)	Examples: X1 X2 X3 X4 X5 mpg murder (List the names of a continuous predictor variables)
Descriptions & Explanation: Input numeric variable names from your dataset you would like to use in discriminant analysis as predictors.	X2-X15 format is also allowed. This is an acceptable format only for discriminant analysis (Macro input 2 = blank). (New feature: Input all continuous variables in the input line 1. Any binary or ordinal variables can be included in the input line 2.)
6. Macro-call parameters:	Options/Examples:
Nonparametric discriminant analysis? (optional statement)	• **Yes:** Canonical discriminant analysis and parametric discriminant function analysis will *not* be performed. Instead, nonparametric discriminant analysis based on the *k*th-nearest-neighbor and kernel density methods will be performed. The probability density in the nearest-neighbor (*k* = 2 to 4) nonparametric discriminant analysis method will be estimated using the *Mahalanobis* distance based on the pooled covariance matrix. Posterior probability estimates in kernel density nonparametric discriminant analysis methods will be computed using these "*kernel=normal r=0.5*" PROC DISCRIM options.
Descriptions & Explanation: Select the type of discriminant. (parametric or nonparametric) analysis.	
	(For details about parametric and nonparametric discriminant analysis options, see SAS online manuals on PROC DISCRIM.[34])
	• **Blank:** Canonical discriminant analysis and parametric discriminant function analysis will be performed assuming all the predictor variables within each group level have multivariate normal distribution.

7. Macro-call parameters:	Options/Examples:
Prior probability options? (required statement)	• **Equal:** To set the prior probabilities equal.
Descriptions & Explanation:	• **Prop:** To set the prior probabilities proportional to the sample sizes.
Input the prior probability option required for computing posterior probability and classification error estimates.	(For details about prior probability options, see the SAS online manuals on PROC DISCRIM.[8])

8. Macro-call parameters:	Options/Examples:
Input ID variable? (optional statement)	• Car
	• id
Descriptions & Explanation:	• model
Input the name of the variable you would like to treat as the ID.	If you leave this field blank, a character variable will be created from the observational number and will be treated as the ID variable.

9. Macro-call parameters:	Options/Explanations:
Input validation dataset name? (optional parameter)	• *diabetic2*
	Input any optional temporary SAS dataset name.
Descriptions & Explanation:	
If you would like to validate the discriminant model obtained from a training dataset by using an independent validation dataset, input the name of the SAS validation dataset.	

10. Macro-call parameters:	Options/Explanations:
A counter value: zth number of run? (required statement)	• 1
	• A
Descriptions & Explanation:	• A1
SAS output files created by the DISCRIM2 macro will be saved by forming a file name from the original SAS dataset name and the counter number provided in the macro input field 10.	Numbers 1 to 10 and any letters are valid.
	For example, if the original SAS dataset name is "fraud" and the counter number included is 1, the SAS output files will be saved as "fraud1.*" in the user-specified folder. By changing the counter numbers, users can avoid replacing the previous SAS output files with the new outputs.

11. Macro-call parameters: Folder to save SAS graphics and output files? (optional statement)	**Options/Explanations:** • *c:\output*—folder name OUTPUT in the C drive • *s:\george*—folder name "George" in the network drive S
Descriptions & Explanation: To save the SAS graphics files in an EMF format suitable for inclusion in PowerPoint presentations, specify the output format as TXT in Version 8.0 or later. In pre-8.0 SAS versions, all graphic format files will be saved in a user-specified folder. Similarly output files in WORD, HTML, PDF, and TXT formats will be saved in the user-specified folder. If this macro field is left blank, the graphics and output files will be saved in the default folder.	Be sure you include the backslash at the end of the folder name. The same imported SAS dataset name will be assigned to the output file. If this field is left blank, the output file will be saved in the default folder.
12. Macro-call parameters: Display SAS output in the output window or save SAS output to a file? (required statement)	**Options/Explanations:** Possible values: • ***DISPLAY***: Output will be displayed in the OUTPUT window. System messages will be displayed in the LOG window.
Descriptions & Explanation: Option for displaying all output files in the OUTPUT window or save as a specific format in a folder specified in option 5.	• ***WORD***: Output will be saved in the user-specified folder and viewed in the RESULTS VIEWER window as a single RTF format (version 8.2 and later). • ***WEB***: Output will be saved in the user-specified folder and viewed in the RESULTS VIEWER window as a single HTML (version 8.2 and later) file. • ***PDF***: Output will be saved in the user-specified folder and viewed in the RESULTS VIEWER window as a single PDF (version 8.2 and later) file. • ***TXT***: Output will be saved as a TXT file in the user-specified folder in all SAS versions. No output will be displayed in the OUTPUT window.

	Note: All system messages will be deleted from the LOG window at the end of macro execution if you do not select DISPLAY as the macro input in #12.
13. Macro-call parameters: Transforming predictor variables? (optional statement)	**Options/Examples:** • **Blank:** No transformation is performed. The original predictor variables will be used in discriminant analysis.
Descriptions & Explanation: You could perform a log-scale or z (0 mean; 1 standard deviation) transformation on all the predictor variables to reduce the impact of between-group unequal variance covariance problem or differential scale of measurement.	• **LOG:** All predictor variables (nonzero values) will be transformed to natural log scale using the SAS LOG function. All types of discriminant analysis will be performed on log-transformed predictor variables. • **STD:** All predictor variables will be standardized to 0 mean and unit standard deviation using the SAS PROC STANDARD. All types of discriminant analysis will be performed on standardized predictor variables.

A.13 Help File for SAS Macro CHAID2

1. Macro-call parameters: Input the temporary SAS dataset name? (required parameter)	**Options/Explanations:** • *Fraud* (temporary SAS dataset called "fraud")
Descriptions & Explanation: Input the temporary SAS dataset name on which to perform a CHAID analysis.	• *Diabet2* It should be in the form of coordinate data (rows = cases and columns = variables).

2. Macro-call parameters: Input categorical group response variable name? (required parameter)	**Examples:** • group (name of a categorical response)
Descriptions & Explanation: Input the categorical group response name from the SAS dataset that you would like to model as the target variable.	
3. Macro-call parameters: Input nominal predictor variables? (optional statement)	**Options/Examples:** • TSTPLGP1 FASTPLGP • **Blank:** Categorical predictors are not used
Descriptions & Explanation: Include categorical variables from the SAS dataset as predictors in CHAID modeling.	
4. Macro-call parameters: Input ordinal predictor variable names? (optional statement)	**Options/Examples:** • X1 X2 X3
Descriptions & Explanation: Include continuous variables from the SAS dataset as predictors in CHAID modeling.	
5. Macro-call parameters: Input validation dataset name? (optional parameter)	**Options/Examples:** • *Diabet2* Temporary SAS dataset: diabetic2 (SAS dataset name)
Descriptions & Explanation: To validate the CHAID model obtained from a training dataset by using an independent validation dataset, input the name of the SAS validation dataset. This macro estimates classification error for the validation dataset using the model estimates derived from the training data.	
6. Macro-call parameters: Input ID variable? (optional statement)	**Options/Examples:** • Car • id • model

Descriptions & Explanation:	If you leave this field blank, a character variable will be created from the observational number and will be treated as the ID variable.
Input the name of the variable you would like to treat as the ID.	

7. Macro-call parameters:	Options/Explanations:
A counter value: zth number of run? (required statement)	• 1
	• A
Descriptions & Explanation:	• A1
SAS output files created by the CHAID2 macro will be saved by forming a file name from the original SAS dataset name and the counter number provided in the macro input field 7.	Numbers 1 to 10 and any letters are valid.
	For example, if the original SAS dataset name is "fraud" and the counter number included is 1, the SAS output files will be saved as "fraud1.*" in the user-specified folder. By changing the counter numbers, users can avoid replacing the previous SAS output files with the new outputs.

8. Macro-call parameters:	Options/Explanations:
Folder to save SAS graphics and output files? (optional statement)	*c:\output* —folder name OUTPUT in the C drive
Descriptions & Explanation:	*s:\george* —folder name "George" in the network drive S
To save the SAS graphics files in an EMF format suitable for inclusion in PowerPoint presentations, specify output format as TXT in Version 8.0 or later. In pre-8.0 SAS versions, all graphic format files will be saved in a user-specified folder. Similarly, output files in WORD, HTML, PDF, and TXT formats will be saved in the user-specified folder. If this macro field is left blank, the graphics and output files will be saved in the default folder.	Be sure you include the backslash at the end of the folder name. The same imported SAS dataset name will be assigned to the output file. If this field is left blank, the output file will be saved in the default folder.

9. Macro-call parameters:	Options/Explanations:
Display SAS output in the output window or save SAS output to a file? (required statement)	Possible values:
Descriptions & Explanation: Option for displaying all output files in the OUTPUT window or saving in a specific format in a folder specified in option 8.	• ***DISPLAY***: Output will be displayed in the OUTPUT window. System messages will be displayed in the LOG window. • ***WORD***: Output will be saved in the user-specified folder and viewed in the RESULTS VIEWER window as a single RTF format (version 8.2 and later). • ***WEB***: Output will be saved in the user-specified folder and viewed in the RESULTS VIEWER window as a single HTML (version 8.2 and later) file. • ***PDF***: Output will be saved in the user-specified folder and viewed in the RESULTS VIEWER window as a single PDF (version 8.2 and later) file. • ***TXT***: Output will be saved as a TXT file in the user-specified folder in all SAS versions. No output will be displayed in the OUTPUT window. Note: All system messages will be deleted from the LOG window at the end of macro execution if you do not select DISPLAY as the macro input in #12.

Appendix III: Instruction for Using the SAS Macros with Enterprise Guide Code Window

SAS Enterprise Guide (EG) is a powerful Microsoft Windows client application that provides a guided user-friendly mechanism to exploit the power of SAS and perform complete data analysis quickly. SAS EG is the front-end applications for SAS learning edition also. However, the SAS macro applications incorporated in the book are not compatible with the SAS EG. To solve this problem, I have developed separate SAS macros and macro-call files that are compatible with the SAS EG code window. These files are already incorporated in the DMSAS2e.zip file in the NODISPLAY folder. Therefore, when you unzip the downloaded zip file you have already installed it in your PC, use the macro-call files that you have saved inside the NODISPLAY folder if you want to use these macros with SAS EG.

```
libname dmsas2nd base "c:\dmsas2e\mcatalog\nodisplay";
options sasmstore=dmsas2nd mstored;
%excelsas(
   /*1. RQ:Input PC  file type?          E.G: excel lotus dbase
                                         & Access TAB CSV*/
      ftype  =  excel
  ,/* 2. RQ:Input PC file folder name ?  E.G: e:\sasdata\ */
      folder =  c:\
  ,/* 3. RQ:Input PC file name(s) ?      E.G: Cars93 diabet */
      file1  =  cars93
  ,/* 4. Optional LIbname ?              E.G: SASUSER mylib */
      lib    =
```

```
,/* 5. Optional data step statements ?   E.G: %str( rename
                                           x1=y2 x4=y3;
                                           logx1=log(yx1)) */
    datstp1 =  %str(rename x1=y1 x2=y2 x3=y3 x4=y4 x5=y5 ;)
,   datstp2 =  %str(ly2=log(x2); sqrty3=sqrt(x3))
,/* 6. Folder to save output ?                 E.G: C:\temp\      */
    output=  c:\
,/* 7. RQ:Display or save SAS output?   E.G. display word web
                                           pdf txt */

    graph =  word


    )
```

See a copy of the EXCLSAS2 macro-call file in the following text. Do not change the syntax. Only input the appropriate macro input in the shaded input area after the "=" sign. If you are not familiar with this type macro-call file, get help from a SAS programmer.

Index

A

ACCESS SAS software, EXLSAS2 macro, 22
Advertising, 68–69, 208–211, 380
AI. *See* Artificial intelligence applications
AIC. *See* Akakike's Information Criterion
Akakike's Information Criterion, 148
Analytical methodology, advancements in, 3–4
ANOVA, 133, 192–193, 308, 331, 334
Apparent error rate, 312
A priori hypothesis, 76, 311
Arithmetic mean, 36–37, 54
Artificial intelligence applications, 3
Artificial neural network methods, 378–379
Assessing model fit. *See* Model fit
Association, market-based analysis, 8
Augmented partial residual, 150, 153–154, 188, 216
Autocorrelation, 154, 167, 198, 224, 226
AUTOREG procedure, 154

B

Backward elimination, 125, 319–320, 324–325
Banking, data mining in, 5
Bar charts, 8, 39, 43–44, 49, 395
BASE SAS software
  CHAID2 macro, 319
  DISCRIM2 macro, 317
  DISJCLS2 macro, 124
  EXLSAS2 macro, 22
  FACTOR2 macro, 83
  FREQ2 macro, 45
  LIFT2 macro, 169
  LOGIST2 macro, 172

  LSCORE2 macro, 173
  RANSPLIT2 macro, 26
  REGDIAG2 macro, 168
  RSCORE2 macro, 171
  UNIVAR2 macro, 51
Binary categorical variable, 416
Binary logistic regression modeling
  assumptions, violations of, 164–165
  binary logistic regression, 239–260
  data descriptions, 234–260
  diagnostic plots, 237–239
  exploratory analysis, 162–165, 237–239
  false negative, 249
  false positive, 249
  gain chart, 252, 257–258
  if-then analysis, 258–260
  influential outlier, 164–165
  interpretation, 163
  LIFT chart, 252, 257–258
  model selection, 161–162, 235–237
  model specification error, 164
  multicollinearity, 165
  overdispersion, 165
  Pearson chi-square, 243–244
  specificity, 249
  two-factor interaction plots, 164
  validation, 252
Biplot display, 82–83, 91, 138, 310, 317
Box plot, 7, 40–41
BOXPLOT SAS procedure, 82, 123, 167, 316
Branches, 315
Breiman studies, 306
Breusch-Pagan test, 206–207, 228
Brier score, 161, 172, 257, 271, 278–279
Bundling of products, market basket association analysis, 380